Patterns of Development in Conidial Fungi

Garry T Cole
University of Texas at Austin

Robert A Samson
Centraalbureau voor Schimmelcultures, Baarn

Patterns of Development in Conidial Fungi

Pitman
London · San Francisco · Melbourne

Pitman Publishing Limited
39 Parker Street, London WC2B 5PB

Fearon Pitman Publishers Inc.
6 Davis Drive, Belmont, California 94002, USA

Associated Companies
Copp Clark Pitman, Toronto
Pitman Publishing New Zealand Ltd, Wellington
Pitman Publishing Pty Ltd, Melbourne

First published 1979

British Library Cataloguing in Publication Data

Cole, Garry T
Patterns of development in conidial fungi.
1. Fungi imperfecti
I. Title II. Samson, Robert A
589′.24′043 QK625.A1

ISBN 0–273–08407–0

Text set in 10/11 pt Photon Times, at The Pitman Press, Bath
and printed and bound at
William Clowes & Sons Limited, Beccles and London

Dedicated to
Judy and Gisela

Contents

Foreword

I want to tell you how this book came to be written—and I don't mean simply the rationale consciously espoused by the authors. I'd like to go back to their roots. Of course to reveal the full story would involve delving into the early days of mycology—back beyond von Höhnel and Lindau and Saccardo, beyond Preuss and Corda and Fries to Persoon (and what, you might ask, about Aristotle?). But I can bear true witness only to the period spanned by my own experience, the last twenty-five years. These have been eventful years, replete with mycological advances, of which students of the Fungi Imperfecti have contributed more than their share.

As a graduate student immersed in fungal ecology under the aegis of Alan Burges and Dennis Parkinson at Liverpool, I worked on the succession of fungi involved in the decomposition of pine needles. I found some fungi I couldn't identify (what a universal experience that is!). They were Hyphomycetes, that much I *could* tell, and I trundled off to the Commonwealth Mycological Institute at Kew to check them out. To my surprise and delight they appeared to be new—undescribed. Martin Ellis, then principal mycologist at C.M.I., advised me to concentrate on the Hyphomycetes. He pointed out that, unconstrained by the need to sporulate inside a fruit body, they can be positively protean; evolutionary pressures can and do produce the most fanciful configurations.

At that time there was no authoritative manual on the group; the literature was still fragmentary and widely scattered. There *was* an opportunity, and I took Martin's advice. As a start on this project I published three of my new fungi. One has since gone into synonymy, one has moved to another genus, and one has managed to hang onto its good name. This, believe it or not, is an average track record for Hyphomycete taxa, as you could quickly tell from looking at Kendrick and Carmichael's compilation of Hyphomycete genera (Chapter 10 in *The Fungi* Vol. IVA—Academic Press). I like to think that this happened because generic concepts in the Hyphomycetes were not as well defined as they are today.

The so-called classification of the Hyphomycetes (not to mention the unspeakably mysterious Coelomycetes) employs 'form' genera, which were then based on such features as the shape and septation of conidia, and the aggregation or enclosure of conidiophores. These characters were hardly intellectually satisfying once one got a feel for the group, and there were no acceptable suprageneric categories until you arrived at the 'form Class'—a condition that persists to this day.

At that time Stan Hughes, another British-born, C.M.I.-trained mycologist, had relatively recently become part of the brain-drain. He did not have a doctorate when he came to Canada, and he decided to remedy this by obtaining a D.Sc. He was able to do this relatively quickly with the aid of a profusion of publications arising from his years at C.M.I., and by producing a synthesis of his many penetrating observations of Hyphomycetes, their taxonomy and their development. You who are considering reading this book (or even just glancing through the pictures) must know that it derives whatever authority it has from that synthesis. 'Conidiophores, conidia and classification' appeared in the *Canadian Journal of Botany* in 1953. That paper, although it, too, had antecedents in the works of Mason and Vuillemin, was the contemporary stimulus needed to set mycologists thinking, working, writing: as they say, 'feverishly,' adding a new developmental dimension to the morphological framework of Hyphomycete taxonomy.

In 1958 I came to Canada as a raw post-doctoral fellow, still wet behind the scientific ears, to work with the selfsame Stan Hughes.

Several years later, when I joined the Faculty at Waterloo, I was ready to test and expand upon some of Hughes's ideas. In due course I received a letter from a disillusioned young school teacher named Garry Cole who had decided that he wanted a second crack at the academic life. He was to be my first graduate student. I guided his steps into the study of Hyphomycetes, and watched the fascination of these organisms weave its spell. Soon we were watching them grow in an ultra-thin culture chamber, and recording their ontogenetic secrets by time-lapse photomicrography. This work led to a satisfying thesis, and many were the novel phenomena we saw and documented *en route*. A post-doctoral year with Henry Aldrich in Florida sharpened Garry's technical skills and set the stage for his current electron microscope studies. Readers of this book will appreciate the quality of the electronmicrographs. But they will also realize that Garry has steadfastly pursued the study of conidiogenesis from time-lapse to T.E.M., from phase contrast to freeze-etching, for ten years. Garry's part in this book is a monument to the single-minded pursuit of an idea.

My connection with, and knowledge of, Rob Samson is not nearly so intimate. But I know enough to say something about his background and his contribution to the making of this book.

He began his mycological career in 1967 as a student at the world-renowned Centraalbureau voor Schimmelcultures in Baarn. By 1969 he had been appointed to the staff, and began his work on phialidic Hyphomycetes. Hughes's ideas were influential in Holland, as elsewhere. Like Garry Cole, Samson found electron-microscopy a useful approach. When Cole visited Baarn in 1974 it was natural for the two to find many interests in common, and so the collaboration began. It was continued and amplified during Samson's subsequent visit to Texas. Samson had experience with the Zygomycetes, and this led to the inclusion of a section on the so-called 'conidial' Zygomycetes. Here, in one of the most basic and valuable contributions of this book, we may find incontrovertible evidence for the morphologically minute but developmentally vital differences between the processes by which the anamorphs of Zygomycetes produce their sporangiospores, and the anamorphs of Ascomycetes and Basidiomycetes, their conidia.

Garry Cole is largely responsible for the electron-microscopy in this book. Rob Samson is largely responsible for the line drawings, nomenclature and taxonomy. But their contributions are so well interwoven that it would be very difficult to separate them.

It is 25 years since Hughes advanced his hypothesis. It is 9 years since the first Kananaskis Conference conclusively demonstrated that the mycological world was taking Hughes's ideas seriously. The intervening years have seen a spate of publications devoted to the subject of development, at both the microscopic and ultrastructural levels. So this book will come as a welcome compilation and digest of all that activity. But it is more than that. The careful reader will find sections of the book that contain much new information, and even the most casual browser will discover a fine collection of drawings and micrographs that fully illustrate the salient features of the text.

This book is timely, as I have just pointed out. But it is also timely in a way its authors may not have intended. It is now clear that there will be no natural classification of the Fungi Imperfecti based on conidium ontogeny. This perhaps regrettable fact is fully documented elsewhere*. This book may, in one way, be a 'dernier cri,' but it is a loud, clear and well-reasoned one. It will serve admirably, if not to lead us into the promised land, at least to consolidate the gains in our knowledge of how the anamorphs of Fungi produce non-motile, asexual spores.

I congratulate the authors on seeing this project through—we are indebted to them for their dedication and persistence.

* in:

(1) *Biosystematics in Agriculture—Beltsville Symposia in Agricultural Research. 2.* Romberger, Foote, Knutson and Lentz (Eds). Allanheld, Osmun, Montclair, N.J., 1977.

(2) *The Whole Fungus—Proceedings of Kananaskis II,* W. B. Kendrick (Ed,). National Museums of Canada, Ottawa, 1978.

Bryce Kendrick,
Department of Biology,
University of Waterloo,
Waterloo, Ontario,
Canada.

March 1978.

Acknowledgements

The senior author gratefully acknowledges the National Science Foundation, Pan American Health Organization and Research Corporation for grants which supported portions of this work. Dr. Samson was the recipient of a ZWO–NATO Fellowship (1975–1976) which permitted him to work with Dr. Cole in Texas. The authors thank Dr. J. A. von Arx, Director of the Centraalbureau voor Schimmelcultures, for his generosity in providing many of the fungal cultures used in this study. Fungi were also provided by Drs. L. Ajello, R. K. Benjamin, C. W. Hesseltine, E. S. Luttrell, A. Padhye, K. Pirozynski, E. Simmons and P. Szaniszlo. The following individuals contributed micrographs which appear in this book and have been acknowledged in the text; G. Benny, R. Campbell, D. H. Ellis, D. A. Griffiths, T. M. Hammill, R. T. Hanlin, T. Hashimoto and K. L. O'Donnell. The authors thank these workers for their cooperation. Finally, sincere appreciation is directed to those colleagues who provided valuable criticisms and suggestions during the preparation of the illustrations and manuscript. They were R. K. Benjamin, W. Gams, W. B. Kendrick, E. S. Luttrell, K. Pirozynski and C. T. Rogerson.

1 Introduction

The Fungi Imperfecti, or Deuteromycotina (Ainsworth, 1971), comprises a group of terrestrial and aquatic fungi which reproduce by asexual means. Although for some species the perfect state has been discovered (most have ascomycetous affinities, a few are Basidiomycetes), the majority of the more than 2,000 genera have no recognized sexual state. With concentrated efforts in establishing sexual-asexual connections, many species of Deuteromycetes will be aligned with appropriate ascomycetous or basidiomycetous genera (Kendrick and Watling, 1977; von Arx, 1977). However, a 'permanent residue of imperfect forms will remain' (Ainsworth, 1971) in which sexuality has been lost and genetic variability is dependent primarily on the parasexual cycle (Pontecorvo *et al.*, 1953).

Deuteromycetes are easily overlooked in nature, even by the trained biologist, but they can be ferreted out by looking for such signs as powdery, woolly, or blue-stained patches on the surface of wood, darkly stained regions of leaves, tiny, dark, spherical or flattened structures on woody and herbaceous plants, foam in streams, rivers and along the shorelines of lakes and oceans, localized areas of coarse, irritated skin on the family dog or cat and even on the next door neighbour. However, the best source of these fungi is soil, whether derived from Africa or Antarctica. Virtually any soil sample suspended in sterile water and plated out on laboratory media containing bacterial antibiotics will yield at least one Deuteromycete. Whenever an undescribed species is found, attempts are made to isolate and grow it in pure culture. Many such axenic cultures are maintained (von Arx, 1970, 1974) in national collections from which subcultures are readily available for study. W. B. Grove (1937) claimed that too many investigators suffer from 'petripatellism' which he described 'as the state of mind of a mycologist who studies his fungus in a laboratory, on agar slants or Petri dishes without paying equal regard to what the fungus can do out of doors in the wide and untrammelled field'. We have been afflicted with the same disease, bowing to this necessity in our wish to examine the reproductive cycle of conidial fungi.

The Form-Subdivision Deuteromycotina is broken up into three Form-Classes; Blastomycetes, Coelomycetes and Hyphomycetes. The first class includes the imperfect yeasts and yeast-like fungi (Fig. 1.1) and is given little consideration in this book. The Coelomycetes (W. B. Grove, 1935; Sutton, 1973) produce conidia in, or on specialized fructifications called pycnidia (Figs. 1.2, 1.3), acervuli and stromata. In order to elucidate their asexual reproductive cycles the tiny fruiting bodies must be carefully broken open or sectioned at various stages of development without excessively damaging their contents. As one would expect, this is extremely tedious and requires masochistic tendencies in the investigator. However, such studies are necessary if we are ever to arrive at a comprehensive developmental scheme for the Deuteromycetes.

This leads us finally to the Hyphomycetes, the Form-Class with which we are mainly concerned here. This group produces loose, floccose mycelia, bearing reproductive cells on hyphae or aggregations of hyphae but never within sporocarps.

Conidia are asexual, nonmotile and usually deciduous propagules produced by the Deuteromycetes. Studies of conidial formation are relevant to several biological disciplines, such as medical mycology and plant pathology. Many features of conidiogenesis revealed by time-lapse light microscopy and electron-microscopic techniques are apparently unique among eukaryotes, and have drawn the attention of developmental biologists and cytologists. Taxonomic significance has also been attributed to characters of conidiogenesis. S. J. Hughes (1953) wrote: 'The need for further studies on the precise method of conidium development is stressed because this promises to provide the

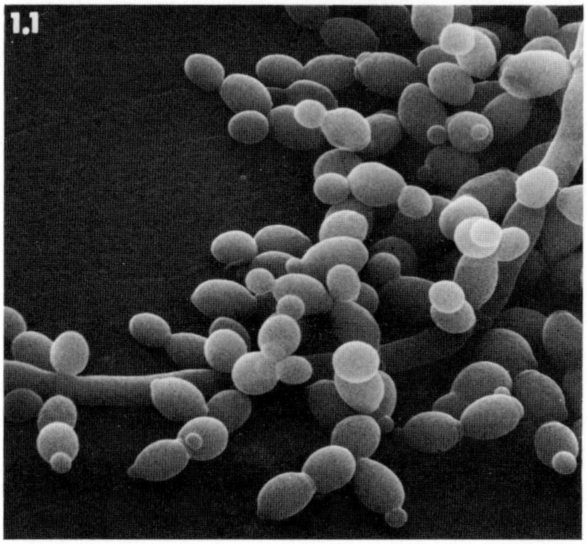

Fig. 1.1 *Candida albicans.* Chains of budding cells and vegetative hyphae. ×2,000.

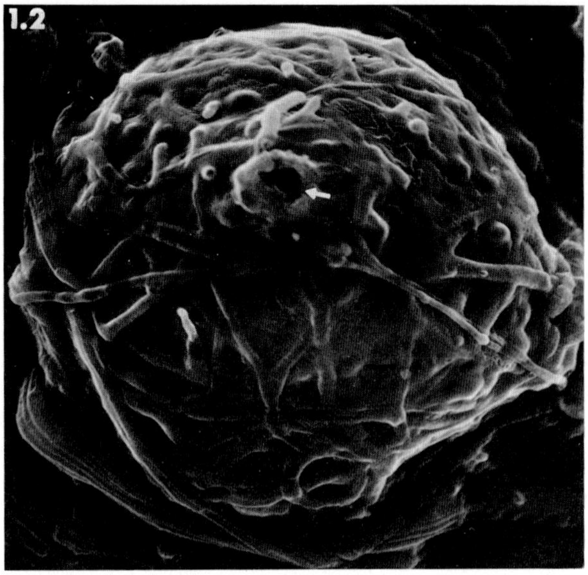

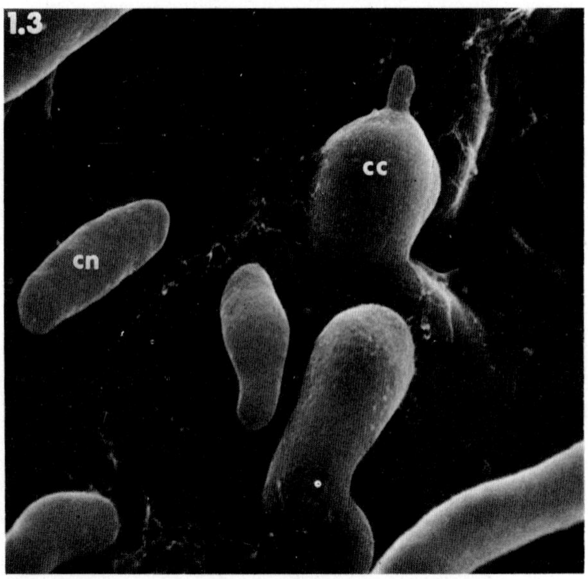

Figs. 1.2–1.3 *Phoma* sp. Pycnidium showing apical pore (arrow in 1.2) and enclosed conidia (cn) and conidiogenous cells (cc) revealed after pycnidium is cryofractured (1.3). × 757 and 4,950.

most stable character for the classification of Fungi Imperfecti as a whole'.

The main impetus for continued examinations of conidiogenesis has been provided by the systematists, and during the years since the publication of Hughes's pioneering paper, many developmental studies have appeared. Several workers have presented revised schemes for classifying the different kinds of conidial ontogeny, and inevitably a surge of new terminology has been generated. The following references include most attempts to synthesize the new developmental information: Goos, 1956; Tubaki, 1958, 1963; Subramanian, 1956, 1962a, b, 1965, 1971a, 1972a, b, 1973, 1978; Madelin, 1966; Simmons, 1966; Barron, 1968; Chesters, 1969; Cole, 1969; Kendrick, 1971a; Reisinger, 1972; Kendrick and Carmichael, 1973; Sutton, 1973; Cooke, 1974; Gams *et al.*, 1975.

During the last five years, much new information has been made available by electron microscopy. This has necessitated revisions or complete abandonment of some earlier concepts while reinforcing others. One of the functions of this book is to assimilate information provided by these investigations and re-evaluate current developmental concepts. In addition, previously unreported ultrastructural and developmental features of conidia and conidiogenous cells in a large number of imperfect fungi are presented. Many light-microscopic studies could not resolve the subtleties of developmental processes. We have attempted to clarify these ambiguities and draw correlations, whenever possible, between information provided by light- and electron-microscopic examinations.

We begin our discussion of patterns of conidial development with some of the concepts generated by participants in the Kananaskis Conference on Taxonomy of Fungi Imperfecti (in Kendrick, 1971a). One of the most thought-provoking ideas to surface during this conference was that two basic modes of conidiogenesis exist in the Deuteromycotina. These were referred to as *blastic* and *thallic* development. The blastic process involved a 'marked enlargement of a recognizable conidium initial *before* the initial is delimited by a septum' and it was suggested that the 'conidium differentiates from *part of* the cell'. On the other hand, in the thallic process, 'any enlargement of the recognizable conidium initial . . . occurs only *after* the initial has been delimited by a septum or septa' and the 'conidium differentiates from a *whole* cell'. A diagrammatic interpretation of these two developmental processes is presented in Fig. 1.4. Although this dichotomy represents the most recent, and probably most successful attempt so far to simplify the classification of the various modes of conidial ontogeny, the definitions still require some refinement. The critical aspect of the timing of septation, implicit in both the blastic and thallic concepts, is difficult to document either by light or electron microscopy. As we shall see in Chapter 3, initiation of septa at the base of the young conidium may occur very early in development, perhaps before there is recognizable enlargement. In addition, delimitation of a conidial initial from the adjacent conidiogenous cell does not necessarily mean that cytoplasmic continuity between the cells has been interrupted. So a clear differentiation of blastic and thallic development based on septation alone is not possible. The real difference appears to lie in the initial contrast between apical growth involving a limited amount of wall area—essentially only the apical dome at first (i.e., blastic)—and a combination of growth and conversion involving significant lengths of extant hypha, including the periclinal wall (i.e., thallic). A discussion of the significance of this distinction must involve consideration of the mechanism and regulation of hyphal tip growth since, as pointed out by Da Riva Ricci and Kendrick (1972), 'it is often the hyphal apex that gives rise to the conidia of Hyphomycetes'.

Fig. 1.4A–C Blastic development: a conidium differentiates by blowing-out of a part of the fertile hypha, or, conidiogenous cell.

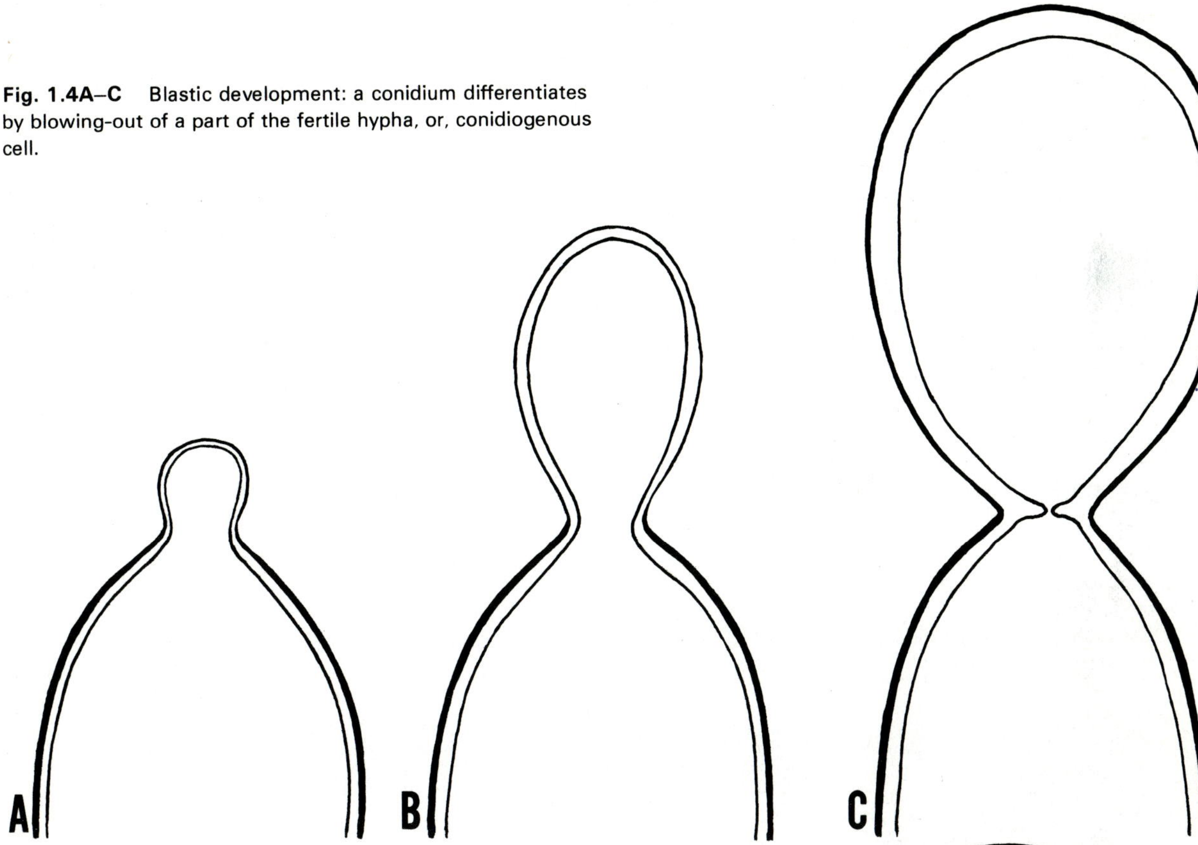

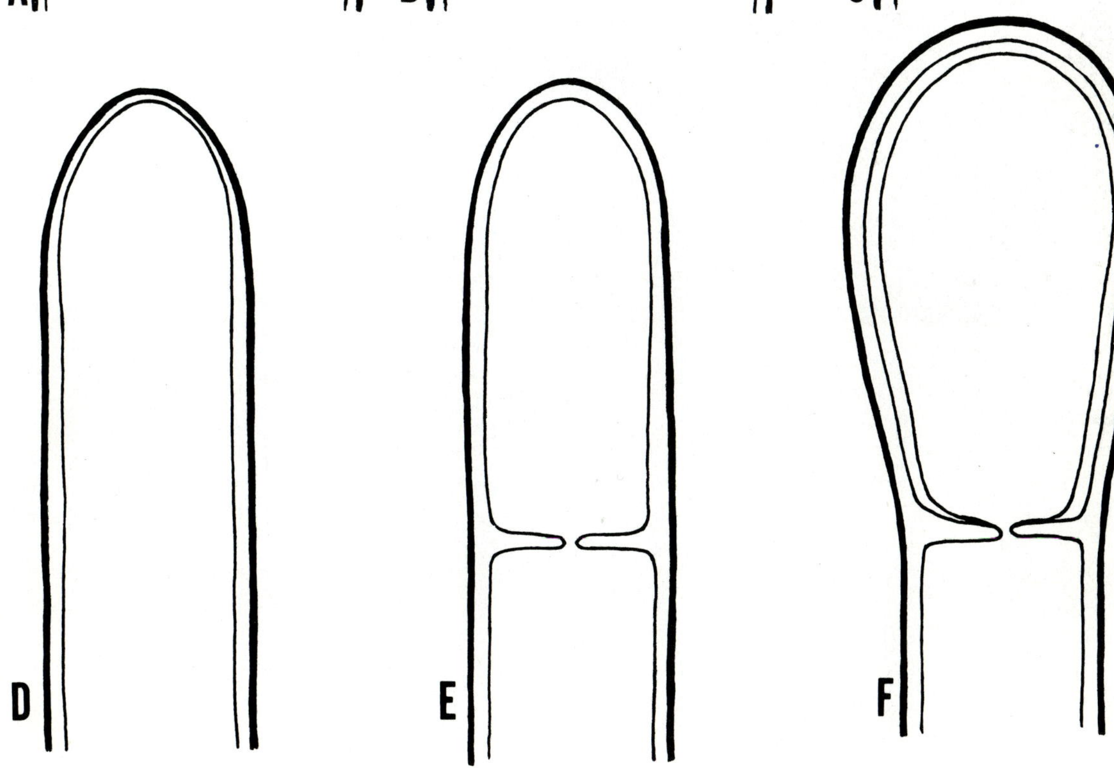

Fig. 1.4D–F Thallic development: a conidium differentiates by conversion and disarticulation of a pre-existing segment of the fertile hypha.

Blastic development

Correlation with hyphal tip growth

Protoplasmic differentiation associated with blastic conidial development resembles that of hyphal tip growth. The conidial initial of *Scopulariopsis brevicaulis* (Sacc.) Bain. shown in Fig. 1.5 has arisen by the blowing-out of a conidiogenous cell apex, increasing the volume of cytoplasm in the initial and the area of the encompassing wall layer. A new electron-transparent wall layer (2°W) has already differentiated inside the primary wall (1°W). The latter is continuous with the outer layer of the conidial initial. The cytoplasm contains several rings of ER cisternae (CR), some in a stage of blebbing or vesiculation (arrow in Fig. 1.5). These organelles have been likened to Golgi apparatuses (Bracker, 1968; Cole and Aldrich, 1971a). Although morphologically distinct dictyosomes have not been found in the Deuteromycetes, a functional equivalent of these organelles must assume a secretory role in these fungi (Hawker, 1965; Mollenhauer and Morré, 1966; Bracker, 1967, 1968; Girbardt, 1969; Oláh, 1970, 1971, 1973, 1974; Moore, 1971). The vesicles adjacent to the cisternal rings in Fig. 1.5 and many juxtaposed to the plasma membrane arise by proliferation of cisternae. The vesicles may also contain wall precursors that are released when these organelles fuse with the cell membrane. The hyphal tip of *Drechslera sorokiniana* (Sacc.) Subram. & Jain (Fig. 1.6) shows a cluster of vesicles at the apical pole where most incorporation of new wall material takes place (Bartnicki-Garcia and Lippman, 1969; Katz and Rosenberger, 1970, 1971; Gooday, 1971; Bartnicki-Garcia, 1973). Vesicles found at the growing tips of vegetative hyphae and germ tubes in a wide range of fungi support the contention that these organelles possess a secretory function and contribute to wall growth and cell extension (R. Marchant *et al.*, 1967; Girbardt, 1969, 1973; S. N. Grove and Bracker, 1970; S. N. Grove *et al.*, 1968, 1970, 1973; Heath and Greenwood, 1970; Bracker, 1971; Bracker *et al.*, 1971; Heath *et al.*, 1971; S. N. Grove, 1972; McCully and Bracker, 1972; Bartnicki-Garcia, 1973; Hess, 1973, 1975; Hess and Weber, 1973; Hunsley, 1973; Syrop, 1973; Barstow and Lovett, 1974; Murray and Maxwell, 1974; Cook, 1972, 1974; Collinge and Trinci, 1974; Hemmes and Wong, 1975; Beckett, 1976; Smith *et al.*, 1976). Although some evidence exists that the contents of apical vesicles are

Fig. 1.5 *Scopulariopsis brevicaulis.* Thin section of conidial initial which reveals several cisternal ring complexes (CR). 1°W, primary wall; 2°W, secondary wall; N, nucleus; mV, multivesicular body. Arrows locate large secretory vesicles; arrowheads indicate small secretory vesicles. ×32,000.

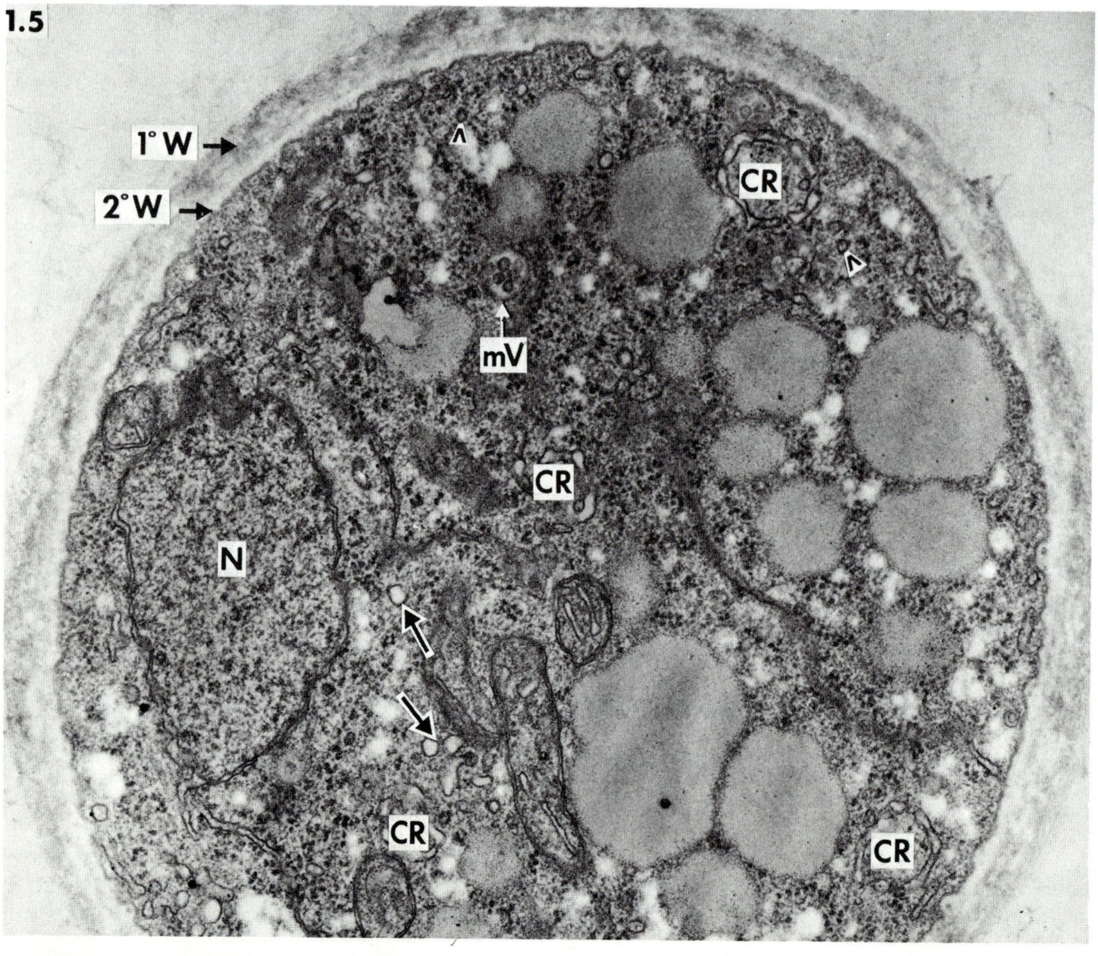

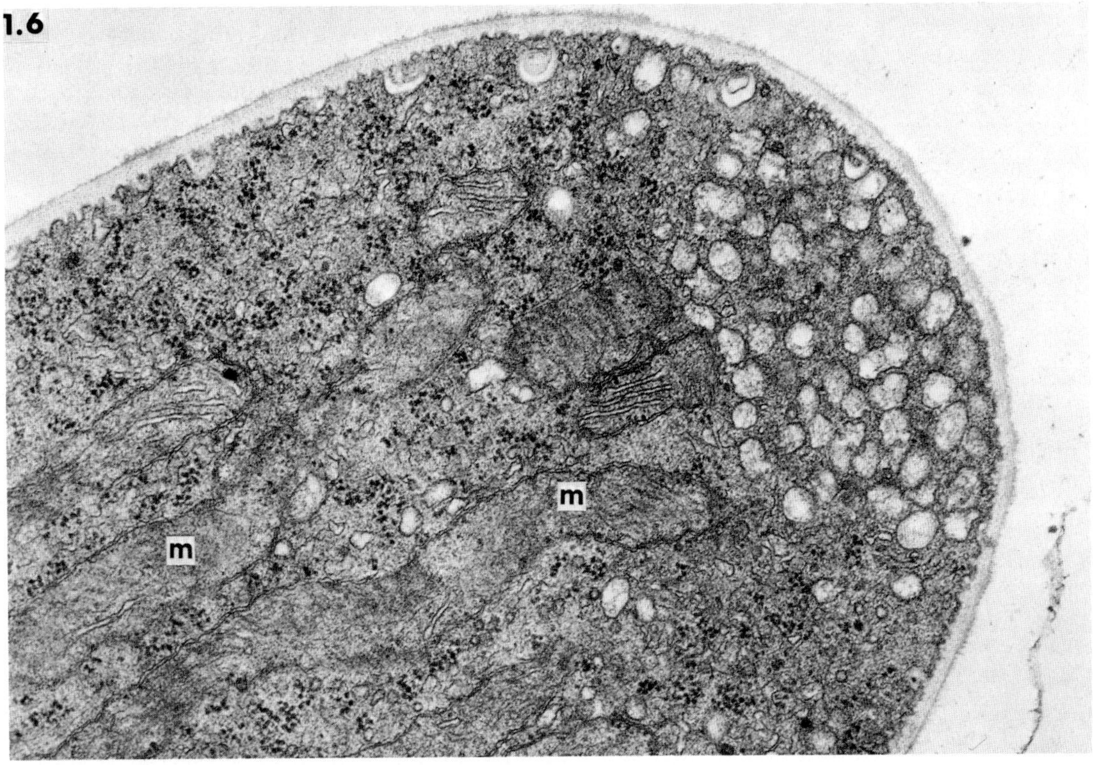

Fig. 1.6 *Drechslera sorokiniana.* Vegetative hyphal tip which contains a polar cluster of secretory vesicles and subapical mitochondria (m). ×39,900.

similar to hyphal wall components (S. N. Grove *et al.*, 1972), the involvement of the vesicles in wall building is not yet understood (Ruiz-Herrera *et al.*, 1975). Freeze-etching shows that these vesicles fuse with the plasmalemma (Cortat, 1971; Vian, 1972) enabling it to expand during extension growth. No apical concentration of dictyosome-derived vesicles can be seen in the young conidium in Fig. 1.5. We will return to the significance of this later. Collinge and Trinci (1974) concluded that the rate of extension of a hypha largely depends on the rate of supply of vesicles to its apex, and that the vesicles may originate several μm back from the tip and then migrate forward. In Fig. 1.7, subapical vesicles like those of the hyphal tip (arrows) are shown in a vegetative hypha of *D. sorokiniana* (cf. Figs. 1.6, 1.9). Microtubules (arrowheads in Fig. 1.7) lie adjacent to the vesicles and more or less parallel to the long axis of the hypha. Microtubules, microfilaments and microfibrils probably play important roles in movement of organelles through fungal hyphae (Girbardt, 1969; Cole, 1972; McKeen, 1972; Raudaskoski, 1972; Raudaskoski and Koltin, 1973; Gull, 1975) and may thus be involved in the migration of secretory vesicles towards the hyphal tip. Microtubules also occur subapically in conidiogenous cells where they may perform a similar function (Carroll and Carroll, 1974a, b; Hanlin, 1976; Hammill, 1977c; Fig. 4.74, Chapter 4 shows microtubules in a conidiogenous cell of *Scopulariopsis brevicaulis*). The secretory vesicles in vegetative hyphae and conidial initials

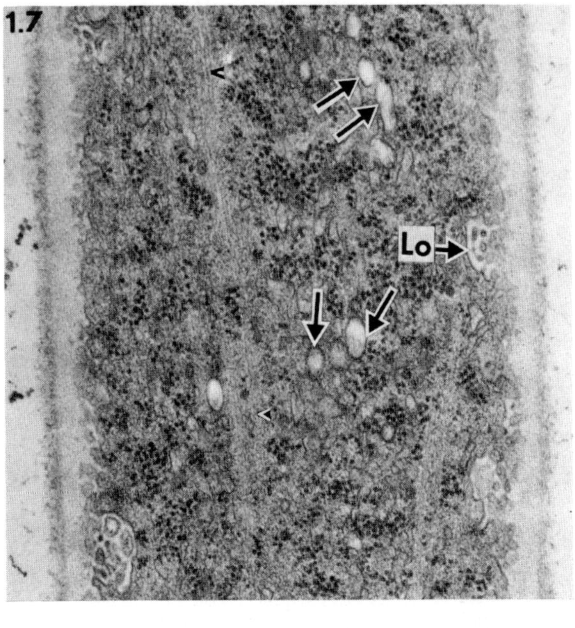

Fig. 1.7 *Drechslera sorokiniana.* Parallel clusters of microtubules (arrowheads) and secretory vesicles (arrows) in the subapical region of a vegetative hypha. Lo, lomasome. ×36,400.

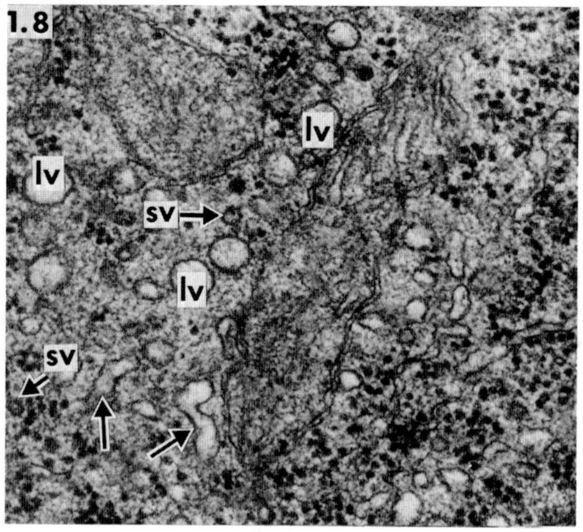

Fig. 1.8 *Drechslera sorokiniana.* Cisternae (arrows), which are apparently in the process of blebbing, and large secretory vesicles (lv) in the subapical region of a vegetative hypha. sv, small secretory vesicles. ×74,000.

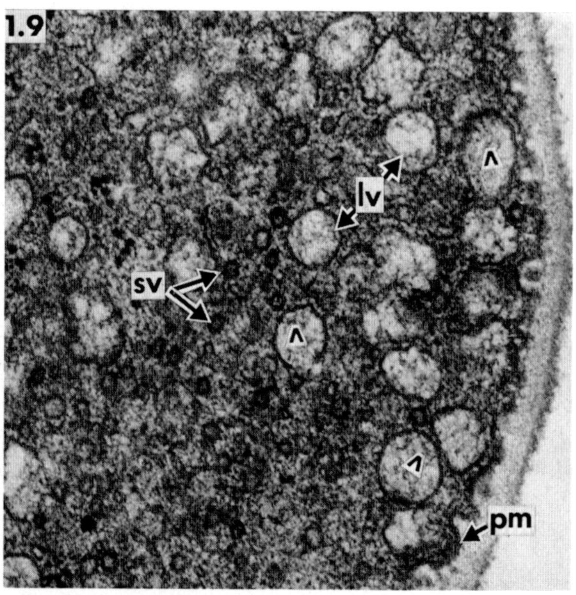

Fig. 1.9 *Drechslera sorokiniana.* Hyphal tip containing large vesicles (lv) and small vesicles (sv), both suggested to have a secretory function. pm, plasma membrane. Arrowheads indicate wall-like material contained by large secretory vesicles. ×71,400.

appear to originate in dictyosome-like bodies. Subapical cisternae (arrows in Fig. 1.8) are often observed near vesicles (lv) similar to those at the hyphal tip (cf. Fig. 1.6). These vesicles may arise by *blebbing* of cisternae, but other origins are possible. Cook (1974) proposed that vesicles in hyphal tips and conidial initials are released from multivesicular bodies (mV, Fig. 1.5). The nuclear membrane is also capable of blebbing to produce vesicles (Carroll, 1967; Reeves, 1967; Oso, 1969) or dictyosomes (Morré *et al.*, 1971). However, Hyphomycetes have not been shown to have an elaborate endomembrane system like that in some species of *Pythium* Prings. (Oomycetes) (S. N. Grove *et al.*, 1967), where the 'nuclear envelope, endoplasmic reticulum, Golgi apparatus and secretory vesicles form a structural and functional unit' (Bracker *et al.*, 1971).

Two populations of vesicles are recognized in the hyphal tip of *Drechslera sorokiniana*: large vesicles (lv) which contain fibrous material (arrowheads in Fig. 1.9) similar to wall microfibrils, and small vesicles (sv) whose contents are homogeneous. The large vesicles are bounded by a unit membrane resembling the plasmalemma (pm). The small vesicles are bounded by a tripartite, *membrane-like shell*, the nature of which is still unknown. Both vesicles are also observed in the subapical region of the vegetative hypha (Fig. 1.8). S. N. Grove and Bracker (1970) demonstrated that the smaller vesicles in *Aspergillus niger* van Tieghem aggregate in a dense cluster at the hyphal tip among the larger vesicles. A similar phenomenon occurs at the hyphal tips of many septate fungi, although in some species this central region within the cluster of larger vesicles is occupied by an electron-translucent core which is vesicle-free (Bartnicki-Garcia, 1973). In either case, this central region at the hyphal tip of septate fungi is probably synonymous with Girbardt's (1957) 'Spitzenkörper'. Small vesicles are also observed adjacent to the cisternal rings and plasmalemma in young, blastic conidia of *S. brevicaulis* (arrowheads in Fig. 1.5), and have been demonstrated in conidial initials of *Stemphylium botryosum* Wallroth (Carroll, 1972), *Stachybotrys atra* Corda (Campbell, 1972), *Stilbothamnium nudipes* Haum. (Roquebert and Abadie, 1973), *Ulocladium atrum* Preuss (Carroll and Carroll, 1974a) and *Aspergillus clavatus* Desm. (Hanlin, 1976). Cook (1972, 1974) has shown small vesicles at the tips of conidial initials in *Oedocephalum roseum* Cook. Each initial later swells apically and develops into a blastic conidium. The author suggested that the small vesicles contain wall-softening enzymes, while the larger vesicles contain wall-building materials. Although no direct evidence exists for this differentiation, it agrees with Bartnicki-Garcia's (1973) concept that a delicate balance between wall synthesis and wall lysis exists at the hyphal tip. He stated that 'a harmonious balance must exist between the processes of wall synthesis and wall lysis to assure the orderly growth of the cell wall. Too much synthesis would cause thickening of the apical wall and arrestment of elongation; too much lysis would lead to bursting (or perhaps swelling) of the apex'. This change from *polarized* to *depolarized* growth is pivotal in differentiation of a fertile hyphal tip into a blastic conidium.

1.10

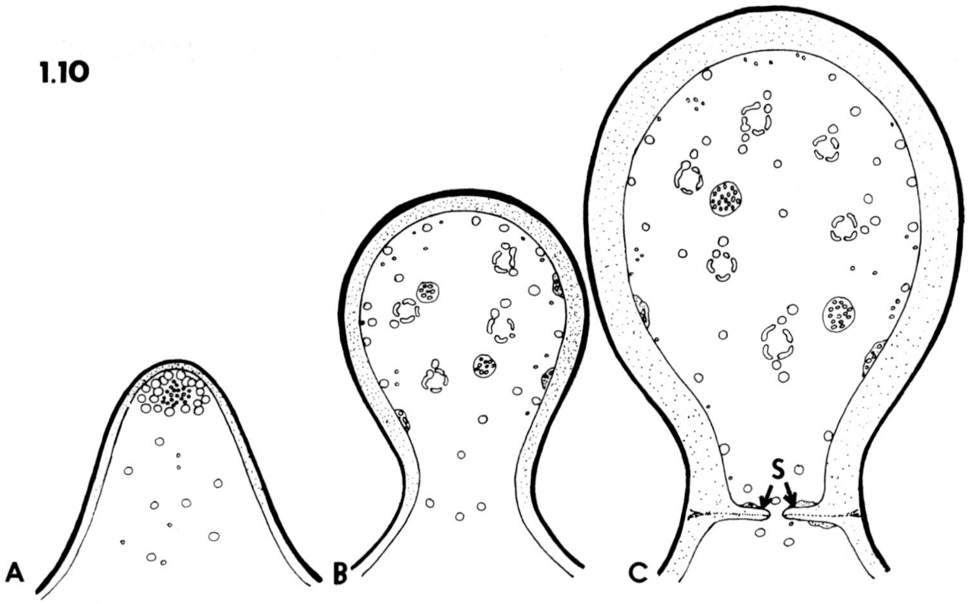

Fig. 1.10A–C Interpretation of protoplasmic differentiation during blastic conidial formation. S, septum.

Fig. 1.11 *Scopulariopsis brevicaulis.* Mature, blastic conidium attached to apex of conidiogenous cell. A transverse septum separates the cytoplasm of the two cells. ×13,230

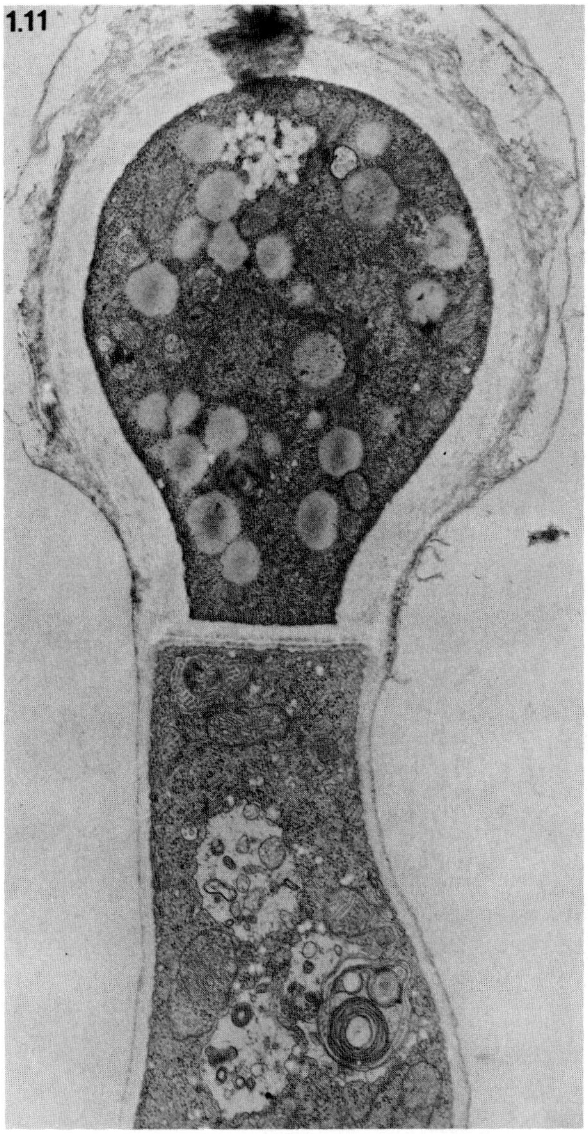

Bracker (1977) described microvesicular structures found in budding yeast and tips of growing hyphae of several species. He called these organelles 'chitosomes'. The minute vesicles are 400–700 Å in diameter and resemble the small vesicles shown in Fig. 1.9. It has been suggested (Bracker *et al.*, 1976; Bartnicki-Garcia *et al.*, 1977) that these structures contain the chitin synthetase zymogen and are responsible for transport of this proenzyme from its point of formation (i.e., rough endoplasmic reticulum) to the plasma membrane (i.e., point of operation of the zymogen on the cell surface). Studies of mural enzyme activity in *Saprolegnia monoica* Prings., presented by Févre (1977) and Févre and Dumas (1977), support this conclusion.

Figure 1.10A–C is a diagrammatic interpretation of three stages of blastic conidial formation. In Fig. 1.10A, which is comparable to the early stage illustrated by Cook (1974) in *O. roseum*, a cluster of small vesicles surrounded by larger vesicles is present at the growing apex. Protoplasmic differentiation at this stage of conidial development resembles that of a hyphal tip. However, the vesicles soon disperse as the apex of the papilla swells and growth involves the entire surface of the initial (Fig. 1.10B). Also, as noted in Fig. 1.5, dictyosome-like bodies now appear which were not present earlier and are absent from a hyphal tip undergoing normal extension growth (Fig. 1.6). The inner wall layer (2°W) begins to thicken at this stage (Fig. 1.10B) and concomitantly, centripetal growth of a septum begins at the base of the young conidium (Fig. 1.10C). Secretory vesicles arising from the cisternal rings fuse with the plasma membrane and contribute to secondary wall development. Cytoplasmic continuity through the septal pore is maintained for some time (Fig. 1.10C). As the new cell expands the 2° wall extends and

1.12

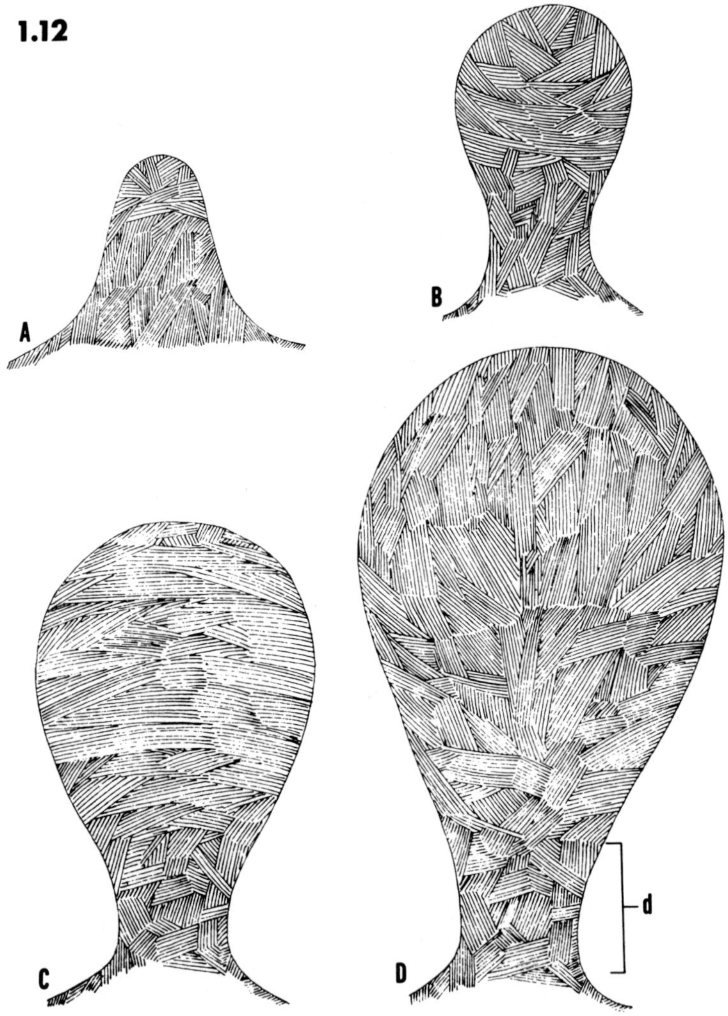

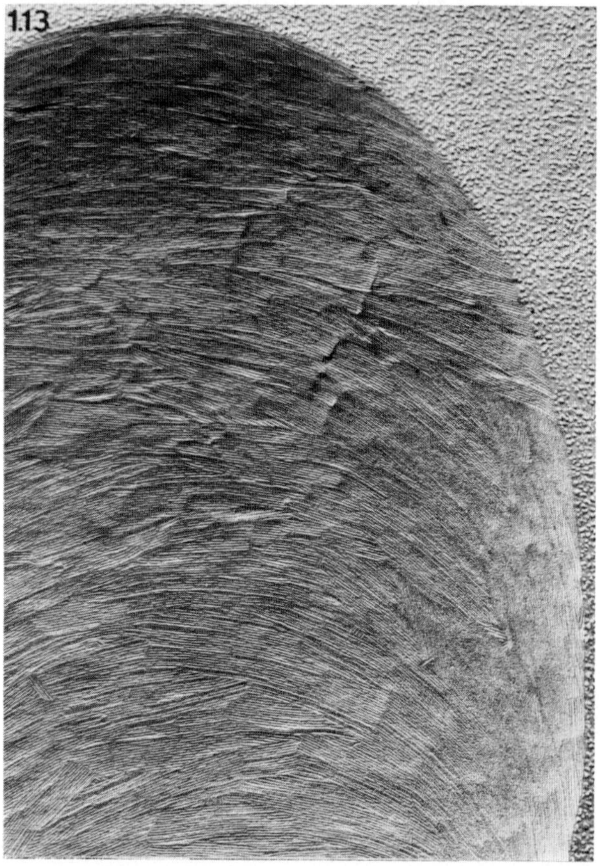

Fig. 1.12A–D Interpretation of changes in *rodlet* orientation during blastic conidial development in *Gonatobotryum apiculatum.*

Figs. 1.13–1.14 *Gonatobotryum apiculatum.* Freeze-fractures of a conidial initial (1.13) and mature conidium (1.14) revealing different orientations of rodlet fascicles. Arrows in 1.14 indicate radiating rodlet fascicles. ×52,200 and 55,000, respectively.

thickens. We will return to this when discussing rodlet orientation and conidiogenesis (Figs. 1.12–1.18). A mature, thick-walled, blastic conidium of *Scopulariopsis brevicaulis* is shown in Fig. 1.11, delimited from the conidiogenous cell by a transverse septum at its base.

Several models of hyphal tip growth in fungi have been proposed (e.g., Robertson, 1965a, b; Park and Robinson, 1966; Burnett, 1968; Da Riva Ricci and Kendrick, 1972; Trinci and Saunders, 1977), each based on a similar theme; that the hyphal tip is differentiated into an apical zone where wall intussusception occurs, and a subapical zone where extension growth, thickening and progressive rigidification occur. Robertson (1965a, b, 1968), and later Kendrick

(1971c) suggested that different parts of the wall may be either in a *set* or an *unset* condition. Kendrick (1971c) defined the *set* wall as 'inelastic and incapable of further deformation except by rupture due to trauma or to secondary enzymatic softening', whereas the *unset* wall 'is elastic and can yield and stretch in response to pressure differentials'. A gradient was presumed to exist between these two conditions.

The concept of set and unset walls was supported by freeze-fracturing examinations of conidiogenesis (Cole, 1973c). Surface fractures of the conidia of several Deuteromycetes have revealed the presence of *rodlets* (Hess *et al.*, 1968) which are components of the outer wall layer. Hashimoto *et al.* (1976) identified the rodlet layer in the hyphomycetous fungus *Trichophyton mentagrophytes* (Robin) Blanchard as a protein-glucomannan complex with a crystalline structure. The rodlets of Hyphomycetes commonly lie just beneath an outermost, hydrophobic layer of undetermined composition. It seems that the rodlet layer itself is only a few hundred angstroms thick. Changes in rodlet orientation on the outer wall surface of freeze-etched conidia appear to be correlated with changes in cell shape during conidium differentiation (Cole, 1973c). This was demonstrated by conidial fractures of *Gonatobotryum apiculatum* (Peck) Hughes (Figs. 1.12–1.14). During the initial expansion period a circumferential pattern of rodlet fascicles is revealed (Fig. 1.13). Later, during cell elongation, the fascicles are arranged approximately parallel to the long axis of the conidium (Fig. 1.14). These changes are summarized diagrammatically in Fig. 1.12A–D. Changes in rodlet arrangement during blastic conidiogenesis may reflect changing stresses applied to the outer wall layer. Roelofsen (1959) pointed out in his 'multinet growth hypothesis', which was an attempt to explain the mechanism of growth in cellulosic walls of higher plants, that 'the orientation of the microfibrils' (cellulosic) 'proves to be parallel to the direction in which tension reaches its highest value, no matter what the direction of growth may be'. This may be a reasonable explanation for the circumferential orientation of rodlets in Fig. 1.12B, C. Although the cells are still elongating, the surface tensions associated with expansion are highest. In contrast, Roelofsen pointed out that if 'tension is equal in all directions, the microfibrils appear to be oriented parallel to the direction in which the cell shows its strongest growth'. In Fig. 1.12D, tensions resulting from widening and elongation may be almost equal. At this stage elongation predominates, and rodlet fascicles are mostly parallel to the long axis. However, because some rodlet fascicles later radiate from a longitudinal to a transverse axis (arrows in Fig. 1.14), some diametric expansion must be continuing. This pattern was seen in all mature, seceded conidia of *G. apiculatum* examined, and in freeze-fractures of blastic conidia in *Scopulariopsis brevicaulis* (Cole and Aldrich, 1971a).

Variation in shape of blastic conidia

Although *G. apiculatum* and *S. brevicaulis* both form ellipsoidal to spherical conidia, many conidial shapes occur in the Deuteromycetes and any general concept of blastic development must explain the ontogeny of ovoid, helicoid, biconical and other shapes. In each case, the early developmental stages are as reported above. Subsequently, the shapes of conidia are determined by the following interrelated factors:

1. the relative rates of wall setting (rigidification) and wall plasticization (blowing-out) ((Kendrick, 1971d)), and
2. local changes in the supply of wall precursors.

This latter process may depend on the arrangement of secretory vesicles.

In order to demonstrate the significance of these factors, we will follow the formation of the medially constricted, ellipsoidal conidia of *Cladobotryum varium* Nees, and the helicoid conidia of *Zalerion maritimum* Linder (Anastasiou). Replicas of surface-fractured conidia of *C. varium* (Figs. 1.16–1.18) show changes in rodlet orientation which indicate

1.15

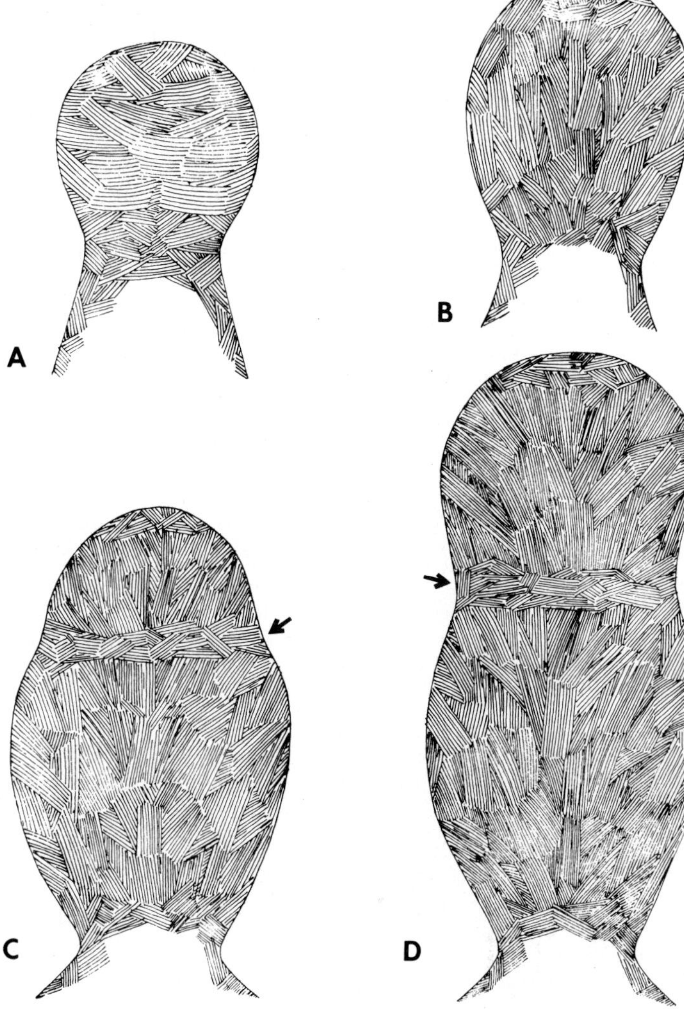

Fig. 1.15A–D Interpretation of changes in rodlet orientation during blastic conidial development in *Cladobotryum varium*. Arrows in (C) and (D) indicate medial constriction of conidium.

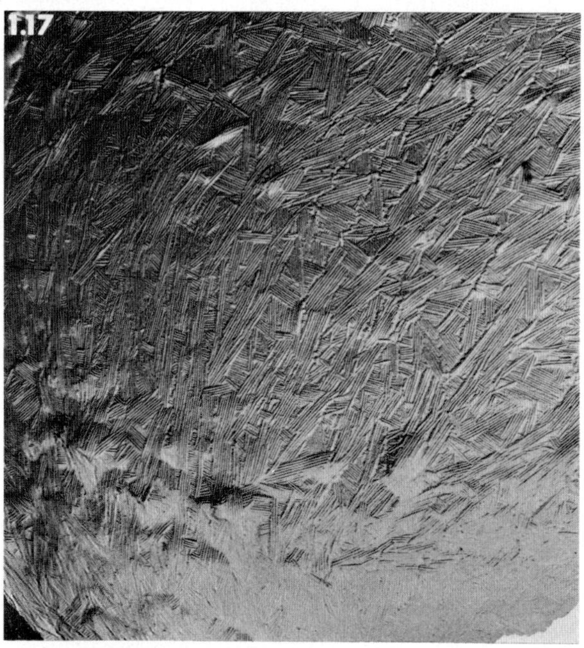

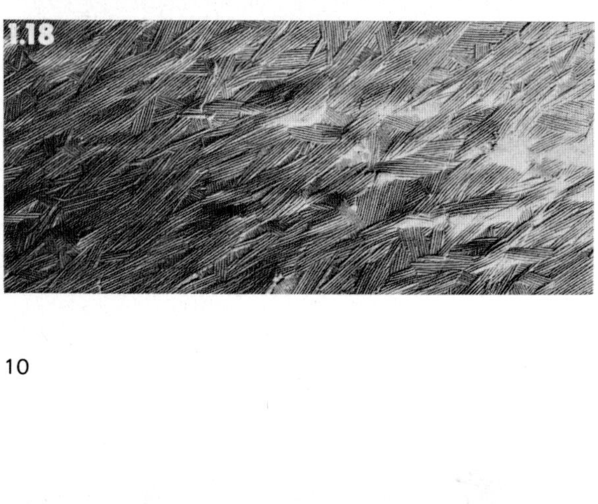

Figs. 1.16–1.18 *Cladobotryum varium.* Freeze-fracture of mature, seceded conidium (1.16), revealing more or less parallel arrangement of rodlet fascicles toward base and apex (1.17), and random arrangement at medial constriction (1.18). ×9,380, 32,000, and 32,000, respectively.

Fig. 1.19A–D Interpretation of relationship between wall differentiation and cell growth during blastic conidial development in *Cladobotryum varium.* Solid lines represent *set* wall layers, dotted lines represent *unset* wall layers. Arrows indicate general direction of cell growth.

1.19

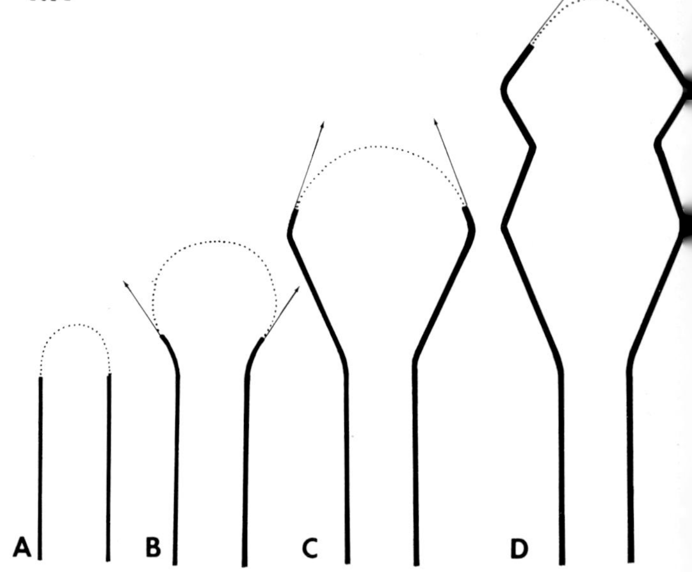

the pattern of wall development. In our interpretation of these changes shown in Fig. 1.15A–D, the early developmental stages are comparable to those in *G. apiculatum* and *S. brevicaulis.* The circumferential orientation of rodlets reflects widening of the initial (Fig. 1.15A). Later, elongation predominates (Fig. 1.15B), though some widening continues. Kendrick (1971d) pointed out that during the initial stages of blastic development, 'wall setting lags behind blowing out by a slight but relatively constant amount. There will thus be more than a hemisphere of unset wall' (Fig. 1.19A, B), 'and setting will thus move along the tangent drawn in the figure. Later, the rate of wall setting overtakes the rate of blowing out, and exceeds the latter by a slight but relatively constant amount. Now there will be less than a hemisphere of unset wall' (Fig. 1.19C), 'and setting will move along the tangent drawn in the figure, steadily narrowing the conidium at its apex'. We suggest that if the original relationship of the rates of 'wall setting' and 'blowing out' were then re-established, the developing conidium would assume the shape shown in Fig. 1.19D, which approximates the shape of the mature conidia of *C. varium* (Figs. 1.15D, 1.16). Returning to rodlet orientation, we find that it is generally parallel to the long axis of the cell where elongation predominates (Figs. 1.15C, 1.17), but more random at the medial constriction (arrows in Figs. 1.15C, D, 1.18) indicating retarded growth.

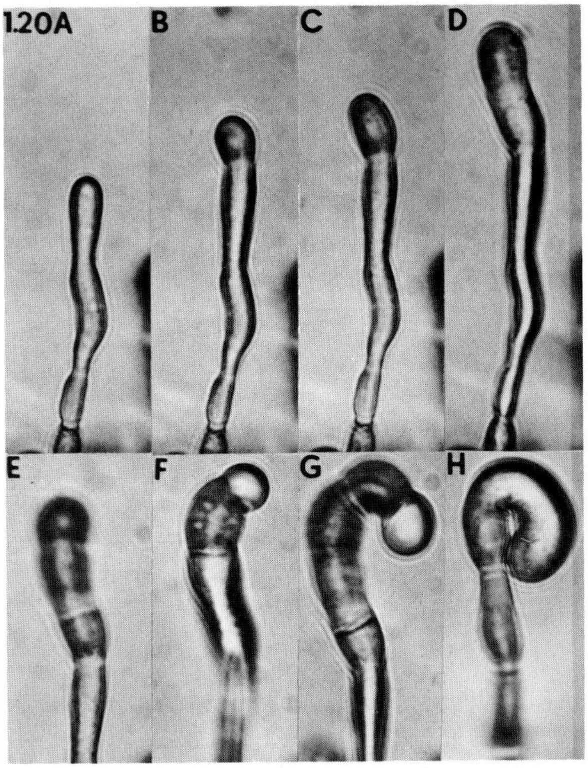

1.21

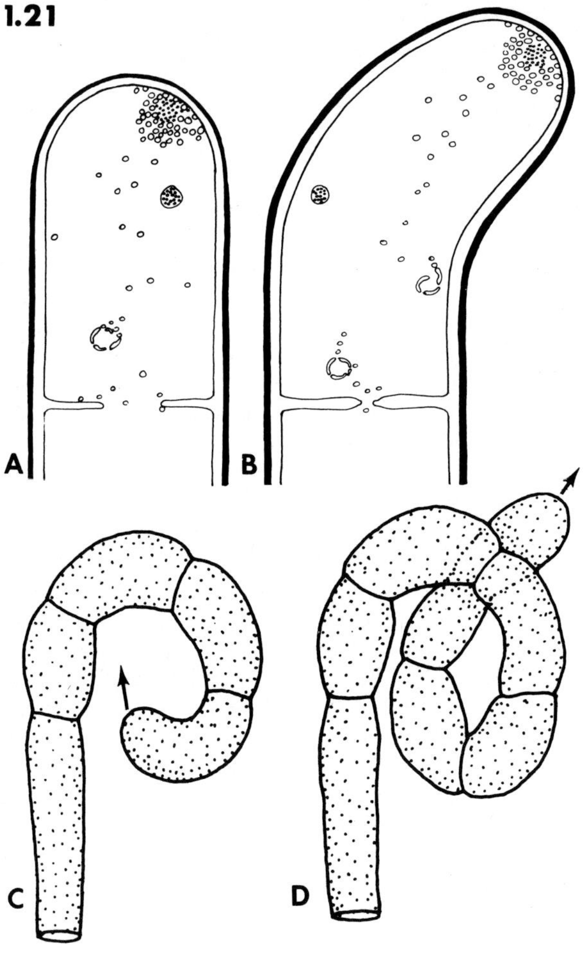

Fig. 1.20A–H *Zalerion maritimum.* Time-lapse, photomicrographic sequences (A–E, F–G) of helicoid conidial development. T = 0, 30, 45, 70, 100 minutes and T = 0, 175 minutes respectively. ×1,248.

Fig. 1.21A–D Diagrammatic interpretation of changes in position of apical cluster of secretory vesicles (A, B) during helical growth of the blastic conidium of *Zalerion maritimum* (C, D). Arrows indicate direction of extension growth.

Zalerion maritimum is a helicosporous Hyphomycete. In the light-microscopic, time-lapse sequence of conidiogenesis (Fig. 1.20A–H) the fertile cell first appears as a short, erect hypha (Fig. 1.20A) which swells apically and elongates (Fig. 1.20B, C). Septa are formed subapically as apical growth continues. A change in the direction of elongation of the young conidium from left (Fig. 1.20B, C) to right (Fig. 1.20D, E) subsequently occurs and is even more evident in Fig. 1.20F–H. We have interpreted this as resulting from an altered distribution of the secretory vesicles which contribute to wall intussusception at the cell tip (Fig. 1.21A–D). Girbardt (1957, 1973) observed that changes in the direction of hyphal growth were consistently preceded by changes in the position of the Spitzenkörper. Trinci and Collinge (1974a) also concluded that the precise location of a new branch arising from a hyphal compartment is decided by 'some event in the cytoplasm (e.g., vesicle accumulation) . . .'. Although the Spitzenkörper has only been seen in the tips of actively growing hyphae (Bartnicki-Garcia, 1973), a similar cluster of vesicles may also exist in the tip of developing helicoid conidia. Other factors, such as osmotic sensitivity at the growing apex (Robertson, 1958, 1959, 1965a, b; Park and Robinson, 1966; Hunsley, 1973; Bartnicki-Garcia, 1973), probably contribute to the shape of the mature conidium and require elucidation.

Considerable taxonomic use has been made of conidial shape in the Deuteromycetes (Talbot, 1971). Blastic conidia show the greatest variation in form and Ainsworth (1971) reviewed the descriptive terminology used for them. The shapes of thallic conidia are less variable, because of their more restricted modes of development, and primarily include globose to obovoid shapes (terminal and intercalary, holothallic conidia) and cylindrical cells (thallic-arthric conidia). Conidial appendages and surface ornamentations, as in Chapter 2, Fig. 2.14, provide additional variations in conidial form.

Karyological features of blastic development

The formation of blastic conidia can also be characterized by karyological events resembling those of hyphal tip growth. Early in development, when swelling and rapid synthesis of new wall materials occurs, the nuclei are located in a subapical region of the conidiogenous cell and not within the conidial initial. A similar phenomenon occurs in the elongating vegetative hypha. The nuclei later migrate into the newly formed cell at about the same time as the septum at the base of the conidium is initiated. These events have been reported by several authors (Kendrick and Chang, 1971;

Hashmi *et al.*, 1972; Cole, 1973a, b; Hammill, 1977a, c) and are demonstrated in *Cladobotryum varium* (Chapter 5, Fig. 5.14A–D).

The cytoplasm in the hyphal tip, and in blastic conidial initials, is also devoid of mitochondria and ER (Zalokar, 1959; McClure *et al.*, 1968; Girbardt, 1968; S. N. Grove *et al.*, 1970; S. N. Grove and Bracker, 1970). Bartnicki-Garcia (1973) suggested that the absence of these organelles indicates that their functions are 'of secondary importance to the main problem at hand: to establish a pattern of wall construction that will continuously transform the hemispherical surface of the apical dome wall into a cylindrical surface'.

Holoblastic and enteroblastic development

The members of the Kananaskis Conference (Kendrick, 1971a) also concluded that two kinds of blastic ontogeny occur in the Deuteromycetes. Either the outer layer(s) of the conidiogenous cell wall remain intact encompassing the conidium until it finally secedes (i.e., *holoblastic* ontogeny), or the conidiogenous cell wall ruptures at some stage during conidial development (i.e., *enteroblastic* ontogeny). Unfortunately, the distinction between these two mechanisms is not as obvious as originally suggested. For example, M. B. Ellis (1971a) thought that conidiogenesis in *Helminthosporium velutinum* Link ex Ficinus & Schubert, *Corynespora cassiicola* (Berk. & Curt.) Wei, and *Alternaria brassicae* (Berk.) Sacc. represents a special kind of enteroblastic development in which the new, innermost wall of the conidiogenous cell grows through a channel in the outer wall layer(s) and subsequently blows-out to form the outer wall of the conidium. Ellis (1971a, b) referred to this process of development as 'tretic'. However, Brotzman *et al.* (1975) concluded, on the basis of their ultrastructural examination of *Helminthosporium maydis* Nisikado & Miyake, that conidiogenesis in this species is *holoblastic*. A more detailed discussion of the controversy surrounding this mode of conidiogenesis is presented in Chapter 3. Confusion between holoblastic and enteroblastic development also arises in a discussion of phialidic and annellidic ontogeny (Chapter 4). Earlier investigations have consistently identified the formation of conidia from phialides as enteroblastic (Subramanian, 1971b) and from annellides as holoblastic (Cole and Aldrich, 1971a). However, certain inconsistencies exist in this distinction, which are discussed in Chapters 4 and 5.

Summary of features of blastic development

The concept of blastic development in conidial fungi clearly overlaps with that of hyphal tip growth. The conidiogenous cell is itself a differentiated part of a hypha, arising either terminally or laterally from the vegetative filament. The formation of each blastic conidium demonstrates a number of apparently common features.

1. *Blowing-out* of part of the conidiogenous cell (formation of successive conidia may involve percurrent and sympodial proliferation, retrogressive and basauxic development—Chapters 3–6).
2. The polarized growth and cytodifferentiation of the fertile region (conidiogenous locus) of the conidiogenous cell is similar to that of a hyphal tip.
3. Subsequent growth of the conidium may
   (a) be totally depolarized, producing a globose cell (i.e., involve the entire surface area of the conidial initial), or
   (b) remain polarized, producing a cylindrical cell (e.g., *Zalerion maritimum*), or
   (c) vary between the depolarized and polarized condition producing a myriad of conidial shapes.

In each case, early growth involves differentiation of a new wall layer (2°W) adjacent to the plasmalemma and inside the outer wall (1°W) which is continuous with the outer conidiogenous cell wall. The latter may remain intact and encompass the mature conidium (i.e., *holoblastic* conidial formation), or may rupture (i.e., *enteroblastic* conidial formation). However, the distinction between these two modes of development is often rather obscure.

4. Conidia are delimited by a septum at their base, but may go on growing—even after secession. Additional septa may be laid down above the basal septum.
5. Conidial secession occurs either by schizolytic splitting of the basal septum or rhexolytic rupture of a supporting cell (S. J. Hughes, 1971b, c; Chapter 2).

The remainder of this discussion comprises descriptions of categories accommodating developmentally similar conidial fungi. In Chapter 3, we examine the formation of solitary, botryose and catenulate conidia. An analysis of ultrastructural and developmental aspects of two specialized kinds of blastic conidiogenous cells, the phialide and the annellide, is then presented in Chapter 4. Similar investigations of retrogressive and basauxic conidiogenous cell development are presented in Chapters 6 and 7, respectively.

Thallic development

Conversion of hyphal elements into conidia

The term *thallic* is derived from Vuillemin's (1910b) reference to 'thallospores', which he defined as 'partie de l'appareil végétatif ou thalle; elles s'adaptent aux fonctions disséminatrices et conservatrices d'une façon secondaire et parfois incomplète'. In another paper Vuillemin (1910a) designated those conidia arising from the fragmentation of a hypha as 'arthrospores' and described the developmental process as follows: 'Coupés carrément à l'origine, ces tronçons peuvent s'arrondir et épaissir leurs parois. On peut voir une variété d'arthrospores . . . qui sans s'isoler, se distinguent des éléments actifs par un phénomène d'enkystement et d'accumulation de réserves nutritives. Leur formation épuise le filament délicat auquel elles survivent'. S. J. Hughes (1953) considered distinctive those fungi in which the conidia 'arise from the fragmentation of conidiophores of determinate length'. The participants in the Kananaskis Conference (Kendrick, 1971a) introduced the term 'thallic' to describe the development of what are now generally known as arthroconidia (Cole and Kendrick, 1969c), as well as certain kinds of solitary conidia. These solitary conidia have, unfortunately, given rise to some confusion in the concept of thallic development.

Formation of holothallic conidia

Carmichael (1971) pointed out that 'some conidia which fall into' the thallic 'group may swell after the septa are laid down and become large and multicellular. Single terminal conidia may also be produced in this way'. Most single, solitary conidia formed at the ends of fertile hyphae fit our concept of blastic development. However, in some cases, the formation of such terminal conidia, unlike blastic development, involves growth and conversion of cytoplasm and wall material of a large segment of the pre-existing hyphal branch (e.g., species of *Microsporum* Gruby, *Trichophyton* Malmsten, and *Epidermophyton* Sabour.; Chapter 7, Figs. 7.1–7.6). Similarly, formation of intercalary propagules of conidial fungi (e.g., intercalary *chlamydospores*) involve growth as well as conversion. The overriding developmental feature in thallic development is the differentiation of a *whole cell*, or hyphal element, into a conidium—not part of a cell, as in blastic ontogeny. There should be concomitant differences in protoplasmic organization and wall growth to distinguish the thallic and blastic modes. However, few ultrastructural data are available on thallic development, and we hope the following discussion will stimulate further investigation.

The developmental features evident during initiation of blastic conidia are not apparent during differentiation of thallic conidia. A diagrammatic interpretation of wall development during thallic conidial formation is presented in Fig. 1.22A–E. Cessation of apical growth probably signals initiation of a terminal holothallic conidium (Fig. 1.22A, B). This is followed by thickening of the inner wall layer (2°W) at the hyphal tip (Fig. 1.22C). The preceding events are unique to thallic ontogeny. Although subapical wall thickening occurs both in vegetative hyphae and during thallic conidiogenesis, the behaviour of the hyphal tip is diagnostic. Little is known of the cytological events associated with subapical deposition of wall precursors but evidence for its occurrence has been provided by light microscopy, autoradiography (Bartnicki-Garcia and Lippman, 1969; Katz and Rosenberger, 1970, 1971; Gooday, 1971; Gooday and Hunsley, 1971; Hunsley and Gooday, 1974), wall chemistry (Hunsley and Burnett, 1968; Hunsley, 1973) and electron microscopy (Heath and Greenwood, 1970; Hunsley and Burnett, 1970; Hunsley and Gooday, 1974). Hunsley and Burnett (1968) observed a subapical increase in the diameters of microfibrillar aggregates comprising hyphal walls of several fungi. They suggested 'that unit microfibrils are added during growth by a process of secondary intussusception' rather than by apposition of new microfibrillar lamellae as previously proposed (Roelofsen, 1950). This implies that the subapical region of the hypha is capable of some extension growth. Several authors have suggested that lomasomes (Moore and McAlear, 1961; Moore, 1965), multivesicular structures occurring between the plasmalemma and wall, may be associated with the synthesis and deposition of wall precursors (Wilsenach and Kessel, 1965b; R. Marchant *et al.*, 1967; R. Marchant and Robards, 1968; Heath and Greenwood, 1970; Weisburg and Turian, 1971; R. Marchant and Moore, 1973; Lin *et al.*, 1975; Hanlin, 1976; Happ *et al.*, 1976). Lomasomes frequently occur in

Fig. 1.22A–E Interpretation of protoplasmic and wall differentiation during formation of holothallic, terminal conidia (A–C) and intercalary conidia (D, E).

1.22

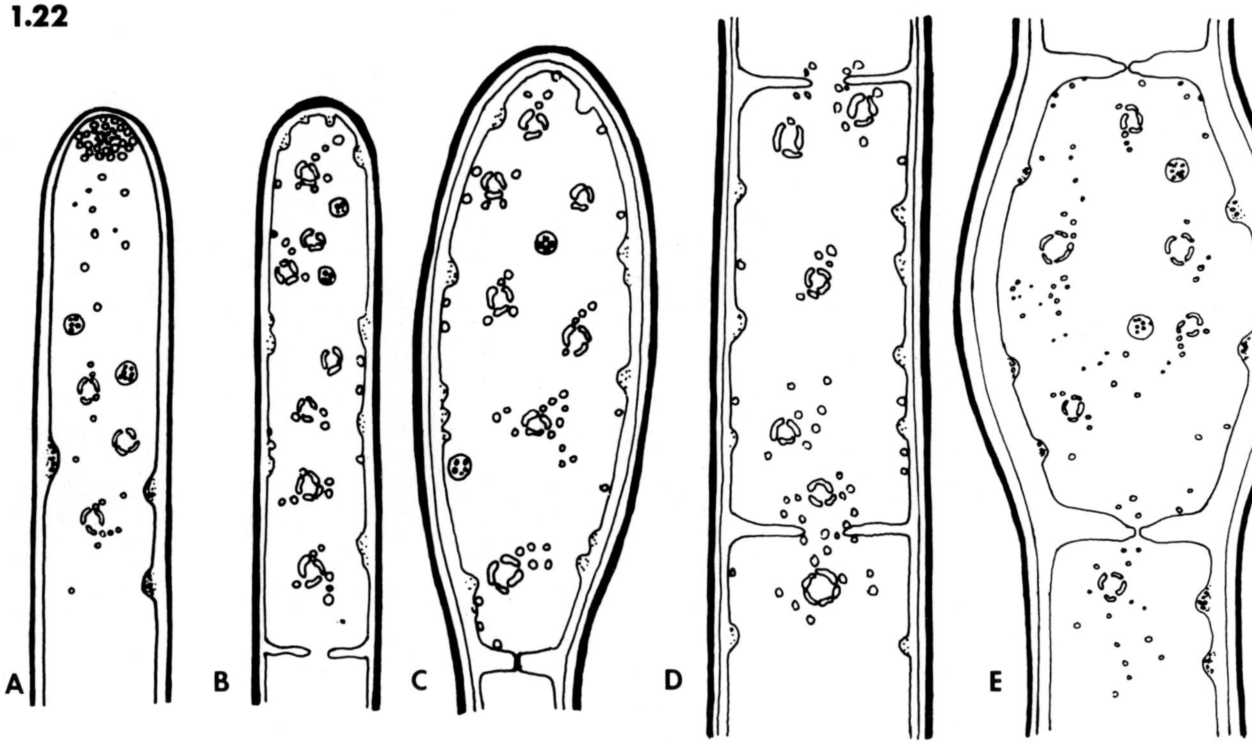

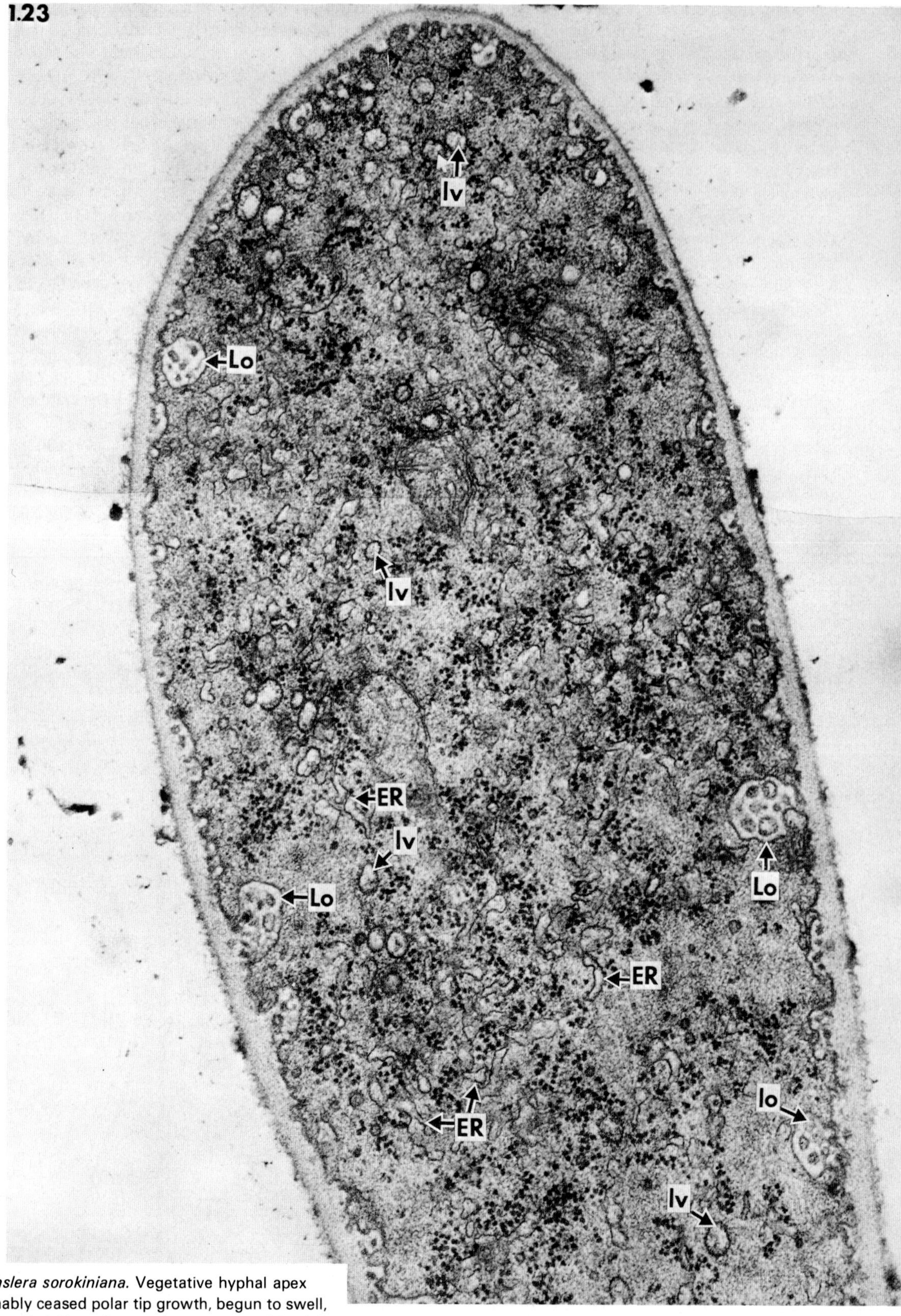

Fig. 1.23 *Drechslera sorokiniana.* Vegetative hyphal apex which had presumably ceased polar tip growth, begun to swell, and become thick-walled. Note dispersed secretory vesicles (lv), profuse ER cisternae (ER), and large number of lomasomes (Lo). ×48,000.

subapical regions of vegetative and fertile hyphae. However, it is now recognized that all paramural bodies in hyphae and conidia are not homologous (Moore, 1965), and the function of these structures should be interpreted cautiously (Heath and Greenwood, 1970; Brushaber and Jenkins, 1971; O'Brien, 1972). We suggest that the hyphal apex of *Drechslera sorokiniana* in Fig. 1.23 had ceased tip growth before fixation, since the apical vesicles have dispersed and several lomasomes are visible near the tip. The protoplasmic differentiation shown in this aborted vegetative hypha is comparable to that of a holothallic initial (Fig. 1.22B). The wall at the apex of a young conidium would later thicken rapidly by secondary intussusception (Fig. 1.22C). Similar events may accompany the development of an intercalary, holothallic conidium, or an intercalary chlamydospore (Fig. 1.22D, E).

Formation of thallic-arthric conidia

Another easily recognized mode of thallic development is fragmentation and disarticulation of determinate fertile hyphae into chains of conidia (Fig. 1.24), which have been termed 'arthric' (Kendrick, 1971b). After hyphal tip growth stops (Fig. 1.21A, B), the innermost wall layer thickens while transverse septa are formed (Fig. 1.24B). These septa become the end walls of the arthric conidia.

Two kinds of thallic-arthric development may be distinguished. We propose the term *holoarthric* for that in which all wall layers of the conidiogenous cell are involved in the formation of the conidial wall (Fig. 1.24A, B, F), and the term *enteroarthric* for that in which the outer wall of the conidiogenous cell usually separates from the newly formed wall of the conidium (Fig. 1.24A–E). Further discussion of thallic-arthric conidiogenesis is presented in Chapter 7.

Summary of features of thallic development

One aim of these introductory remarks is to establish that two main categories of conidium ontogeny, blastic and thallic, exist in the deuteromycetous fungi. Examples of each process are provided in the appropriate chapters. However, at this point it is useful to summarize data presented in Chapter 7 on thallic development, for comparison with the synopsis of blastic development given on p. 12.

Fig. 1.24A–F Interpretation of wall differentiation during formation of holoarthric (A–C, F) and enteroarthric conidia (A–E).

1.24

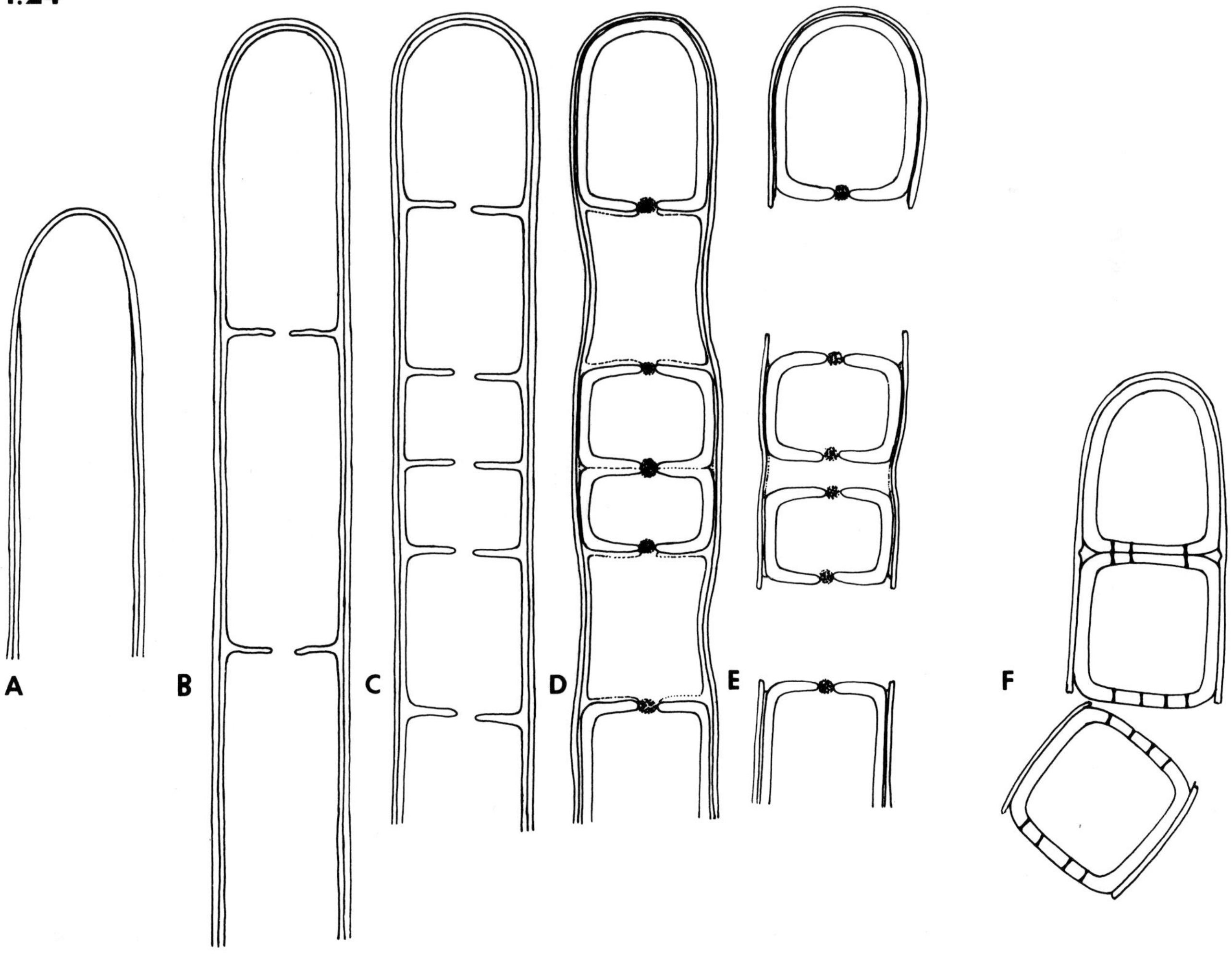

1. Conversion and disarticulation of whole segments of determinate fertile hyphae producing a solitary conidium (terminal or intercalary) or chains of conidia.

2. Conidiogenesis is signalled by (a) cessation of 'polar growth' and (b) thickening of the inner wall and the development of the conidial septum (septa). Differentiation of solitary conidia involves all wall layers of the fertile cell (*holothallic*). Formation of chains of conidia (*thallic-arthric* mode) may be *holoarthric* or *enteroarthric* and septation may be acropetal, basipetal or random.

3. Wall thickening may occur uniformly along the hypha during conidium differentiation, (i.e., holoarthric ontogeny; Fig. 1.24A, B, F), or a variation in wall thickness may appear between cells of the chain and the thin-walled cells degenerate while thick-walled cells continue to develop (i.e., enteroarthric ontogeny; Fig. 1.24A–E). All cells delimited by septation of the fertile hypha do not necessarily become conidia.

4. Delimited segments of fertile hyphae can become septate even after secession.

5. Variation can occur in the degree of secondary intussusception associated with formation of conidial septa (e.g., *Oidiodendron*).

6. Thallic conidia demonstrate variations in their tendency to secede. Adjacent holoarthric conidia separate by schizolytic splitting of their septa. Secession of holothallic and enteroarthric conidia in many species is rhexolytic but may also be passive.

7. Thallic-arthric conidia in a few species may bud (holoblastic ontogeny) while still part of the conidial chain (e.g., *Trichosporon*) giving rise to a new conidium or hypha (vegetative or fertile).

A critical aspect of both blastic and thallic development is the differentiation of a septum (or septa) which, in most cases, will eventually be involved in secession. Recent interest in the morphogenesis of fungal septa has yielded much new information. Relevant aspects of these studies are presented in Chapter 2.

In concluding these introductory remarks, a diagrammatic summary of the major processes of conidial and conidiogenous cell development demonstrated by the Hyphomycetes and examined in this text is presented in Fig. 1.25. The blastic and thallic modes of conidiogenesis have been separated into developmental categories based on differences in (1) details of wall differentiation during conidium ontogeny, (2) order of production and (3) aspects of conidiogenous cell proliferation. The adjectival terms which are used to denote each ontogenetic mechanism are defined in the following chapters. These have replaced earlier terms which, in many cases, did not adequately describe the developmental processes.

Fig. 1.25 Diagrammatic summary of modes of conidial and conidiogenous cell development examined in text. The following symbols are used to denote the three major criteria which distinguish the developmental categories: CW, conidial wall differentiation; OA, order of production and arrangement of conidia on conidiogenous cell; CC, process of conidiogenous cell proliferation. Classification of ontogenetic processes based on modes of conidial wall differentiation (CW) under blastic and thallic development is indicated on the diagram. Additional subdivisions and a list of the representative species illustrated in Fig. 1.25 are presented below.

CW: HOLOBLASTIC

(A, B) *Nigrospora sphaerica.* OA: Terminal, Solitary; CC: Determinate.
(C–F) *Tritirachium oryzae.* OA: Terminal, Solitary; CC: Proliferous (Sympodial).
(G, H) *Botrytis cinerea.* OA: Synchronous, Botryose; CC: Determinate.
(I, J) *Gonatobotryum apiculatum.* OA: 1° Conidia Synchronous, Botryose; 2° Conidia Asynchronous, Catenulate; CC: Proliferous (Percurrent).
(K–M) *Deightoniella torulosa.* OA: Terminal, Solitary; CC: Porogenous.
(N–P) *Spegazzinia tessarthra.* OA: Terminal, Solitary; CC: Basauxic.

CW: ENTEROBLASTIC

(A–C) *Phialophora lagerbergii.* OA: Basipetal, Solitary; CC: Phialidic.
(D–G) *Scopulariopsis brevicaulis.* OA: Basipetal, Solitary; CC: Proliferous (Annellidic).
(H–K) *Cladobotryum varium.* OA: Basipetal, Solitary; CC: Retrogressive.

CW: HOLOTHALLIC

(A–C) *Microsporum gypseum.* OA: Terminal, Solitary; CC: Determinate.

CW: HOLOARTHRIC

(D–F) *Geotrichum candidum.* OA: Random, Chains; CC: Determinate.

CW: ENTEROARTHRIC

(G–J) *Sporendonema purpurascens.* OA: Random, Chains; CC: Determinate.

1.25

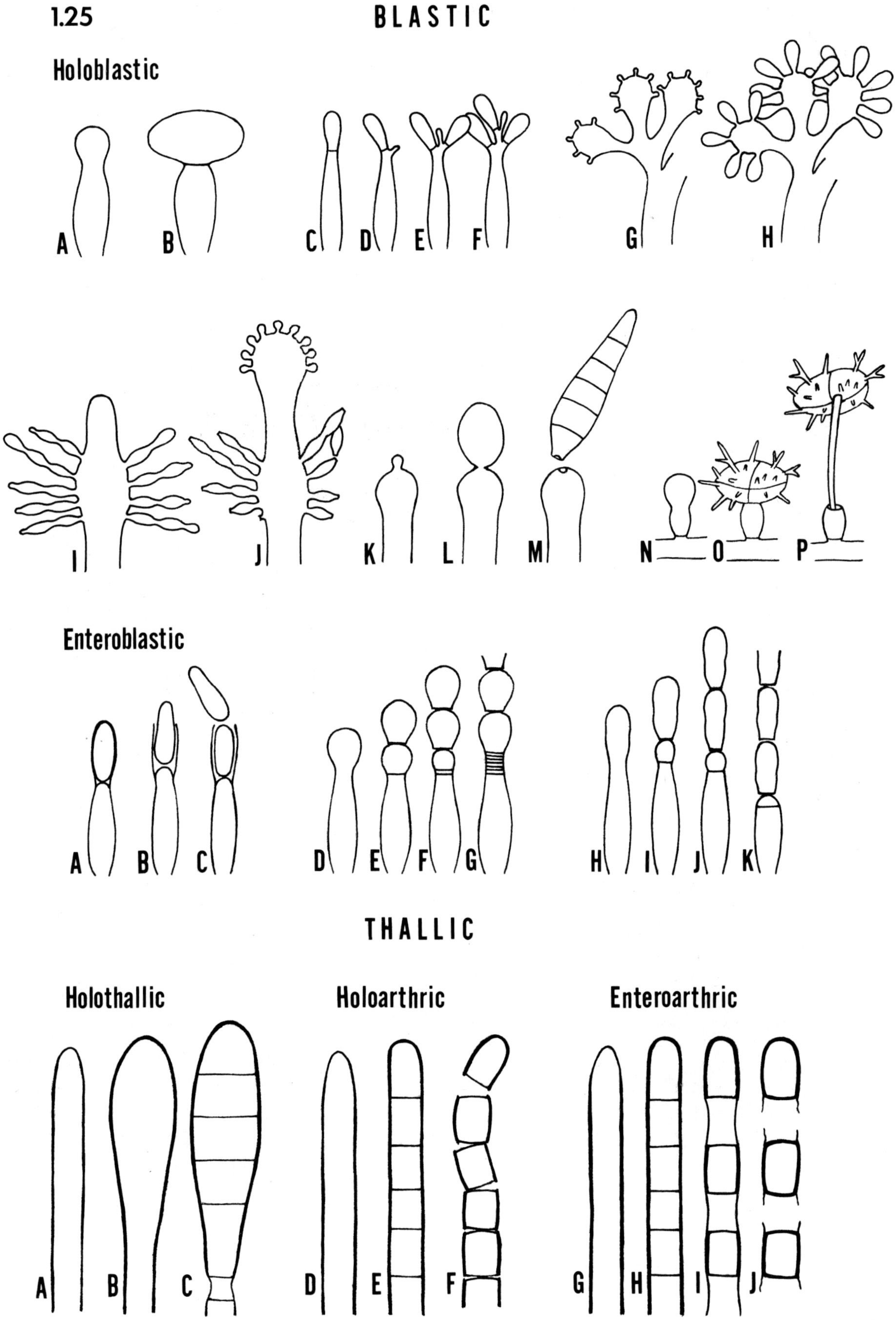
BLASTIC
Holoblastic
A
B
C
D
E
F
G
H
I
J
K
L
M
N
O
P
Enteroblastic
A
B
C
D
E
F
G
H
I
J
K
THALLIC
Holothallic
Holoarthric
Enteroarthric
A
B
C
D
E
F
G
H
I
J

2 Conidial Septation and Secession

Present concepts of conidiogenesis are based largely on aspects of wall differentiation revealed by thin-section examinations. These studies have also demonstrated that different patterns of conidium ontogeny are associated with distinct processes of septum formation. Bracker and Butler (1963) pointed out: 'Knowledge of the structure of fungous septa is a prerequisite for complete understanding of certain biological phenomena in the septate fungi . . . the mechanisms of nuclear migration, or . . . the translocation of protoplasm and nutrients from one part of a fungous thallus to another . . . the nature of the septum forms a basis for the concept of the cell in fungi, and may aid in developing theories of phylogeny'.

Initiation of septa

It has been suggested that septum formation in all taxonomic groups of fungi has basic similarities (Buller, 1933; Burnett, 1968; Hunsley and Gooday, 1974). The development of septa in conidia is probably similar to that in vegetative hyphae, about which more is known because they are easier to prepare for thin-sectioning. The first indication of septation is circumferential invagination of the plasmalemma and thickening of the adjacent inner wall layer (Fig. 2.1). In some

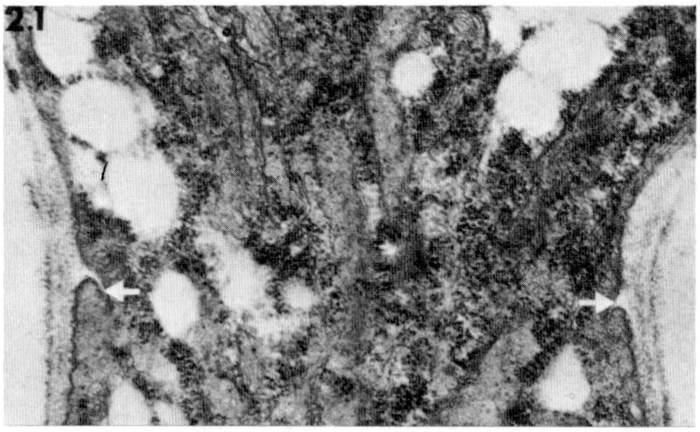

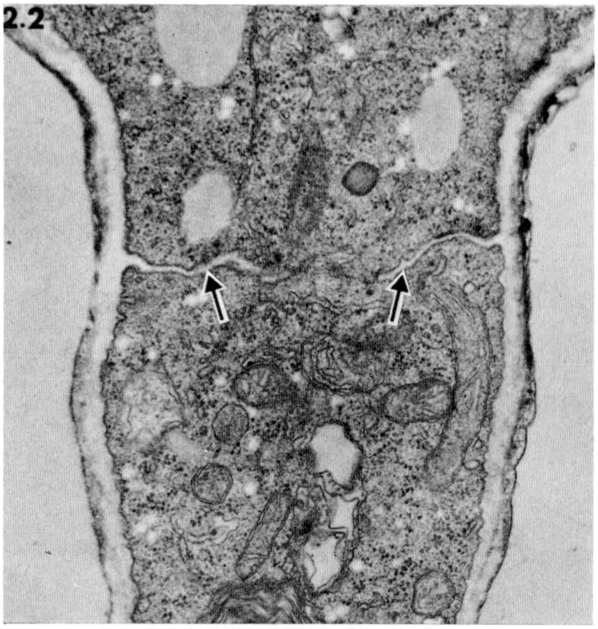

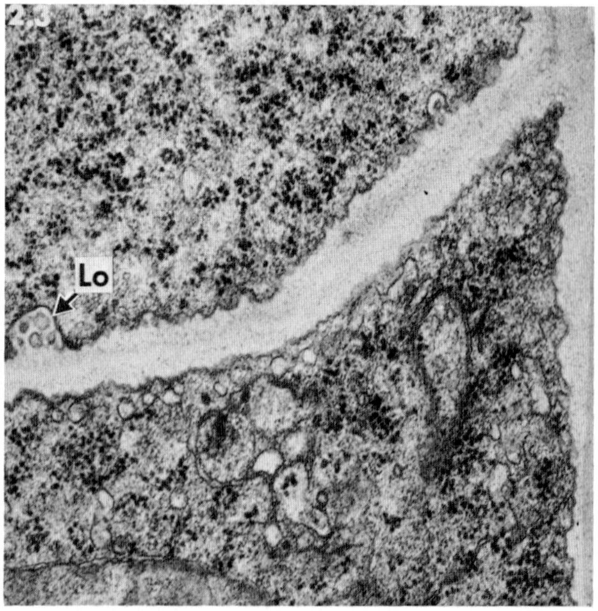

Figs. 2.1–2.3 Stages of septal development. 2.1. Invagination of plasmalemma (arrows) in *Cladobotryum varium* signals initiation of septation. 2.2. Thin, undulated cross-wall (arrows) at base of a conidium in *Scopulariopsis brevicaulis*, typical of early stage of septal development. 2.3. Later stage demonstrated by *Drechslera sorokiniana*. Note lomasomes (Lo) and vesicles adjacent to plasma membrane. ×26,000, 29,400 and 42,000 respectively.

fixation procedures, thin fibrils resembling microfilaments (Allen and Kimiya, 1964; Parthasarathy and Mühlethaler, 1972) are found in the cytoplasm near the infolded cell membrane (Hunsley and Gooday, 1974; Benny and Aldrich, 1975) and may be instrumental in its invagination. Although vesicles are sometimes observed near developing septa (Ehrlich *et al.*, 1968; Cortat, 1971; Heath *et al.*, 1971; Hashimoto *et al.*, 1973), their involvement in septal growth has not been confirmed. The mode of formation of the thin, centripetally developing, undulated diaphragm (Fig. 2.2), which will become the septum, is unknown. However, it appears that centripetal growth is completed before thickening by apposition occurs. This second developmental state (Fig. 2.3) probably involves lomasomes such as that shown in Fig. 2.3 (Happ *et al.*, 1976).

Kinds of septa

From an ultrastructural perspective, three groups of fungal crosswalls are recognized: partial, complete and perforate (uniperforate and multiperforate) septa. Partial septa are simply considered as thickenings of the lateral walls (Bracker, 1967) and may, in some cases, represent abortive crosswalls. In the case of complete septa, centripetal growth continues until the ingrowing walls meet, leaving only the finest of pores, too narrow even to permit the passage of organelles (Figs. 3.55, 4.24). The plane of sectioning is critical in demonstrating the presence of this channel with the electron microscope. The fate of the plasmalemma at the centre of this kind of septum is unclear (Kreger-van Rij and Veenhuis, 1974). Complete septa occur at the base of phialoconidia (Chapter 4) in several Deuteromycetes with ascomycetous affinities (Fletcher, 1971; Hammill, 1972a; Carroll and Carroll, 1974b; R. Marchant, 1975; Hanlin, 1976). On the other hand, septa at the base of the conidiogenous cells (phialides) and in the vegetative hyphae of these same species are perforated by a much wider central pore (Cooper *et al.*, 1973; Hanlin, 1976). Representative asexual states in the Uredinales produce complete septa as well as perforate septa (Moore, 1963; Boyer and Isaac, 1964; Bracker, 1967; Manocha and Shaw, 1967; Ehrlich *et al.*, 1968; Coffey *et al.*, 1972; Ehrlich and Ehrlich, 1969; Jones, 1973; Littlefield and Bracker, 1971; Robb *et al.*, 1973). It is interesting and perhaps phylogenetically significant that phialoconidia have been reported only in the Ascomycetes and Uredinales (S. J. Hughes, 1970). Complete septa also develop at the base of holoblastic conidia formed at successive apices of sympodially proliferating conidiogenous cells (Hammill, 1973b; Chapter 3). This mode of septation may also be associated with other kinds of conidiogenesis and has been reported among the Hemiascomycetes (Bracker, 1967), since there is a limited number of ways in which septa can be laid down. Similarities even exist between the formation of complete septa in fungi and certain groups of bacteria (Moore, 1965; Burdett and Murray, 1974; Decker and Maier, 1975; Gilleland and Murray, 1975; Stevens, 1975).

However, much more information is available from studies of uniperforate septa in ascomycetous and basidiomycetous

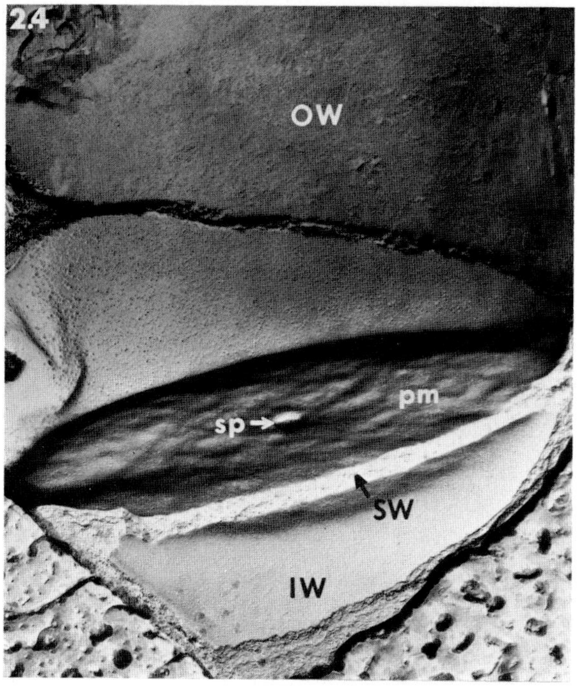

Fig. 2.4 *Scopulariopsis brevicaulis.* Freeze-fracture of vegetative hypha revealing outer surface (OW) and inner surface (IW) of lateral wall, broken portion of septal wall (SW), and plasma membrane (pm) which is perforated by a centrally located septal pore (sp). ×24,000.

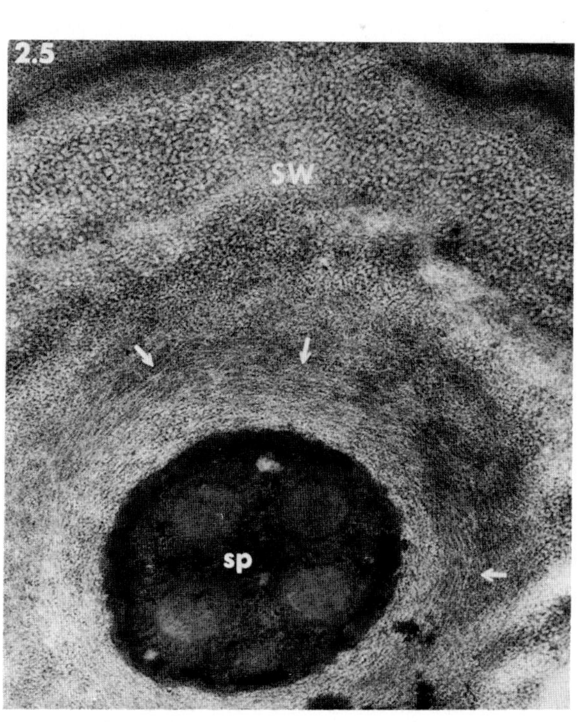

Fig. 2.5 *Drechslera sorokiniana.* Thin section through septal pore (sp) of vegetative hypha showing microfibrillar *collar* (arrows). ×30,400.

fungi, and these data have been frequently used to reinforce concepts of fungal phylogeny (Moore and McAlear, 1962; Moore, 1965, 1971, 1972; Demoulin, 1974; Kohlmeyer, 1975). The typical septum of Ascomycetes and related deuteromycetous fungi is a simple crosswall with one central pore (Figs. 2.4 and 2.5). The freeze-etch micrograph of a fracture through a vegetative hypha of *Scopulariopsis brevicaulis* (Sacc.) Bain. in Fig. 2.4 reveals the cell membrane with a central pore (most of the septum has broken away). In Fig. 2.5, concentric rings of microfibrillar material are visible surrounding the pore of *Drechslera sorokiniana* (Sacc.) Subram. & Jain. Studies of isolated septa have revealed a 'collar' around the pore (Hunsley and Gooday, 1974) composed of an annular arrangement of microfibrils (Kitajima *et al.*, 1976), though it is not yet known whether the collar is present in all ascomycetous septa. Uniperforate septa develop during conidiogenesis in many of the fungi examined in the following chapters. In contrast to the structurally simple crosswall of Ascomycetes, the septum of many Basidiomycetes (Homobasidiomycetidae and Tremellales) is typically more elaborate and has been referred to as the 'dolipore septum' (Girbardt, 1958, 1961, 1973; Moore and McAlear, 1962; Bracker and Butler 1963, 1964; Wells, 1964; Moore, 1965; Willsenach and Kessel, 1965a; Bracker, 1967; Jersild *et al.*, 1967; Niederpruem and Wessels, 1969; T. T. Ellis *et al.*, 1972; Moore and R. Marchant, 1972; Thielke, 1972). However, it is now evident that between the platelike ascomycetous septum and the dolipore septum of basidiomycetous fungi many intermediate configurations exist, some of them demonstrated by primitive Basidiomycetes (Dykstra, 1974b; Heath, 1975), including Uredinales (references cited above) and Ustilaginales (Moore, 1971, 1972). Also, the Ascomycetes *Endomycopsis monospora* Saito and *E. platypodis* Baker & van Rij (Kreger-van Rij and Veenhuis, 1969) have elaborate, centrally thickened septa considered similar to the dolipore septum. These same authors (Kreger-van Rij and Veenhuis, 1972) later illustrated a distinctive multiperforate septum in the Ascomycete *Endomyces* Rees. In this case, the crosswalls show several narrow pores (Wilsenach and Kessel, 1965c; Bracker, 1967; Hashimoto *et al.*, 1973; H. J. Marchant, 1976). Illustrations of multiperforate septa are presented in Chapter 7 (Figs. 7.9–7.11) in connection with the discussion of holoarthric conidiogenesis in *Geotrichum candidum* Link ex Pers. (anamorph of *Endomyces geotrichum* Butler & Petersen). Although septa involved in conidiogenesis in this species are multiperforate, the crosswalls of vegetative hyphae are uniperforate (Figs. 2.6, 2.8, 2.9)—additional evidence for an apparent plasticity of septal development in the fungi. Other species of Deuteromycetes which undergo a similar thallic-arthric mode of conidiogenesis have uniperforate septa (Kreger-van Rij and Veenhuis, 1971a; Cole, 1975a). Although multiperforate septa have been reported in several additional species (Takada *et al.*, 1965; Reichle and Alexander, 1965; Hawker *et al.*, 1966; Wetmore, 1973), they are probably not widespread in the fungi. The lower fungi also demonstrate variations in septal development (Farr and Lichtwardt, 1967; Young, 1969a; Reichle and Lichtwardt, 1972; Lichtwardt, 1973; Benny and Aldrich, 1975).

The phylogenetic implications of septal ultrastructure are uncertain. Heath (1975) suggested that variations in septal structure may simply reflect a 'physiological peculiarity of the species concerned' and warned that phylogenetic conclusions should not be drawn from limited ultrastructural data. However, comparative studies of septa have also led investigators to consider the involvement of crosswalls and pore complexes in hyphal and conidial development, and some of these data are discussed next.

Functions of septal pores

Valves: septal plugging

Undoubtedly one of the principal functions of fungal septa is to prevent catastrophic loss of cytoplasm when the mycelium is damaged. When the hyphal wall is ruptured, the septal pore is quickly plugged, usually by an organelle that is slightly larger than the pore and is termed a Woronin body. These organelles are also found plugging the basal septal pore of mature conidia (Chapter 3, Fig. 3.77; Chapter 4, Figs. 4.82, 4.83; Cole and Aldrich, 1971a; Cole, 1973a, b). Woronin bodies, first seen under the light microscope (Buller, 1933; Carroll, 1966), and later examined with the electron microscope (Bracker, 1967; McKeen, 1971; Hammill, 1974b) are spheroidal, electron-opaque and membrane-bounded (Fig. 2.6). Woronin bodies develop within microbodies as homogeneous, electron-opaque inclusions, then move into evaginations of the latter and are pinched off by constriction and fusion of the encompassing membrane. Stages in this development are shown in Fig. 2.7. Wergin (1973) and Heath (1977) have pointed out that amorphous, paracrystalline and crystalline inclusions of microbodies found in animals, plants (Frederick *et al.*, 1975; Vigil, 1973), and fungi (Avers, 1971) have been credited with enzymatic activity, and Woronin bodies may possess similar activity. We will return to the implications of this proposal later.

A second kind of *septal plug* was seen by Trinci and Collinge (1973) and Hammill (1974b) who described gradual occlusion of septal pores in *Neurospora crassa* Shear & Dodge and *Trichoderma saturnisporum* Hammill, respectively, by deposition of electron-opaque material. Hunsley and Gooday (1974) noted the presence of a fibrous, annular, electron-opaque structure constricting the pore in *N. crassa.* A similar structure seen in *Geotrichum candidum* is shown in Fig. 2.8.

Trinci and Collinge (1974a) described a third kind of septal plug in *N. crassa* in the form of hexagonal crystals. Similar hexagonal, membrane-bounded bodies plug the septal pores of vegetative hyphae in *Whetzelinia sclerotiorum* (Lib.) Korf & Dumont (Hoch and Maxwell, 1974) and *Geotrichum candidum* (Fig. 2.9). Hoch and Maxwell (1974) concluded that the hexagonal inclusions 'are primarily proteinaceous in nature and are similar in composition to the Woronin bodies observed in other Ascomycetes'.

These data support the concept that two basic mechanisms of septal plugging—one gradual and programmed, the other a quick emergency measure—exist in fungi with uniperforate septa (Trinci and Collinge, 1974a). The first process, the gradual occlusion of a septal pore by deposition

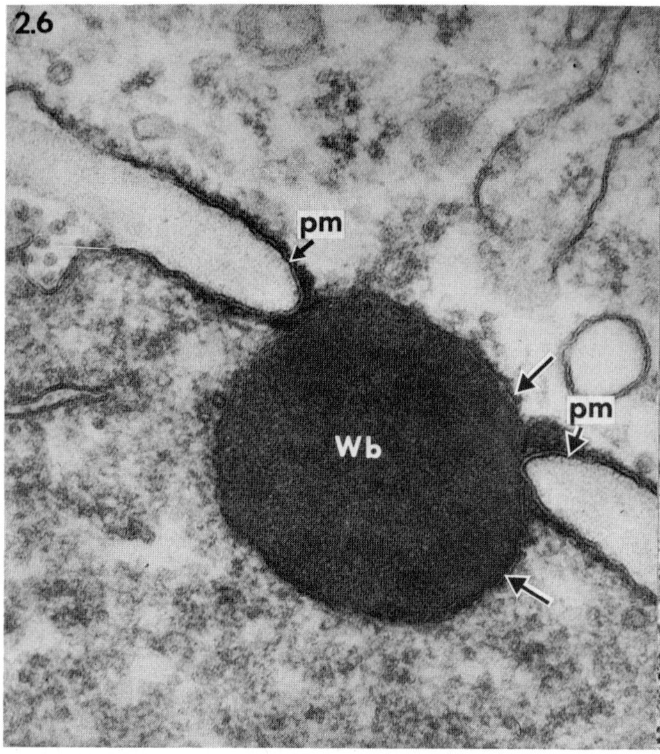

Fig. 2.6 *Geotrichum candidum.* Woronin body (Wb) in the septal pore (sp) of vegetative hypha. Arrows indicate distinct membrane encompassing Woronin body; pm, plasma membrane. ×54,400.

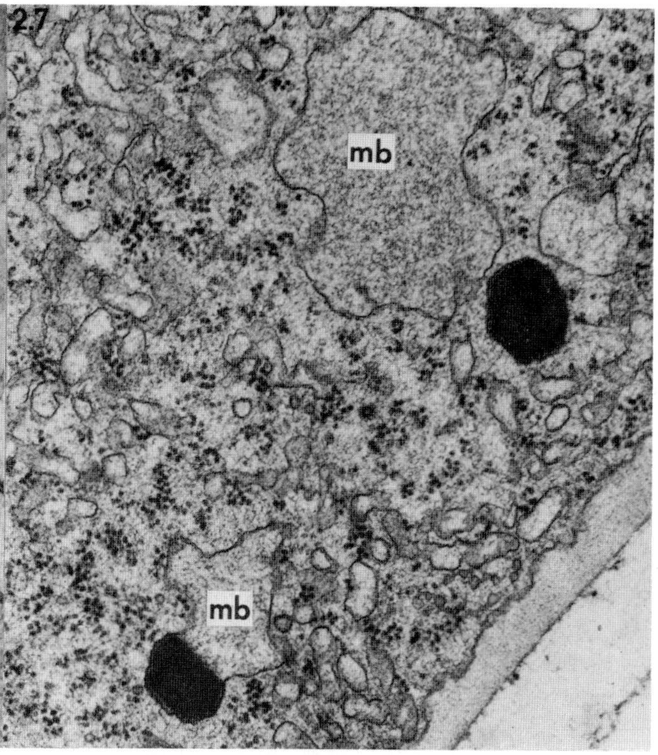

Fig. 2.7 *Drechslera sorokiniana.* Structural association of Woronin bodies and microbodies (mb). ×44,100.

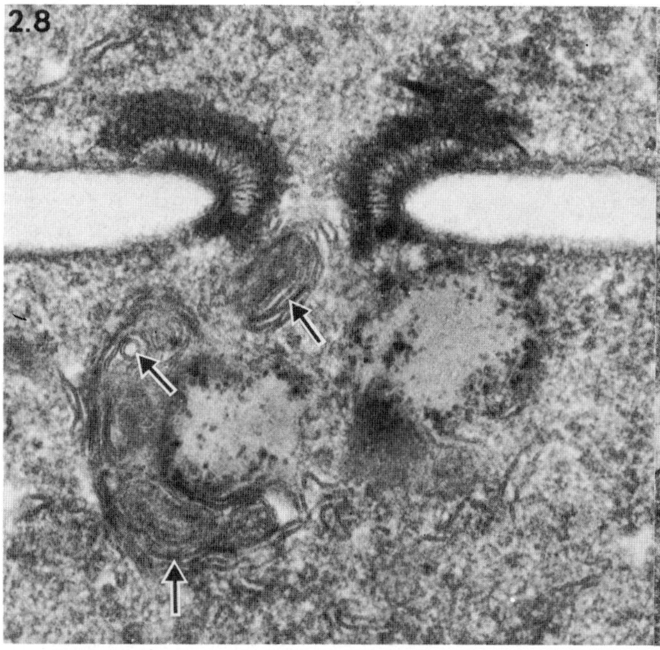

Fig. 2.8 *Geotrichum candidum.* Unusual septal pore complex of vegetative hypha showing electron-opaque pore lining and inner, fibrous-like strands. Arrows point to membranous configurations indicative of cytoplasmic degeneration. ×85,000.

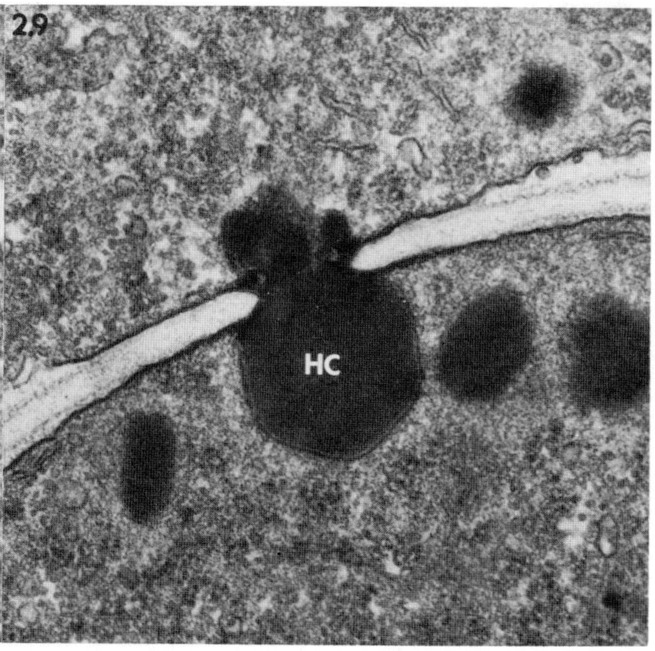

Fig. 2.9 *Geotrichum candidum.* Hexagonal crystal (HC), apparently membrane bounded, has plugged septal pore of vegetative hypha. ×45,000.

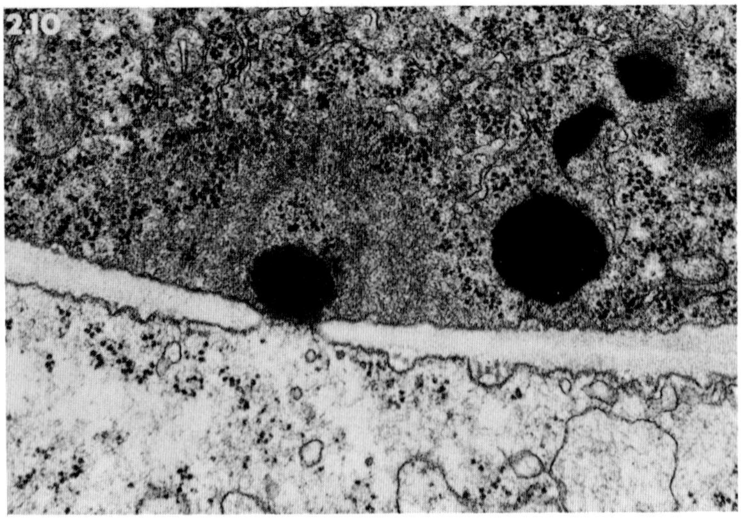

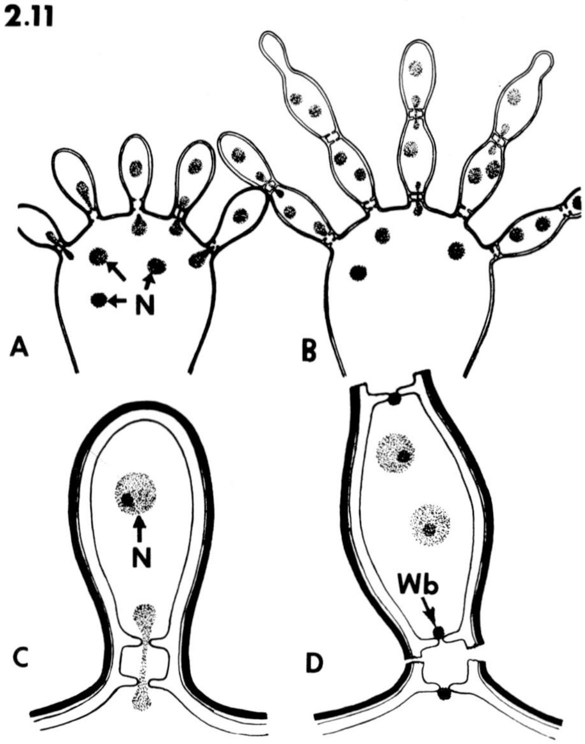

Fig. 2.10 *Drechslera sorokiniana.* Septal pore of vegetative hypha plugged by Woronin body and surrounded by ribosome-free cytoplasm. Note difference in cytoplasmic density on either side of septum. ×39,000.

Fig. 2.11A–D Interpretation of nuclear migration pattern and septal development during formation of acropetal chains of conidia in *Gonatobotryum apiculatum* (A–C), and arrangement of plugs at the time of conidial secession. N, nucleus.

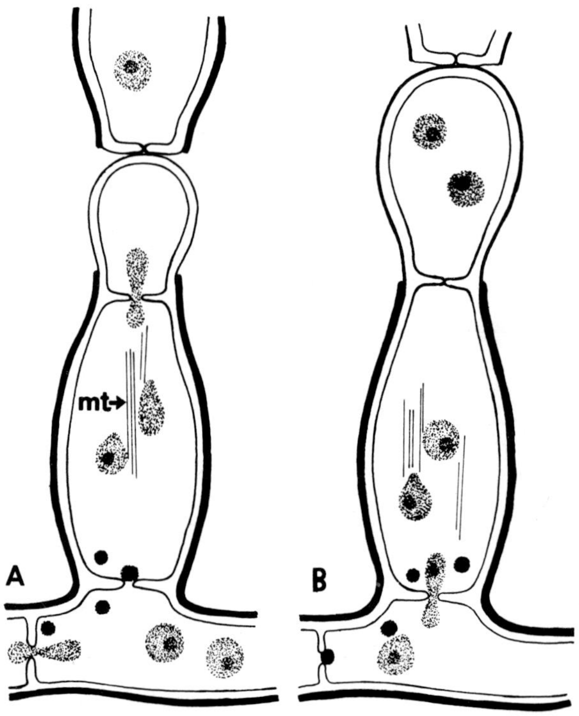

Fig. 2.12A, B Diagrammatic interpretation of nuclear migration pattern and alternate plugging and opening of pore in basal septum of phialide during formation of successive conidia. Mt, microtubules.

of plugging material, results from slow physiological changes which may be caused by senescence or adverse conditions such as nutrient deficiency (in Fig. 2.8, atypical membrane configurations [arrows] indicate senescence—Carroll and Carroll, 1973). In a variation of the above, the septal pore is initially sealed by a Woronin body (or similar-sized organelle) which is later cemented in position by unidentified deposits. Regulation of secondary deposition may involve enzymes released from the Woronin body itself. In Fig. 2.10, a Woronin body has plugged a septum between a normal and a senescent cell. Although the region surrounding the Woronin body is virtually ribosome free, the reason for this is unknown. This kind of gradual occlusion has been demonstrated in seceded conidia (Chapter 4, Fig. 4.82; Campbell, 1970; Cole and Aldrich, 1971a; Hammill, 1972d; Reisinger, 1972) which have a distinct pore in their basal septum. Conidia produced by sympodially proliferating conidiogenous cells have a *complete* basal septum, pierced only by a very fine pore which is also gradually sealed by electron-opaque material in a programmed manner.

The second *emergency* mechanism involves rapid plugging by Woronin bodies, hexagonal crystals and other cytoplasmic inclusions which are presumably thrust into position by the sudden change in cell turgor pressure caused by rupture of the cell wall. This mechanism probably occurs normally in acropetally developing chains of conidia (e.g., *Gonatobotryum apiculatum* [Peck] Hughes, Fig. 2.11A–D) because cytoplasmic continuity must be maintained throughout the growth of such chains and only quick-acting plugs can control the flow of cytoplasm from damaged or disarticulating conidia (Fig. 2.11D). Thin-section examinations of *G. apiculatum* have shown Woronin bodies lodged in the septal pores at the base of seceded conidia (Cole, 1973b; Chapter 3, Fig. 3.77).

Both gradual and rapid plugging mechanisms appear to be associated with thallic development. The formation of arthroconidia in *Geotrichum candidum* involves the differentiation of multiperforate septa and the pores are gradually occluded by electron-opaque material. On the other hand, uniperforate septa of vegetative hyphae of this same species may be quickly plugged by Woronin bodies or hexagonal inclusions (Figs. 2.6 and 2.9, respectively). Little ultrastructural information is available on septal plugging during holoarthric or enteroarthric conidial development. Although the scanning electron microscope has shown conidia of species of *Peniophora* Cooke, *Sporendonema* Desm. ex Fr. and *Coremiella* Bubák & Krieger (Chapter 7) to have uniperforate septa, details of septal occlusion are lacking.

Translocation and organelle migration

Septa and septal pores permit cytoplasmic streaming between cells. Karyological studies of blastic (Chapter 1) and thallic (Chapter 7) ontogeny have indicated that nuclear movement between vegetative hyphae, conidiogenous cells and conidia is largely determined by septal ontogeny. During thallic-arthric conidiogenesis, the timing of septal occlusion is critical in assuring that each conidium receives at least one nucleus before the fertile hypha fragments (Chapter 7, Fig. 7.58). Ultrastructural examinations of blastic conidia show that development of the delimiting septum begins before the nucleus, or nuclei move into the initial. The septal pore thus remains sufficiently large, at least in this stage, to permit nuclear migration. Thin sections of septa showing nuclei in transit (Bourne, 1968; Hunsley and Gooday, 1974) demonstrate the capacity of nuclei to constrict and pass through a narrow pore. Evidence from examinations of basidiomycetous fungi indicates that extensive intercellular migration of nuclei occurs in vegetative hyphae (Giesy and Day, 1965; Snider, 1968; Raudaskoski and Koltin, 1973; Mayfield, 1974). The movement of these and other organelles has been linked to microfilaments containing contractile proteins (Adelman and Taylor, 1969; Nachmias and Ingram, 1970; Nachmias *et al.*, 1970; Raudaskoski, 1970), microfibrils (Cole, 1972; McKeen, 1972) and microtubules (Girbardt, 1968; Hammill, 1977c, d) which are often juxtaposed to the nuclei and parallel to the long axis of the hypha (Chapter 1, Fig. 1.7). Similar structural associations occur in conidiogenous cells producing basipetal successions of blastic conidia at their apices (Fig. 2.12A, B). One anomaly of this process is that the septal pore at the base of aerial fertile cells is frequently plugged, apparently preventing migration of nuclei and other organelles from the vegetative hypha (Fig. 2.12A). This has been demonstrated in species of *Trichoderma* Pers. ex Fr. (Hammill, 1974a, b) and *Aspergillus* Mich. ex. Fr. (Hanlin, 1976). In each case, the cytoplasm of the phialide appeared normal, showing no signs of organelle degeneration. If the fertile apex gives rise to a large number of conidia in rapid succession (e.g., *Aspergillus* spp.), new material must be available to compensate for the loss of cytoplasm as each conidium is delimited. These fertile cells must be supplied through the pore in their basal septum. The septal plug in Fig. 2.12A cannot be fixed in position—it must be periodically dislodged to permit translocation of cytoplasmic material. A simple fluctuation in turgor pressure of adjacent cells may provide a mechanism for opening and closing the septal pore (Fig. 2.12B).

Even when the pore has been firmly sealed by gradual deposition of material, dissolution of the plug may still occur by localized hydrolytic enzyme activity. Septal degradation has been observed in the Basidiomycetes where it is suggested to be associated with nuclear migration in vegetative hyphae (Raudaskoski and Koltin, 1973; Mayfield, 1974). Vesicles which often accumulate near these septa may be the source of hydrolytic enzymes (R. Marchant and Wessels, 1973). A similar phenomenon occurs in some Ascomycetes.

Differentiation of multiseptate conidia

Early stages in development of septa subdividing conidia are as described above. As thickening and pigmentation of these crosswalls occurs, a thin electron-translucent zone is visible between darker staining layers (arrowheads in Fig. 2.13). At the periphery of the septum, adjacent to the outer wall layer(s) of the conidium, a slight broadening of the translucent zone is visible. In darkly pigmented fungi the septal layers on either side of the translucent zone become more heavily pigmented as the conidium matures (Campbell, 1968; White *et al.*, 1973; Harvey 1974a, b; Reisinger and Kiffer, 1974). Reisinger and Guedenet (1968) recognized the region of wall separation adjacent to the outer conidial wall layer(s) as a 'cavité intercellulaire'. This may be the site of lytic enzyme activity in mature propagules. For example, in *Spegazzinia tessarthra* (Berk. & Curt.) Sacc., the outer wall of the young conidium is intact (Chapter 6, Fig. 6.1A, B), but the mature conidium reveals deep cleavage furrows between the four cells (Fig. 2.14). Each cell is capable of forming one or more germ tubes (Cole, 1974), suggesting physiological independence. Cytoplasmic continuity is probably maintained between cells of conidia during early stages of development (Fig. 2.15) but the pores are later sealed by Woronin bodies (Fig. 2.13). On the other hand, the outer wall layer(s) of most multicellular conidia remain intact (Fig. 2.13). Although the intercellular spaces are also present in these latter conidia, any wall degradation which occurs apparently does not affect the outer wall layer.

Luttrell (1963) recognized two major groups of multicellular conidia which he referred to as *euseptate* and *distoseptate.* He stated that 'euseptate conidia are surrounded by a single wall and have true septa formed as inward extensions of the lateral wall'. Our concept of wall differentiation during the formation of a euseptate conidium is illustrated in Fig. 2.16A, B and exemplified by *S. tessarthra* (Fig. 2.14). This differs from Luttrell's (1963) and later interpretations (Mangenot and Reisinger, 1976) in that the conidia have two wall layers when the septa are initiated and only the inner wall layer is continuous with the

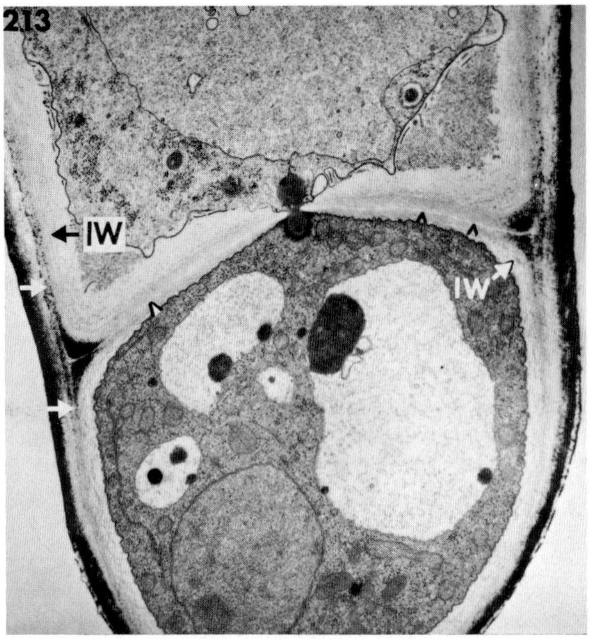

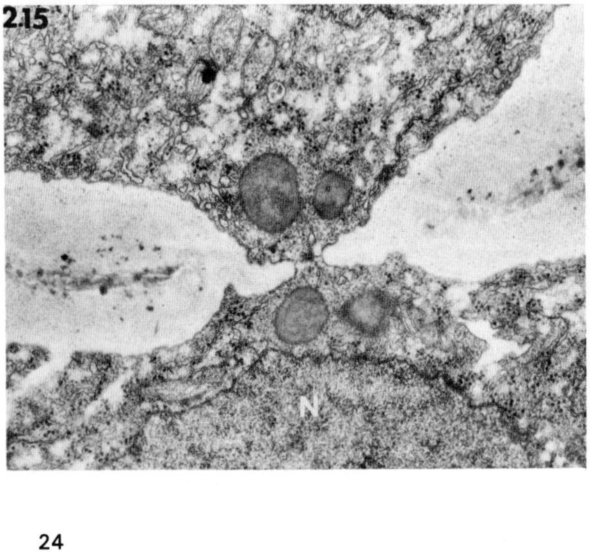

2.16

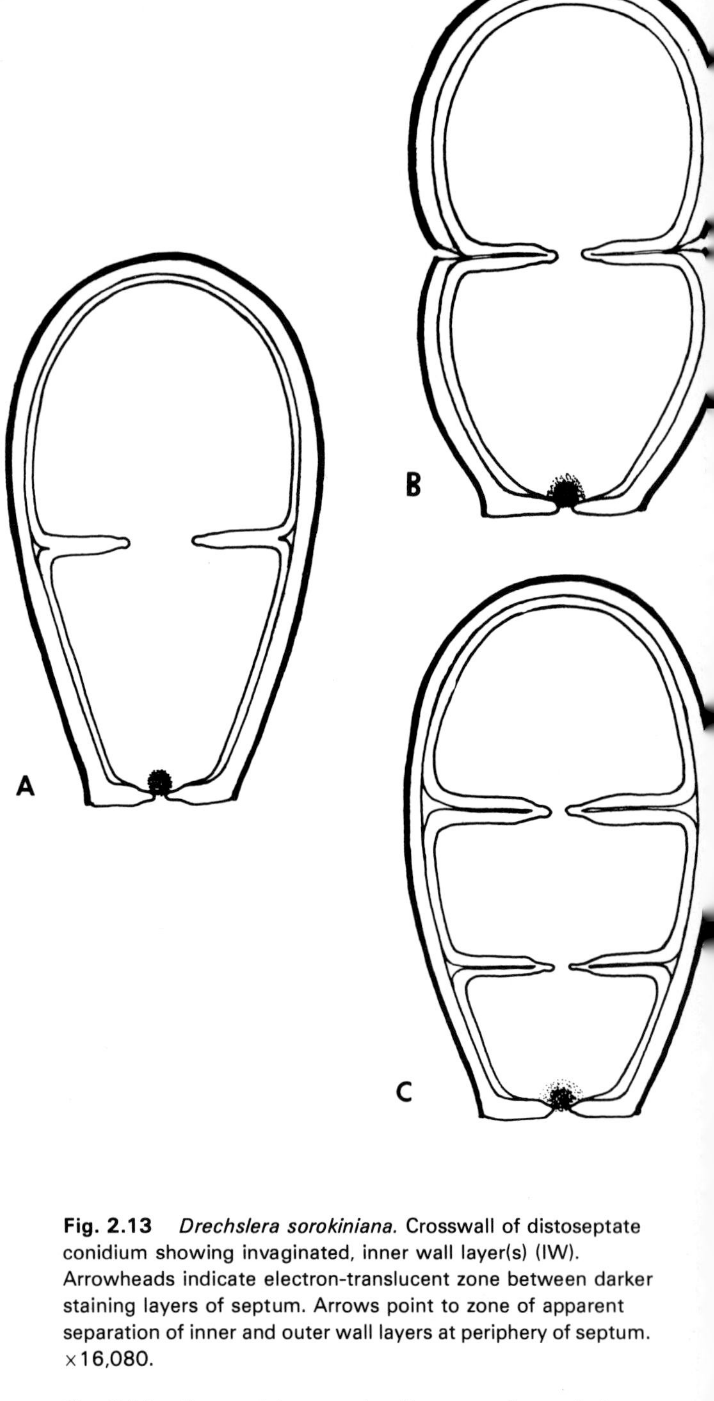

Fig. 2.13 *Drechslera sorokiniana.* Crosswall of distoseptate conidium showing invaginated, inner wall layer(s) (IW). Arrowheads indicate electron-translucent zone between darker staining layers of septum. Arrows point to zone of apparent separation of inner and outer wall layers at periphery of septum. ×16,080.

Fig. 2.14 *Spegazzinia tessarthra.* Euseptate, four-celled conidium showing regions of outer wall discontinuity (arrows). ×2,500.

Fig. 2.15 *Drechslera sorokiniana.* Narrow central pore of conidial septum guarded by several Woronin bodies. N, nucleus. ×26,400.

Fig. 2.16A–C Diagrammatic interpretation of wall differentiation during formation of euseptate (A, B) and distoseptate (A, C) conidia.

developing septum (Fig. 2.16A). Ultrastructural details and biochemical aspects of *euseptation*, however, require further elucidation. Luttrell (1963) distinguished *distoseptate conidia* as having 'a common outer wall enclosing more or less spherical cells each of which is surrounded by an individual wall'. Apparently, the wall of the young conidium is double. Invaginations of the inner wall divide the protoplast into a series of cells resembling 'peas in a pod'. In our diagrammatic interpretation of distoseptation (Fig. 2.16A, C) the initial stages of differentiation are the same as those of euseptate conidia (Fig. 2.16A). The distinction is demonstrated later, during maturation. There is a gradual separation of the inner and outer wall layer(s), first in the region of the intercellular spaces and then radiating from these points (arrows in Fig. 2.13). This occurs without breakdown of the outer wall. As in euseptation, virtually nothing is known of the cytochemistry of this process. One aspect which should be explored is whether distoseptate conidia bear some developmental similarity to enteroarthric conidia (Chapter 7). For example, in *Sporendonema purpurascens* (Bon.) Mason & Hughes (Chapter 7, Figs. 7.38–7.43) differentiation of endogenous conidia within the fertile hypha (enteroarthric development) involves a process of wall ontogeny which resembles distoseptation. Conidiogenesis in *S. purpurascens* (Cole, 1975a) could be used as a model system for studies of septation, with emphasis on attempts to identify specific mural enzyme activities associated with conidial differentiation.

Saccardo (1886) recognized the taxonomic value of conidial septation and devised a terminology for various forms of conidia (e.g., amerospore, didymospore, phragmospore, dictyospore). In some cases, when the shape of the conidium is the most striking feature, Saccardo applied a term to this feature rather than septation (e.g., scolecosporous = filiform conidia; helicosporous = helicoid conidia). Aspects of conidial shape were discussed in the previous chapter.

Conidial secession

This discussion concerns development and ultrastructure of the delimiting septum at the base of the conidium (Figs. 2.17A, B, 2.18). S. J. Hughes (1971b, c) applied the terms *schizolytic* and *rhexolytic* (Correns, 1899) to the two mechanisms of conidial secession which occur in the Deuteromycetes. The schizolytic process is most common and involves a splitting of the delimiting septum so that one half of the crosswall becomes the base of the seceding conidium and the other half covers the apex of the conidiogenous cell (Fig. 2.17A). This process occurs during secession of blastic conidia from annellides (Cole and Aldrich, 1971a; Hammill, 1971, 1972b, 1973a; Chapter 4), phialides (Campbell, 1972; Hammill, 1972a, 1974a, b; Oláh and Reisinger, 1974; Hanlin, 1976; Chapter 4), retrogressive conidiogenous cells (Chapter 5), and secession of most holoarthric conidia (Cole 1975a; Chapter 7). Schizolytic separation probably also liberates solitary, blastic conidia from determinate, sympodially proliferating and basauxic conidiogenous cells (Chapters 3 and 6). In each case, centripetal septum splitting produces a broadening electron-translucent zone between layers of electron-opaque material. The translucent zone represents degradation of wall microfibrils probably due to hydrolytic enzymes. Dictyosome-like organelles, ER and vesicles found near these septa (Cole and Aldrich, 1971a) may be associated with transport and release of such mural enzymes.

The alternative method of secession, referred to as rhexolytic (Correns, 1899), does not involve splitting of the basal septum but rather the circumscissile splitting of the periclinal wall of the cell below the basal septum (Fig. 2.17B). Rhexolytic secession includes all kinds of secession not covered under the heading Schizolytic—the break occurs, not across the basal septum, but across a specialized or unspecialized cell below the conidium. Specialized supporting cells may degenerate enzymatically, may fracture at a built-in zone of weakness, or may be thinner walled than cells above or below them. This kind of secession was originally described by Vuillemin (1911) as being diagnostic of his *Aleuriospores*, and more recently interpreted by Carmichael (1971). Rhexolytic secession is exemplified by formation of solitary blastic conidia in *Camposporium* Harkn. (Fig. 2.18A), botryose blastic conidia in *Gonatobotryum apiculatum* (Peck) Hughes (Chapter 3, Fig. 3.75) and *Peziza quelepidotia* Korf & O'Donnell (Chapter 3, Fig. 3.76) and thallic conidia in *Microsporum* Gruby and *Trichophyton* Malmsten (Fig. 2.18B and C, respectively).

2.17

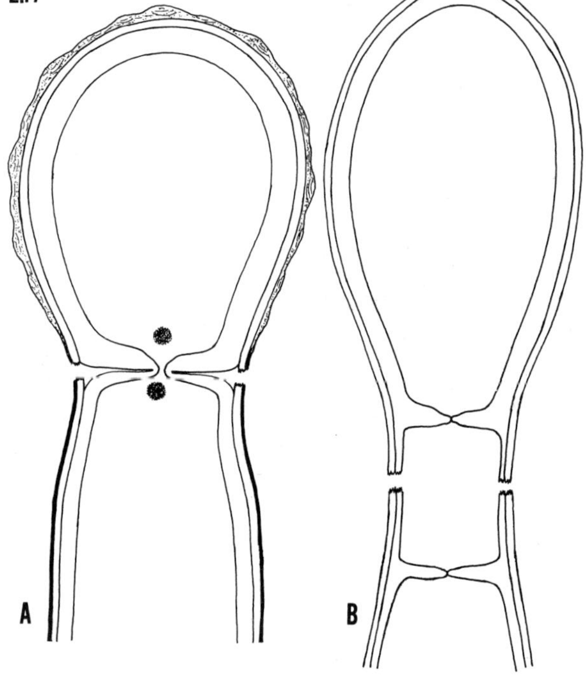

Fig. 2.17A, B Interpretation of conidial and conidiogenous cell wall relations during *schizolytic* (A) and *rhexolytic* (B) separation of cells.

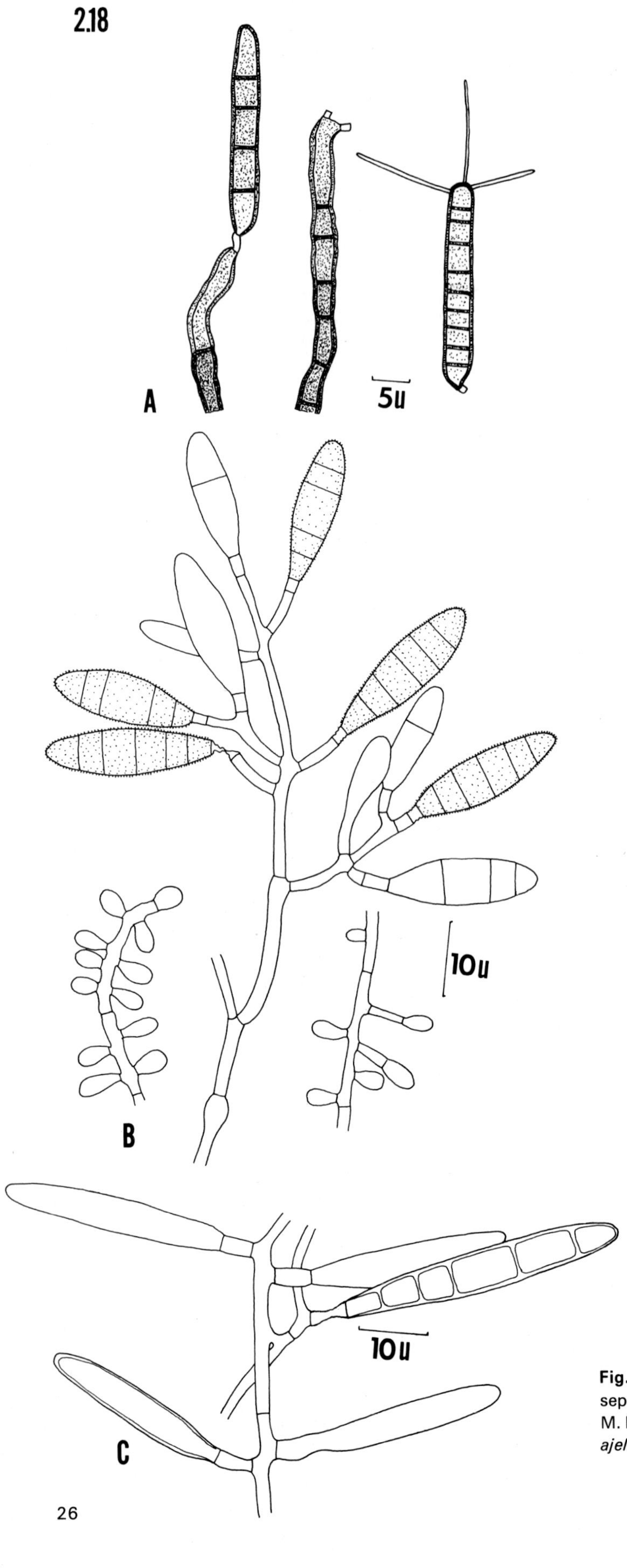

Fig. 2.18A–C. Examples of fungi which undergo rhexolytic separation of conidia; *Camposporium antennatum* (A, after M. B. Ellis, 1971b), *Microsporum gypseum* (B), and *Trichophyton ajelloi* (C).

3 Blastic Conidia

A scheme incorporating the different modes of blastic ontogeny is proposed: it involves aspects of wall differentiation, arrangement of conidia on, and ontogeny of, conidiogenous cells. These categories are not claimed to reflect natural relationships but are simply patterns of development. On the basis of wall differentiation, blastic ontogeny is separable into *holoblastic* and *enteroblastic* modes (Kendrick, 1971b), which were defined in Chapter 1.

Three principal kinds of blastic conidial arrangement are found in the Deuteromycetes:

1. *Solitary*: conidia form singly, terminally and/or laterally from conidiogenous cells, or in basipetal succession from the apex of specially differentiated conidiogenous cells (i.e., phialides, annellides or retrogressive conidiogenous cells; Chapters 4 and 5).
2. *Botryose*: conidia form in clusters, usually from the swollen apex of the conidiogenous cell, and may arise synchronously or asynchronously.
3. *Catenulate*: conidia form in acropetal chains (simple or branched), from the apex and/or sides of conidiogenous cells.

3.1

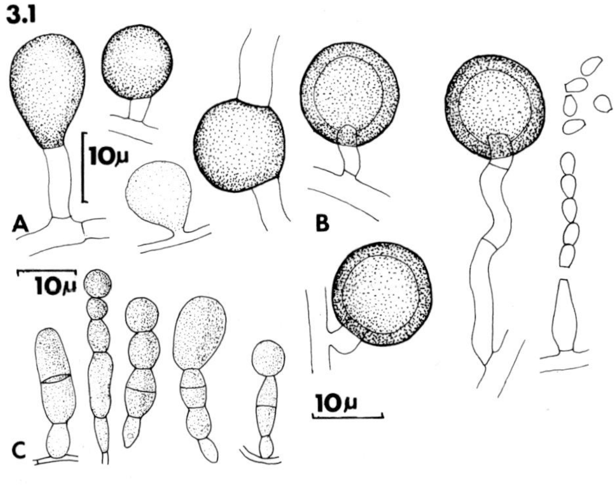

Fig. 3.1A–C *Humicola* spp. Solitary, terminal and intercalary holoblastic conidia produced by *H. grisea* var. *thermoidea* (A) and *H. grisea* (B), and catenulate cell formation in pure cultures of *H. alopallonella* (C). *Humicola grisea* (B) is pleomorphic, forming phialides in addition to holoblastic conidia from determinate conidiogenous cells.

Three modes of associated conidiogenous cell development are recognized:

1. *determinate* (i.e., ceasing extension growth with initiation of the primary, terminal, blastic conidium), or
2. *retrogressive* (i.e., shortening as a result of formation of successive conidia), or
3. *proliferous* (i.e., lengthening during formation of successive conidia).

Using combinations of these three sets of characters, most kinds of blastic conidial development can be characterized. We begin our discussion with an analysis of holoblastic, solitary conidia.

Holoblastic (solitary) conidia

Several earlier terms were coined for this kind of conidium (e.g., *terminus spore*: Mason, 1933; S. J. Hughes, 1953; *gangliospore*: Subramanian, 1962a; *radulaspore*: Tubaki, 1963). Instead of providing clarification, these terms have generally been too imprecise or too narrow to be of much use. Confusion has also arisen in the distinction between *terminal chlamydospores* and solitary, terminal, holoblastic conidia. Members of the Kananaskis Conference (Kendrick, 1971a) opted to reject most of the earlier terminology (with the exception of *chlamydospore*) and adopted concise, descriptive terms which differentiate between conidium formation and conidiogenous cell development. The chlamydospore was considered to be 'a thick-walled, thallic, terminal or intercalary spore'. We concur and deal with the development of this propagule separately in Chapter 7. At this point, we will consider only species producing solitary holoblastic conidia from determinate conidiogenous cells.

Determinate conidiogenous cells

Thin-sections of *Humicola grisea* Traaen (Griffiths, 1974) show that conidia are typically holoblastic (Fig. 3.2). A uniperforate basal septum differentiates and lipid bodies gradually accumulate within the conidial cytoplasm. The thick wall of the mature conidium is differentiated into an outer, electron-opaque (melanized) layer, and an inner, electron-transparent zone (Fig. 3.3). Conidial secession appears to be schizolytic (arrows in Fig. 3.3) and the septal pore is plugged by an electron-opaque deposit (Fig. 3.4). Griffiths (1974) observed that the cytoplasm of the conidiogenous cell degenerates after the conidium has differen-

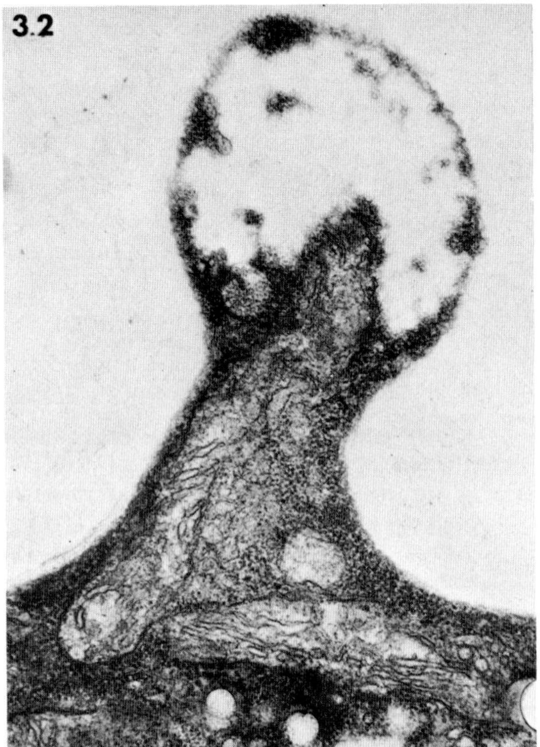

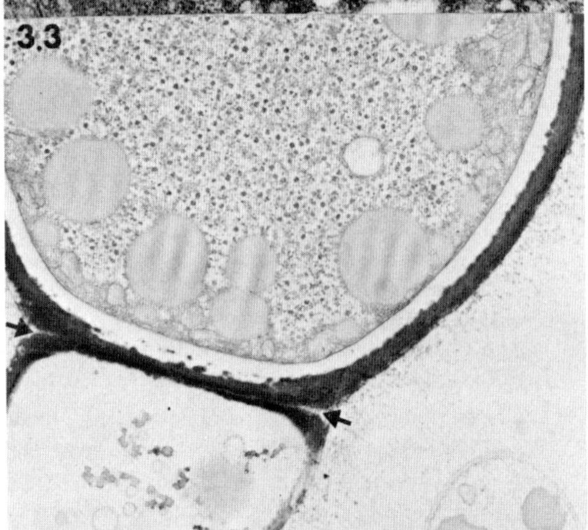

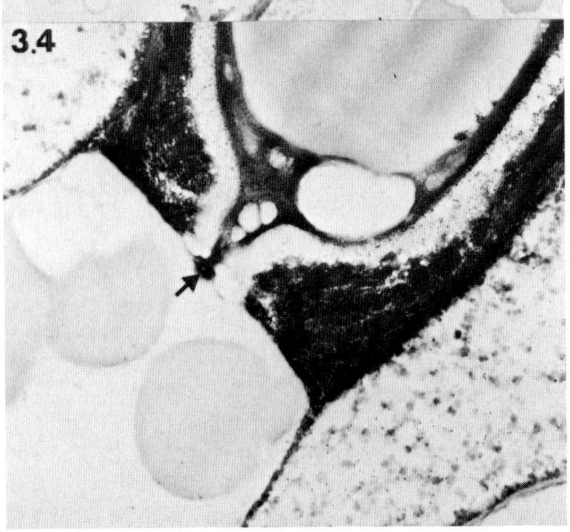

Fig. 3.2 *Humicola grisea.* Holoblastic conidial initial produced laterally from a hypha. ×22,000 (from Griffiths, 1974).

Fig. 3.3 *Humicola grisea.* Mature, holoblastic conidium at apex of conidiogenous cell. Arrows indicate points of initiation of schizolytic secession of conidium. ×5,700 (from Griffiths, 1974).

Fig. 3.4 *Humicola grisea.* Pore through the basal septum of conidium plugged with an electron-opaque deposit (arrow). ×16,000 (from Griffiths, 1974).

tiated, and suggested that this may also facilitate conidial secession (rhexolytic?).

The strain of *Humicola grisea* which Griffiths (1974) examined produces conidia both within the agar and from aerial conidiogenous cells (Fig. 3.1A, B). Some Hyphomycetes (e.g. *Humicola alopallonella* Meyers & Moore, Fig. 3.1C) produce catenulate as well as solitary conidia, and some (e.g., *H. grisea,* Fig. 3.1B; *Botryotrichum piluliferum* Sacc. & March, Fig. 3.5A) form both solitary, holoblastic conidia and basipetal chains of propagules arising from phialides (Fig. 3.1B). This phenomenon, called pleomorphism, is common in the Deuteromycetes and is a strong deterrent to those who would classify this group on the basis of conidial ontogeny alone.

Acremoniella Saccardo (Figs. 3.5B, 3.6–3.9) produces large, ovoid to subglobose conidia at the tapered apices of conidiogenous cells (Fig. 3.6). Less commonly conidiophores produce a terminal, globose vesicle bearing numerous phialides (Fig. 3.5B), or give rise directly to phialides at their apex (M. B. Ellis, 1971b). In *Acremoniella verrucosa* Tongnini the tip of the conidiogenous cell becomes inflated (Fig. 3.7), leaving a narrow, cylindrical isthmus between the conidium and the fertile cell (arrow in Fig. 3.8). Conidial secession occurs by a fracture through the wall of the connective (arrow in Fig. 3.9), and is rhexolytic (Chapter 1).

The determinate fertile cells of *Nigrospora sphaerica* (Sacc.) Mason are typically ampulliform or subspherical, each supporting a single, terminal, holoblastic conidium (Figs. 3.10–3.14). The inflated conidiogenous cell may be associated with the violent discharge of the propagule (i.e., related to increased turgor pressure of the fertile cell), reported by Webster (1952). A globose conidial initial in Fig. 3.10 is attached to an elongated, flask-shaped conidiogenous cell. Subsequent development produces broadly ellipsoidal, mature conidia and swollen, more or less subspherical conidiogenous cells (Figs. 3.11 and 3.13). The scanning electron microscope (SEM) shows that the outer wall layer of mature conidia is not continuous with that of the conidiogenous cell. The membranous layer in Fig. 3.11, which we interpret as the outermost wall layer, is visible at the junction between the conidium and fertile cell on the left (arrows). However, the adjacent conidiogenous cell shows only fragments of this on the surface of the fertile cell (arrowheads) while the schizolytic zone (sz) is clearly shown between the two cells. The apex of a conidiogenous cell which has released its

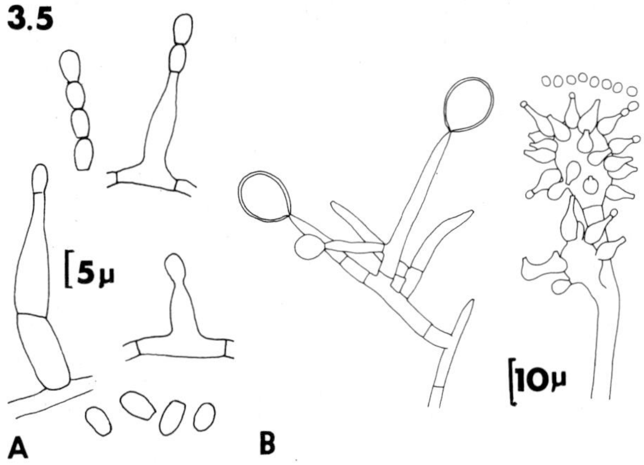

Fig. 3.5A *Botryotrichum piluliferum.* Philalides and chains of phialoconidia.

Fig. 3.5B *Acremoniella atra* Sacc. Pleomorphism, demonstrated by formation of terminal, holoblastic conidia at apices of tapered conidiogenous cells, and successive conidia from clusters of phialides attached to globose vesicles.

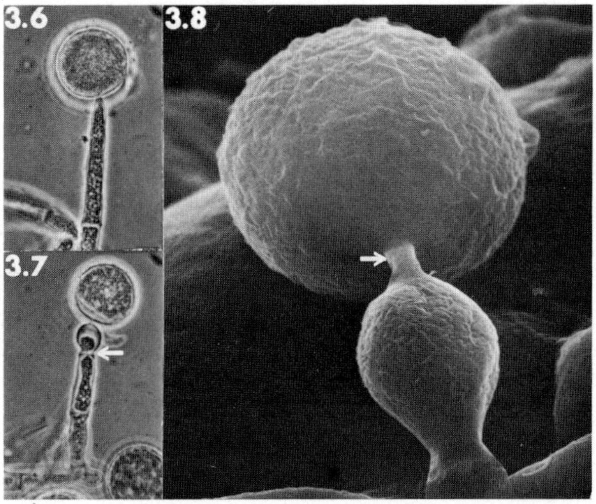

Figs. 3.6–3.9 *Acremoniella verrucosa.* Terminal, holoblastic conidia produced from determinate conidiogenous cells, the apex of which is usually tapered (3.6) but sometimes differentiated into an inflated cell (3.7–3.9). The latter, delimited at its base by a septum (arrow in 3.7), gives rise to the conidium at its apex (3.8). A short, cylindrical isthmus (arrow in 3.8) separates the two cells. Arrow in 3.9 indicates fractured outer wall layer of isthmus after rhexolytic secession of conidium. ×760, 760, 2,975 and 4,400, respectively.

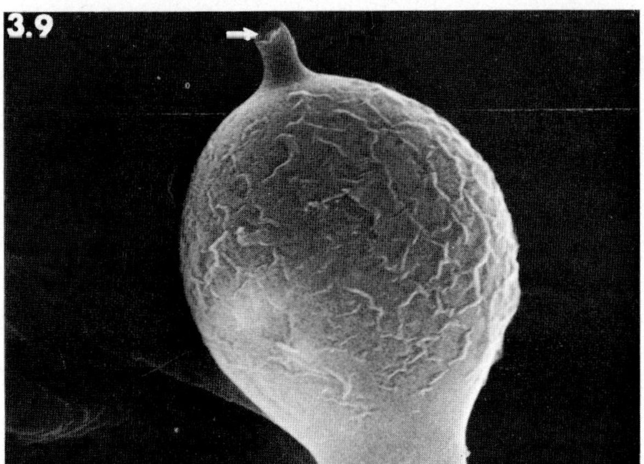

conidium is also shown in Fig. 3.11. After liberating its conidium, a fertile cell occasionally undergoes percurrent proliferation (arrow in Fig. 3.12), and forms a second, terminal conidium at its new apex. Both determinate and proliferous fertile cells may be produced by the same mycelium, again demonstrating the morphogenetic plasticity of the Deuteromycetes. The circumferential fracture (F) though the conidial wall in Fig. 3.14 is partly artefact brought about by our drying technique used in preparation for SEM examination, but is also indicative of a zone of weakness in the wall which may be significant for germ tube emergence. *Thermomyces lanuginosus* Tsiklinsky, which has sometimes been confused with *Humicola* Traaen (Apinis, 1963), is a thermophilic Hyphomycete producing terminal, holoblastic conidia from determinate conidiogenous cells (Figs. 3.15–3.17). The conidium secedes by schizolysis (Fig. 3.17). While the solitary holoblastic conidia of *Humicola, Botryotrichum* Sacc. & March, *Acremoniella, Nigrospora* Zimmermann and *Thermomyces* Tsiklinsky are all nonseptate, those of *Epicoccum purpurascens* Ehrenb. ex Schlecht. (= *E. nigrum* Link) are muriformly septate (Fig. 3.18A–D), though this septation is often obscured by the darkly pigmented wall of mature cells. Development of septation in *E. purpurascens* has been documented by Schol-Schwarz (1959) and Griffiths (1973a). The conidiogenous cells, which arise from sporodochia, are determinate and become septate in later developmental stages (Fig. 3.18A). Conidia are usually terminal but sometimes arise directly from the sides of fertile hyphae (Fig. 3.18B). Conidiogenous cells occasionally form more than one conidium (Fig. 3.18C). Conidial secession is either schizolytic (Griffiths, 1973a) or rhexolytic (Fig. 3.18C and D, respectively). Webster's (1966) suggestion that conidia of *E. purpurascens* are forcibly discharged requires investigation. Solitary, holoblastic conidia produced by the marine Hyphomycete *Monodictys pelagica* (Johnson) Jones (Figs. 3.19, 3.20) are also dictyoseptate. The locations of sep ta are indicated in the SEM by slight grooves in the outer wall. Mature conidia of *Dictyosporium toruloides* (Corda) Guegen (Figs. 3.21–3.22) are 'shaped rather like a hand with fingers held close together' (Ellis, 1971b). A simple conidial initial arises at the end of a determinate fertile hypha (Fig. 3.21), but soon branches and undergoes septation to form 6–8 parallel rows of cells (Figs. 3.22). The wartlike excrescences have not been previously reported.

The equally elaborate, tetraradiate conidia produced by the aquatic Hyphomycete *Tripospermum camelopardus* Ingold, Dann & McDougall (Figs. 3.23A–D, 3.24, 3.25) are holoblastic and arise laterally from vegetative hyphae (Fig. 3.23C). Mature conidia have four divergent, septate arms (Figs. 3.23D, 3.25). Ingold *et al.* (1968) have shown that a

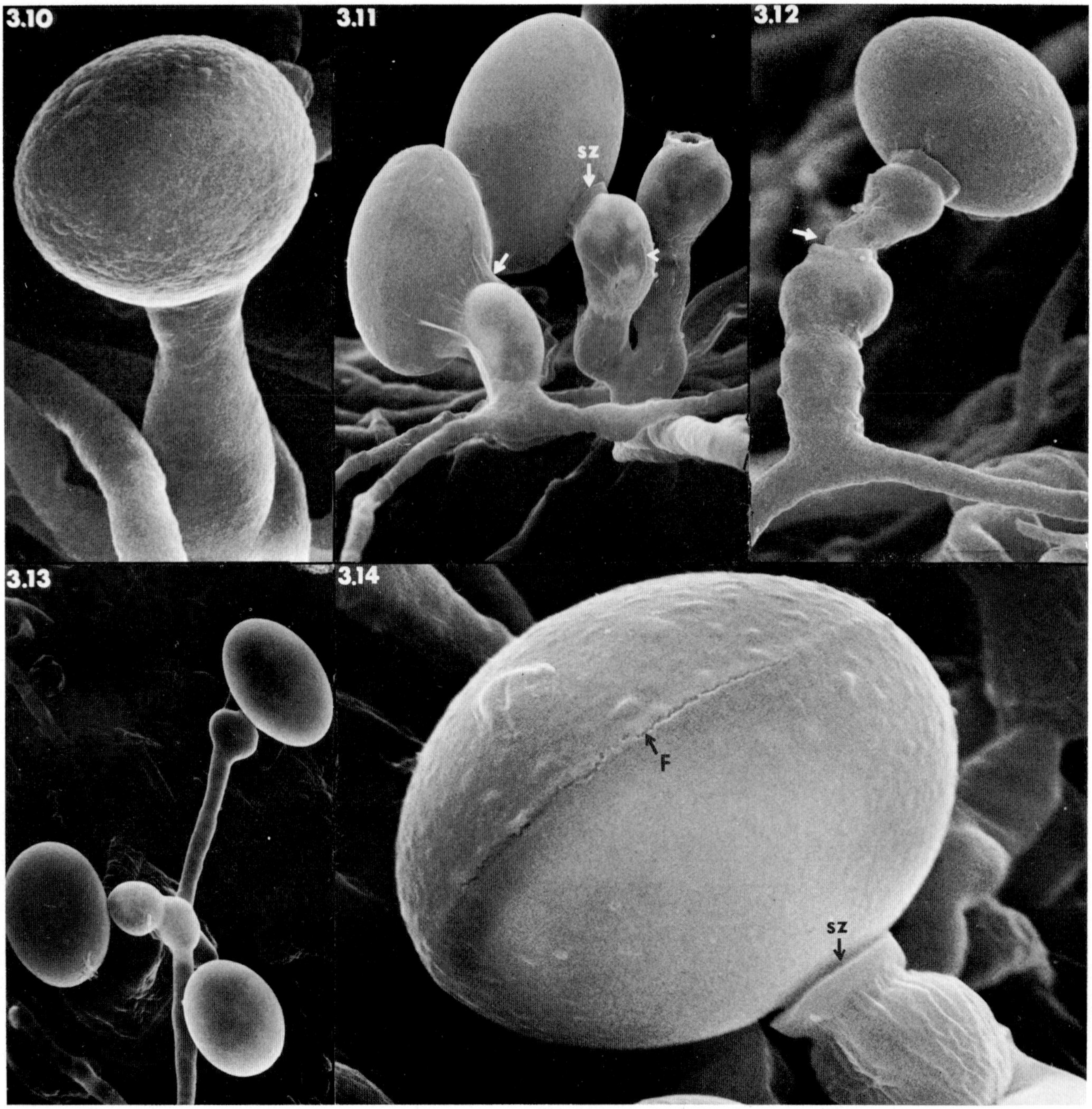

Figs. 3.10–3.14 *Nigrospora sphaerica.* Solitary, terminal, holoblastic conidia produced by determinate (3.10, 3.11, 3.13) and percurrently proliferating conidiogenous cells (arrow in 3.12). Arrow in 3.11 indicates continuous membranous wall layer between conidium and fertile cell, and arrowhead locates fragments of this same wall on the surface of the adjacent conidiogenous cell. The schizolytic zone (sz) between conidium and fertile cell is revealed in 3.11 and 3.14. F, circumferential fracture through conidial wall. ×6,325, 2,210, 2,640, 1,190 and 5,750, respectively.

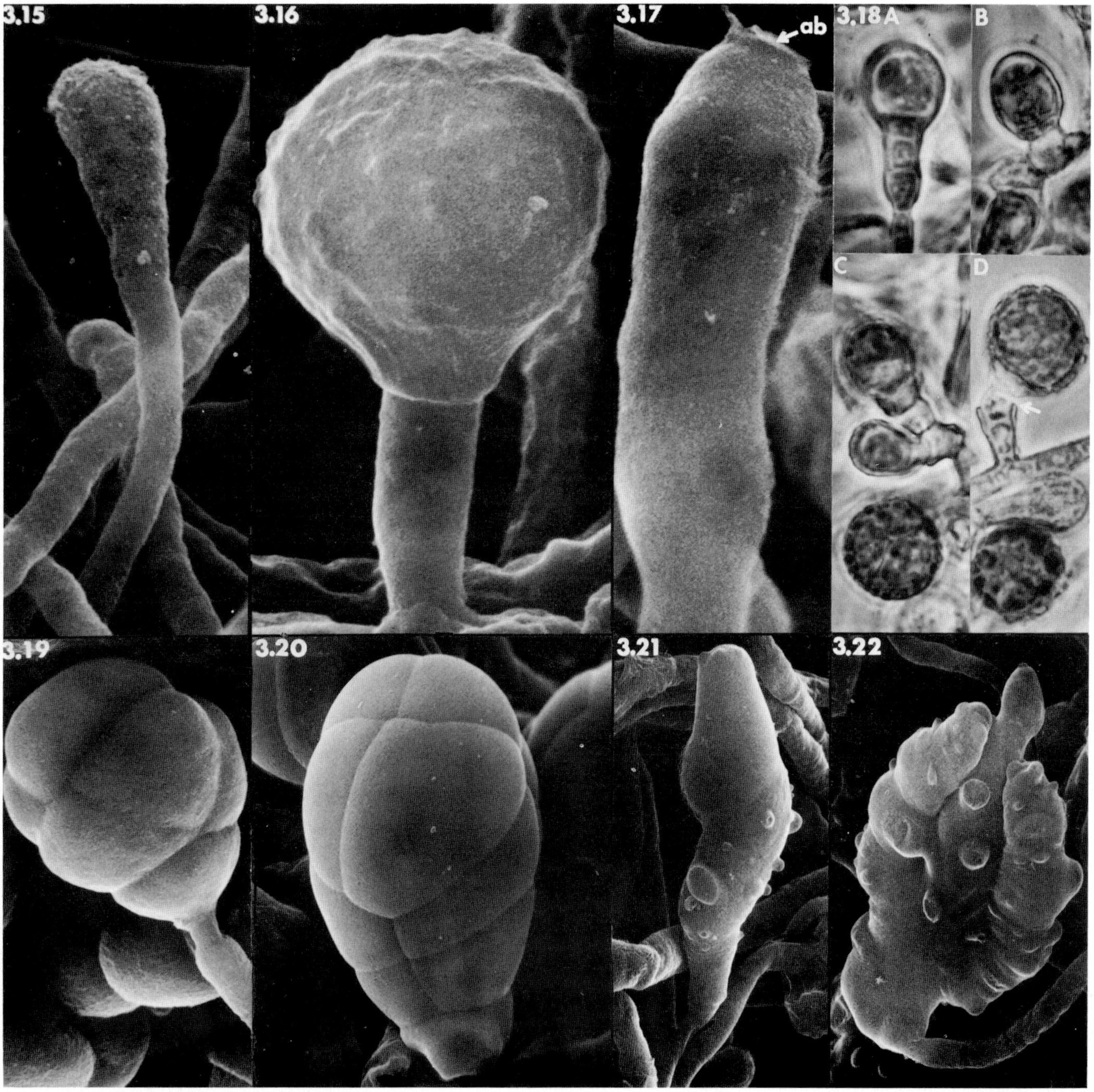

Figs. 3.15–3.17 *Thermomyces lanuginosus.* Stages in development of terminal, holoblastic conidium from a determinate conidiogenous cell (3.15, 3.16); abstriction scar (ab) left at apex of fertile cell (3.17). ×8,880, 9,230 and 13,650, respectively.

Fig. 3.18A–D *Epicoccum purpurascens.* Muriform, holoblastic conidia produced terminally (A) and laterally (B) from fertile cells. Secession of conidia occurs both by schizolytic (C) and rhexolytic (arrow in D) mechanisms. All ×2,000.

Figs. 3.19, 3.20 *Monodictys pelagica.* Muriform, holoblastic conidia. ×3,510 and 3,080, respectively.

Figs. 3.21, 3.22 *Dictyosporium toruloides.* Developmental stages of branched, terminal, holoblastic conidium. ×5,950 and 2,625, respectively.

3.23

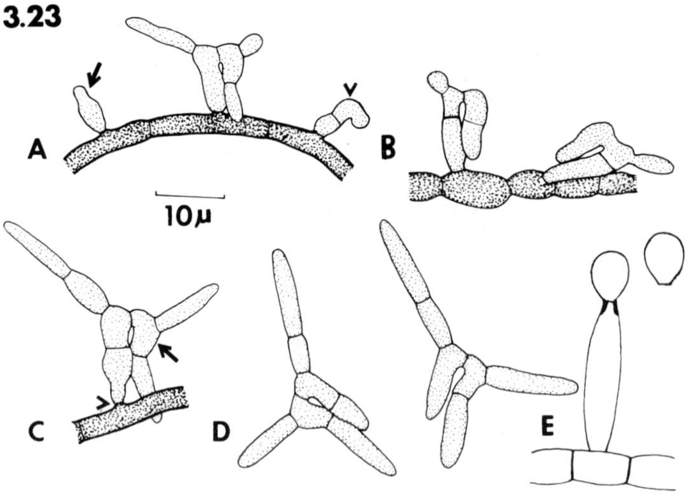

Fig. 3.23A–E *Tripospermum camelopardus.* Developmental stages of holoblastic tetraradiate conidia. Arrow in (A) indicates stalk cell. Arrowhead in (A) points to apical cell which has arisen from stalk cell. Arrow in (C) represents proliferated cell which forms two arms of the conidium. Arrowhead in (C) locates point of conidial secession. Pleomorphism in this species is demonstrated by development of phialides (E).

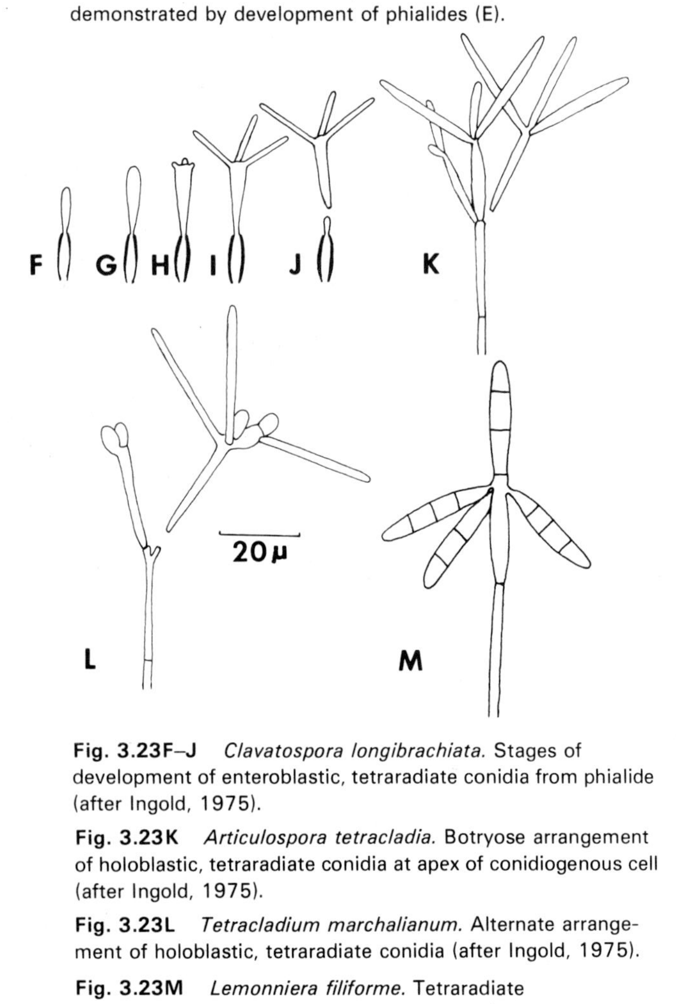

Fig. 3.23F–J *Clavatospora longibrachiata.* Stages of development of enteroblastic, tetraradiate conidia from phialide (after Ingold, 1975).

Fig. 3.23K *Articulospora tetracladia.* Botryose arrangement of holoblastic, tetraradiate conidia at apex of conidiogenous cell (after Ingold, 1975).

Fig. 3.23L *Tetracladium marchalianum.* Alternate arrangement of holoblastic, tetraradiate conidia (after Ingold, 1975).

Fig. 3.23M *Lemonniera filiforme.* Tetraradiate phialoconidium attached to apex of phialide (after Ingold, 1975).

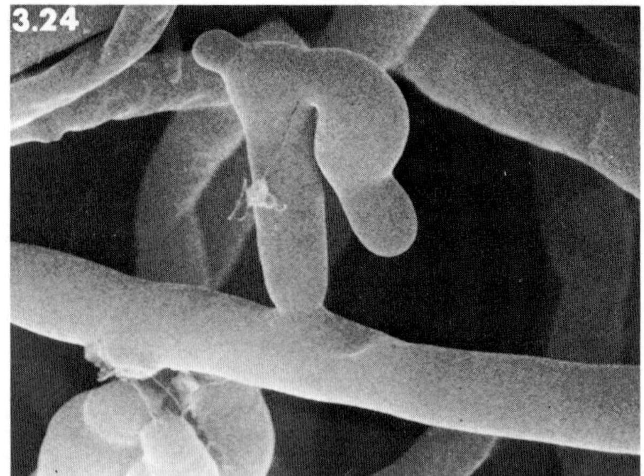

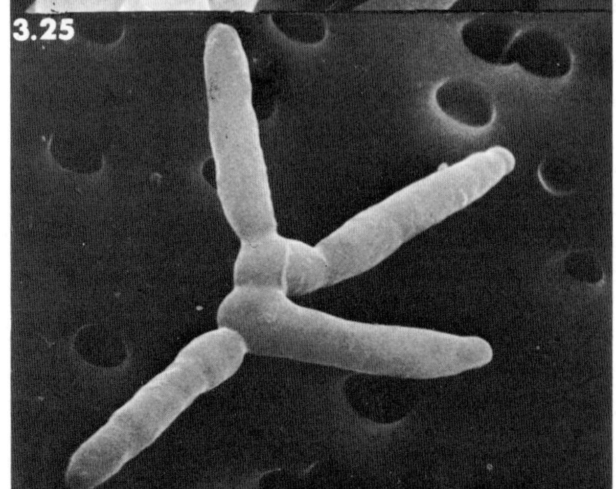

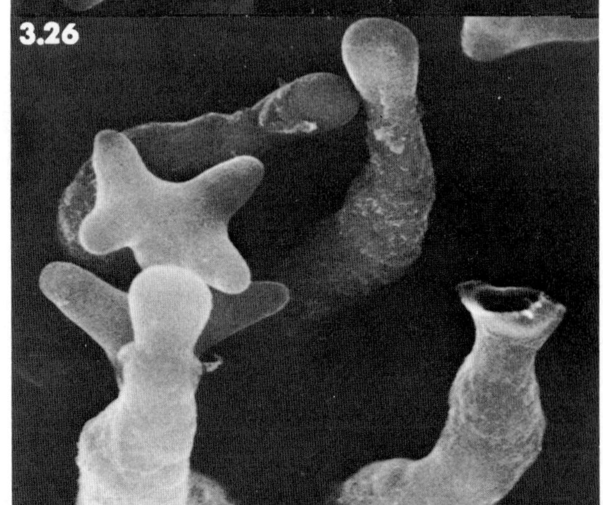

Figs. 3.24, 3.25 *Tripospermum camelopardus.* Holoblastic, tetraradiate conidial initial (3.24, cf. 3.23B) and mature, seceded conidium (3.25). ×4,060 and 2,925, respectively.

Fig. 3.26 *Clavatospora stellatacula.* Various stages of phialoconidium formation. ×5,130

Fig. 3.27 *Zalerion maritimum.* Helicoid conidium at apex of fertile hypha. ×3,230

Fig. 3.28 *Helicosporium phragmitis.* Helicoid conidium attached to conidiogenous cell. ×3,825.

Fig. 3.29A *Helicodendron tubulosum* (Riess) Linder. Helicoid conidia at apices of determinate conidiogenous cells (after Matsushima, 1975).

Fig. 3.29B *Helicoon fuscusporum* Linder. Helicoid conidia produced from septate, fertile hyphae. Note proliferated, secondary conidium (after Matsushima, 1975).

Fig. 3.29C *Helicoma dennisii* M. B. Ellis. Abstriction scars of clusters of helicoid conidia remain on conidiogenous cells (after Matsushima, 1971).

Fig. 3.29D *Xenosporium larvale* (Morgan) Pirozynski. Helicoid conidia composed of two rows of cells except at apex and base (after Matsushima, 1975).

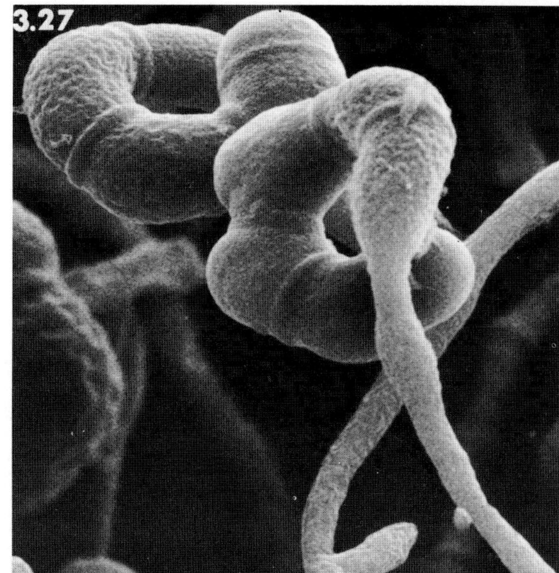

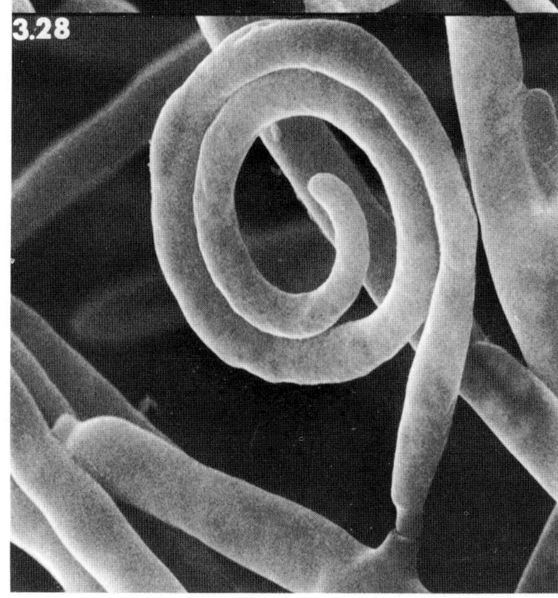

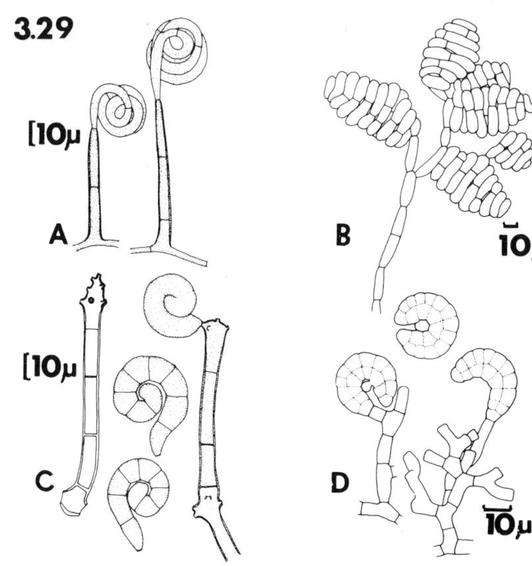

stalk cell develops on a hypha (arrow in Fig. 3.23A) and gives rise to a second apical cell (arrowhead in Fig. 3.23A) which subsequently proliferates and forms two septate arms (Figs. 3.23B, 3.24). The basal cell of one arm proliferates to form a third septate arm (arrow in Fig. 3.23C). Secession occurs at the base of the stalk cell (arrowhead in Fig. 3.23C), and it becomes the fourth arm of the now tetraradiate conidium. A similar developmental scheme was presented by G. C. Hughes (1975). Ingold (1975) discussed the widespread occurrence of tetraradiate conidia in aquatic Hyphomycetes (Figs. 3.23K–M) and logically suggested that their form is important both for dispersal and subsequent adherence to a substrate. *Tripospermum camelopardus* also produces small globose phialoconidia in pure cultures (Fig. 3.23E). Ingold (1975) determined that two kinds of tetraradiate conidia are produced by aquatic Hyphomycetes: terminal thalloconidia (formerly referred to as *aleuriospores*), and phialoconidia. We suggest that the former are actually solitary, holoblastic conidia, exemplified by the propagules of *T. camelopardus*. Formation of tetraradiate phialoconidia is illustrated by *Clavatospora longibrachiata* (Ingold) Nilsson (Fig. 3.23F–J) and *C. stellatacula* Kirk (Fig. 3.26). Ingold (1975) also observed that tetraradiate conidia are attached to the fertile hypha either by one of the four arms (e.g., *T. camelopardus*) or near their point of divergence (e.g., *Lemonniera filiforme* Petersen, Fig. 3.23M). Acropetal chains of tetraradiate conidia have not been reported, and considering their shape, this configuration seems highly improbable.

G. C. Hughes (1975) stated that studies of conidiogenesis in marine Hyphomycetes and Coelomycetes 'are virtually nonexistent', but found that 'in most of the lignicolous dematiaceous Hyphomycete genera having marine representatives; e.g., *Cirrenalia* S. P. Meyer & R. T. Moore, *Dictyosporium, Humicola, Orbimyces* Linder, *Trichocladium* Harz and *Zalerion*, conidia are produced singly or, rarely, in short chains and are usually . . . blown-out ends of . . . conidiogenous cells on poorly developed conidiophores'.

3.30

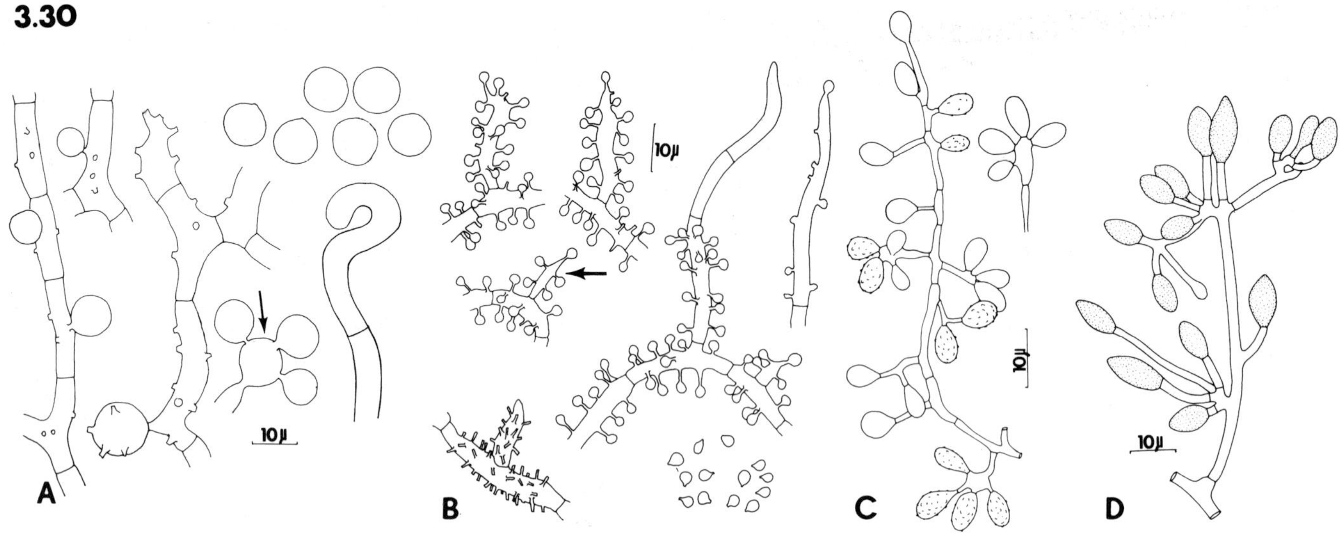

Fig. 3.30A *Beniowskia graminis.* Acropleurogenous arrangement of solitary, holoblastic conidia. Occasionally, botryose clusters of secondary conidia are produced by primary conidia. The arrow indicates a cluster of secondary conidia.

Fig. 3.30B *Myriodontium keratinophilum.* Acropleurogenous arrangement of solitary, holoblastic conidia produced on short denticles. Conidia sometimes form at successive apices of sympodially proliferating, conidiogenous cells (arrow).

Fig. 3.30C, D Acropleurogenous arrangement of conidia in *Myceliophthora thermophila* and *Sporotrichum aureum,* respectively.

Conidial formation in *Zalerion maritimum* has already been described in Chapter 1. The SEM reveals the three-dimensional aspect of the mature, helicoid conidium (Fig. 3.27). The transverse septa, which form successively as the conidium elongates (Chapter 1, Fig. 1.20A–H), are also apparent. The conidium develops from a slender conidiogenous cell (cc) and rhexolytic secession sometimes occurs at the constricted region near the conidial base (arrow in Fig. 3.27), but more often the conidium remains attached to the fertile cell, which secedes schizolytically at its base. The helicoid conidia of *Helicosporium phragmitis* Höhnel (Fig. 3.28) are produced on short denticles and coiled 2–4 times in one plane. Although it is tempting to speculate that a natural relationship exists among this fungal group (Saccardo, 1886; Linder, 1929; Moore 1955, 1957; Cole, 1976b), morphological, developmental and functional differences (Barron, 1961) suggest otherwise (Fig. 3.29).

Solitary, holoblastic conidia may be borne terminally and laterally. *Beniowskia graminis* Rac. produces globose conidia on short denticles at the apex and sides of fertile hyphae (Fig. 3.30A). Occasionally, primary conidia produce a cluster of secondary conidia (arrow in Fig. 3.30A; Fig. 3.31). The fertile hypha may continue to elongate at its tip as conidiogenesis proceeds subapically. *Myriodontium keratinophilum* Samson & Polonelli produces a similar arrangement of conidia on elongated denticles (Figs. 3.30B, 3.32, 3.33) as well as at successive new growing points which arise by sympodial proliferation of a lateral branch of the fertile hypha (arrows in Figs. 3.30B and 3.33). Species of *Sporotrichum* Link and *Myceliophthora* Cost. produce conidia which arise terminally and laterally from poorly differentiated conidiogenous cells (Figs. 3.30C, D, 3.34–3.36). The conidium base in *Sporotrichum aureum* Link is broad, and secession is rhexolytic (arrows in Figs. 3.34, 3.35). Conidia of *Myceliophthora thermophila* (Apinis) v. Oorschot (Fig. 3.30C) and *S. aureum* (Fig. 3.30D) are sometimes arranged alternately at the apex of short fertile branches. This may be the result of sympodial proliferation of the conidiogenous cell, or random formation of propagules.

In other Deuteromycetes which produce solitary, holoblastic conidia, conidiogenesis is generally restricted to the fertile apices of conidiogenous cells, which may be retrogressive or proliferous.

Retrogressive conidiogenous cells

Certain Hyphomycetes produce conidia in basipetal succession from the apex of a conidiogenous cell which undergoes a concomitant shortening (Cole and Kendrick, 1968a, 1971; Kendrick and Cole, 1969). As each conidium secedes, a portion of the fertile cell wall is removed and incorporated into the conidial wall (Fig. 3.37A–G). This mechanism is discussed in Chapter 5.

Proliferous conidiogenous cells

In another group of conidial fungi the fertile cell has a basal meristem which causes it to elongate after differentiation of the apical conidium, an event called *basauxic development* (Fig. 3.37H–M) described in Chapter 6.

Conidiogenous cells may also proliferate sympodially (Fig. 3.37N–T), while conidia arise 'as blown-out ends of successively produced new growing points developing to one side of the previous conidium' (S. J. Hughes, 1953). Hughes formed a separate developmental category (Section II) to accommodate such species. Kendrick (1962) also stated that

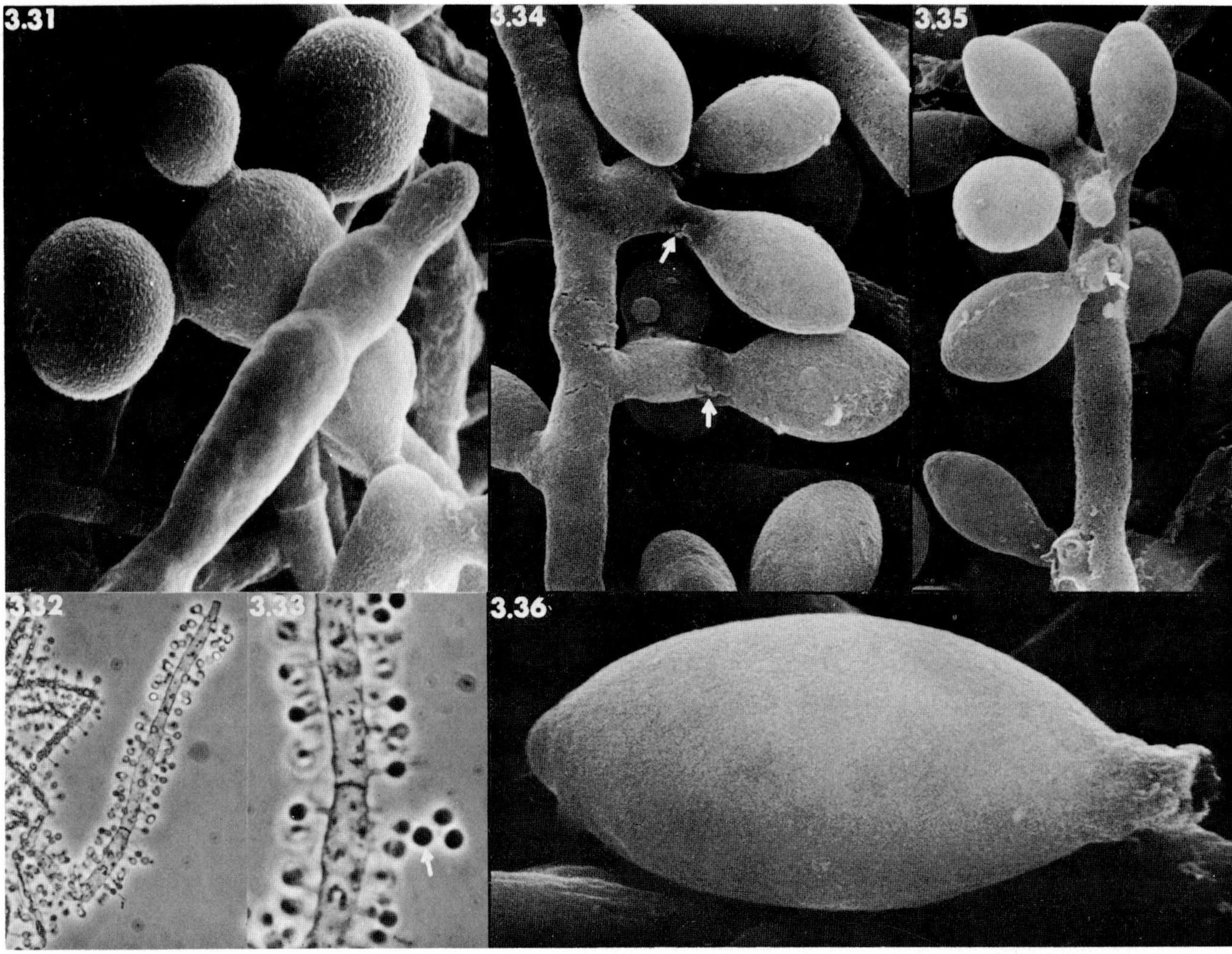

Fig. 3.31 *Beniowskia graminis.* Swollen, primary conidium which has formed at least three secondary conidia. ×3,485.

Figs. 3.32, 3.33. *Myriodontium keratinophilum.* Acropleurogenous and sympodial (arrow 3.33) arrangement of conidia. ×688 and 860, respectively.

Figs. 3.34–3.36 *Sporotrichum aureum.* Terminal and alternate arrangement of conidia on poorly differentiated conidiogenous cells. Arrows in 3.34 and 3.35 indicate cells at the base of conidia which seem to have degenerated (rhexolytic secession). A seceded conidium in 3.36 shows part of the torn outer wall of the degenerate cell attached to its base. ×3,060, 2,430 and 8,320, respectively.

'any term intended to describe the characteristics of Section II should refer to the sporogenous cell, as this and not the spore displays the diagnostic features of the group'. Cole (1971) pointed out that sympodially proliferating conidiogenous cells give rise to several different kinds of conidia (e.g., solitary, holoblastic conidia (Fig. 3.37N–T), catenulate, holoblastic conidia (Fig. 3.37U), botryose, holoblastic conidia (Fig. 3.37V), enteroblastic conidia (Fig. 3.37W), and holoarthric conidia (Fig. 3.37X)). The Kananaskis Conference recognized that both the mode of formation of individual conidia and the behaviour of the conidiogenous cells were important developmental features, but considered that they should be treated as separate, and sometimes independent taxonomic characters.

Tritirachium oryzae (Vincens) de Hoog (= *T. roseum* van Beyma) has sympodially proliferating conidiogenous cells which produce successive, solitary, holoblastic conidia. The time-lapse sequence of photomicrographs (Fig. 3.38A–J) shows that the fertile cell first forms a terminal conidium (Fig. 3.38A, B) and subsequently differentiates a new growing point (Fig. 3.38C) below and to one side of the propagule. A short period of elongation (Fig. 3.38D) is followed by swelling and differentiation of a new terminal, holoblastic conidium (Fig. 3.38E). These events are repeated many times (Fig. 3.38F–J), producing a geniculate extension of the conidiogenous cell, or *rachis* (Ainsworth, 1971). A similar process occurs in *Acrodontium hydnicola* (Peck) de Hoog, the rachis bearing globose to subglobose conidia (Fig. 3.39) on conspicuous denticles (Fig. 3.40). The abstriction scars in *Tritirachium oryzae* (Fig. 3.41) are less distinct, a morphological difference which de Hoog (1972) used to separate these two genera. Hammill (1973b) examined the

3.37

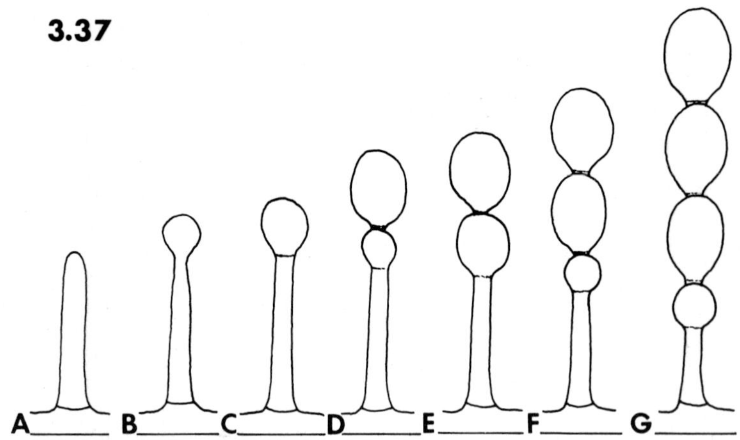

Fig. 3.37A–G Retrogressive conidiogenous cell development.

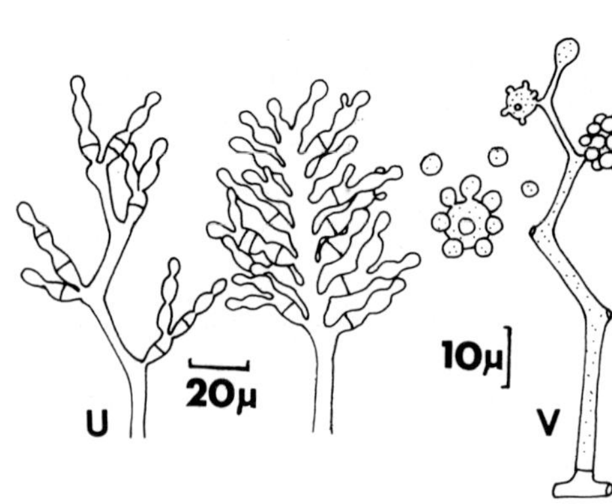

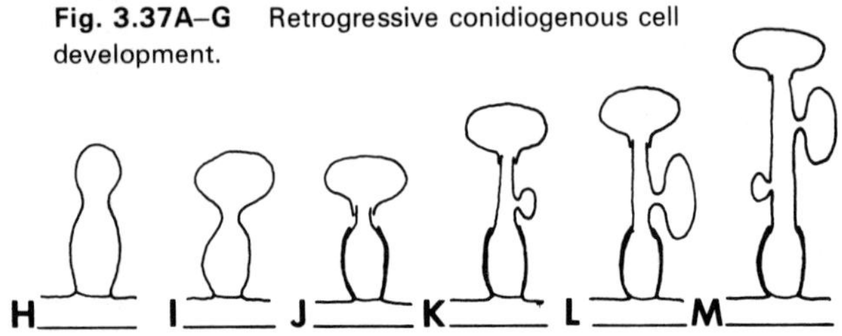

Fig. 3.37H–M Basauxic conidiogenous cell development.

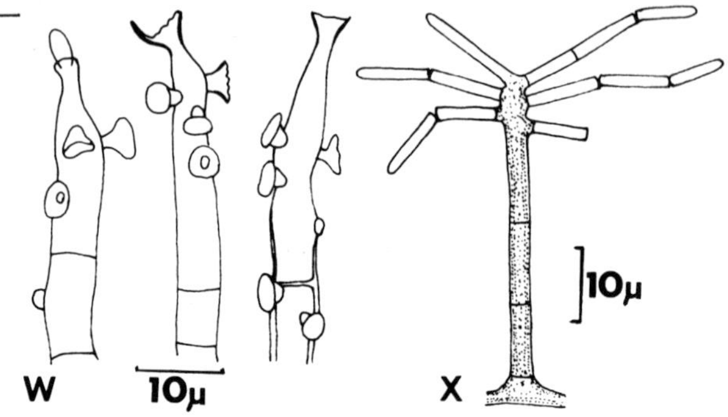

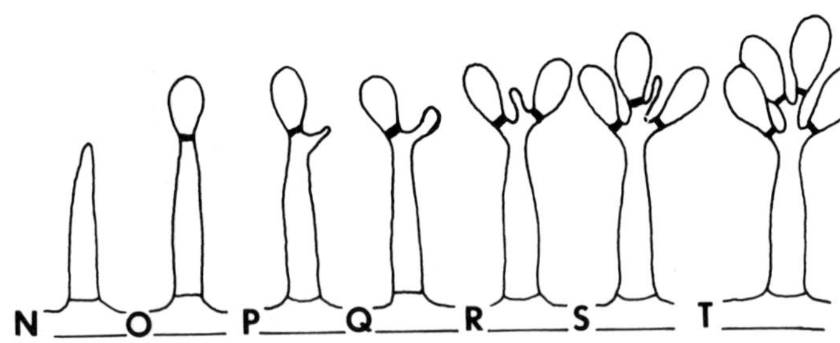

Fig. 3.37N–T Sympodially proliferating conidiogenous cell development.

Fig. 3.37U–X Examples of sympodially proliferated conidiogenous cells which have produced acropetal chains of holoblastic conidia (U, *Cladosporium variabile* (Cooke) de Vries), botryose, holoblastic conidia (V, *Blastobotrys nivea* Klopotek), enteroblastic conidia (W, *Codinaea fertilis* Hughes & Kendrick) and holoarthric conidia (X, *Sympodiella* sp.).

ultrastructure of conidiogenesis in *T. oryzae* and showed that each conidium is delimited by a thickened, basal septum which separates schizolytically. A diagrammatic interpretation of this process is presented in Fig. 3.42A–D. The septum appears to be complete, although median-longitudinal sections are not available. As the developing crosswall thickens, the lytic zone becomes more evident (arrow in Fig. 3.42C). After secession occurs the denticle is represented by a short, exposed cylinder of septal wall material and a frill of the ruptured outer wall is left at the base of the conidium (Fig. 3.42D). The narrow channel through the thick septal wall is later occluded. A comparative thin-section examination of *Acrodontium* de Hoog is desirable in the light of de Hoog's (1972) taxonomic decisions.

Conidium and conidiogenous cell development in the entomogenous Hyphomycete *Beauveria bassiana* (Bals.) Vuill. [=*B. globulifera* (Speg.) Picard] was described by Kendrick and Cole (1968) and is comparable to that in *Tritirachium* Limber. The rachis in Fig. 3.43 has several attached conidia, the youngest at the tip (arrow). Thin-sections of this species (Reisinger and Oláh, 1974) show that septal development at the base of the conidium occurs by the process illustrated in Fig. 3.42.

In the synnematous Hyphomycete *Phaeoisaria clematidis* (Fuckel) Hughes (Fig. 3.44), the conidiogenous cells are not flexuous but still demonstrate an alternate arrangement of conidia and cylindrical denticles (Fig. 3.45). Rarely, the fertile cell may bifurcate (Fig. 3.46).

Sympodial proliferation of fertile cells may also result in formation of widely separated conidia as in *Sympodiophora stereicola* Arnold (Figs. 3.47, 3.48, 3.52A). The fertile branches of this species form transverse septa during proliferation (Fig. 3.52A; arrow in Fig. 3.47). However, the ultrastructural aspects of schizolytic secession (Fig. 3.48) resemble that in *Tritirachium* and *Beauveria* Vuill. (Fig. 3.51).

Dendrosporium lobatum Plakidas & Edgerton ex Crane produces elaborately lobed, 1-septate holoblastic conidia on sympodially proliferating conidiogenous cells (Figs. 3.49, 3.50, 3.52B). The ovoid initial (arrow in Fig. 3.52B)

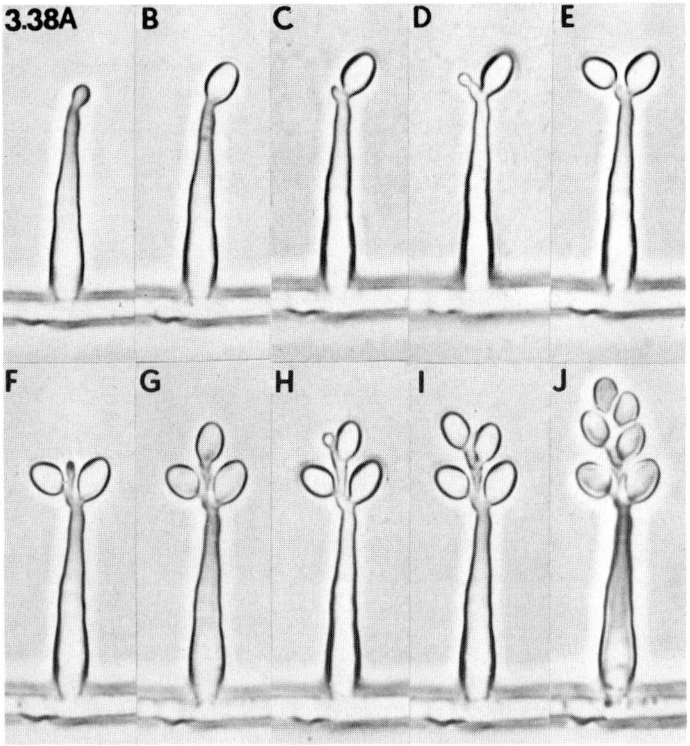

Fig. 3.38A–J *Tritirachium oryzae.* Time-lapse sequence of holoblastic conidial development at successive apices of sympodially proliferating conidiogenous cell. T = 0, $\frac{3}{4}$, $1\frac{1}{4}$, $1\frac{3}{4}$, $2\frac{1}{2}$, $3\frac{1}{4}$, $4\frac{1}{2}$, $5\frac{1}{2}$, $7\frac{1}{4}$ and $12\frac{1}{4}$ hours, respectively. ×3,654.

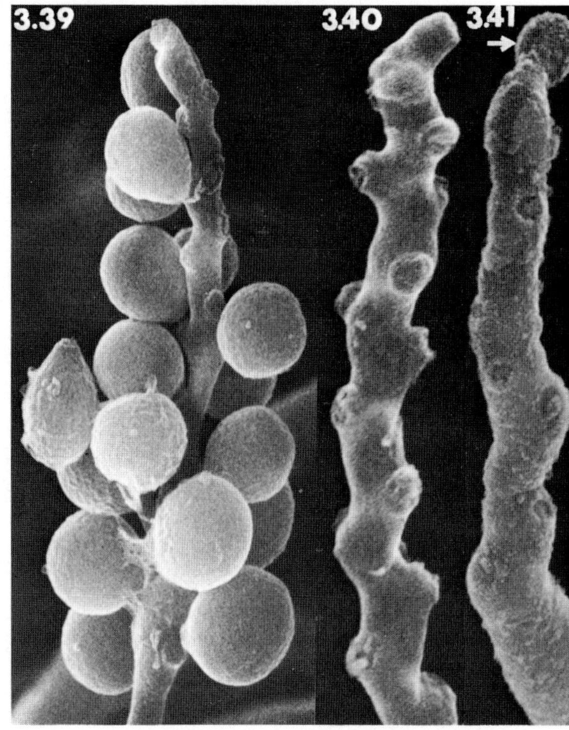

3.42

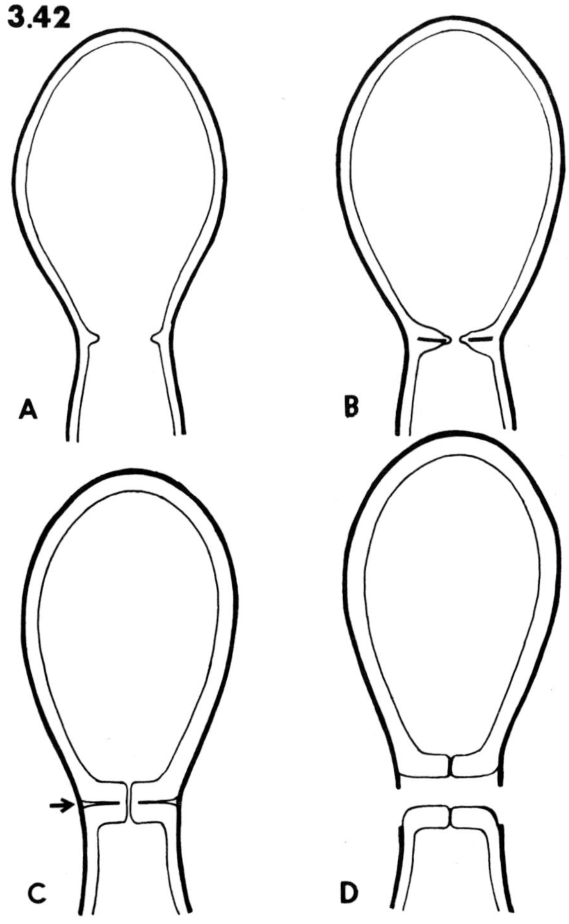

Figs. 3.39, 3.40 *Acrodontium hydnicola.* Alternate arrangement of conidia attached to sympodially proliferated conidiogenous cell (3.39). Rachis of fertile cell after conidial secession (3.40). ×6,700 and 13,000, respectively.

Fig. 3.41 *Tritirachium oryzae.* Rachis of fertile cell. Arrow indicates conidial initial. ×10,640.

Fig. 3.42A–D Interpretation of wall differentiation during holoblastic conidial formation in *Tritirachium oryzae.* Arrow in (C) indicates zone of schizolysis.

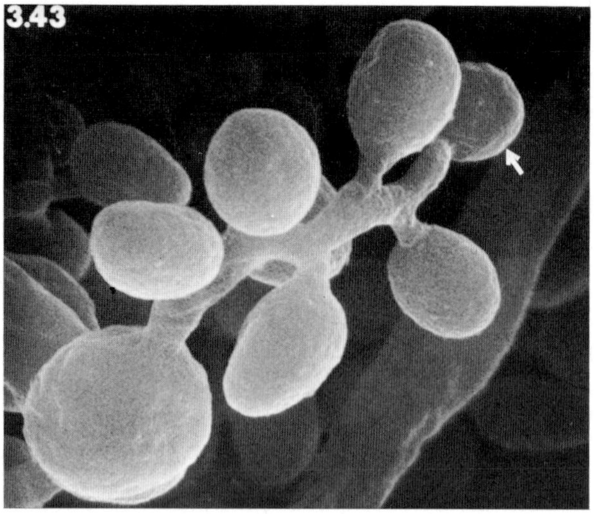

Fig. 3.43 *Beauveria bassiana.* Globose conidiogenous cell, rachis and attached conidia. Arrow indicates youngest, apical conidium. ×8,500.

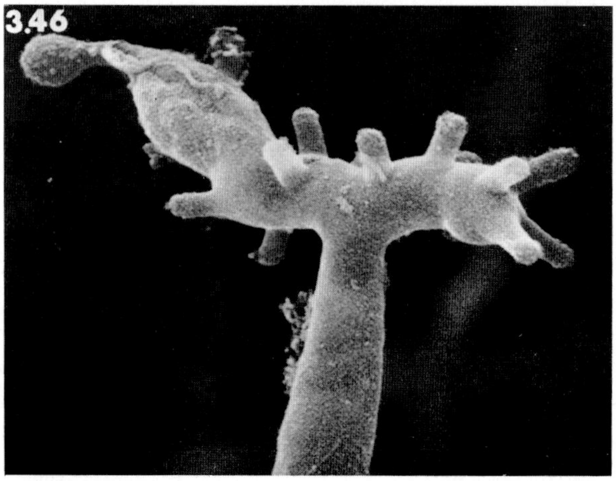

Figs. 3.44–3.46 *Phaeoisaria clematidis.* Synnema (3.44) and conidiogenous cell (3.45) showing alternate arrangement of denticles and terminal conidium. Unusual, bifurcated fertile cell (3.46) showing denticles on each branch. ×110, 4,725 and 7,290, respectively.

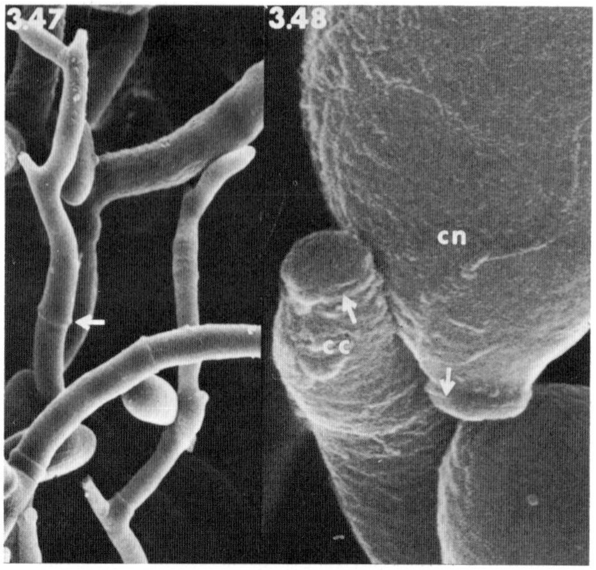

Figs. 3.47, 3.48, 3.52A *Sympodiophora stereicola.* Sympodially proliferated conidiogenous cells (3.47, 3.52A). Arrow in 3.47 indicates transverse septum. A conidium (cn) has seceded from its conidiogenous cell (cc) in 3.48 (arrows indicate rings of fractured, outer wall layer). 3.47, ×1,600; 3.48, ×12,600.

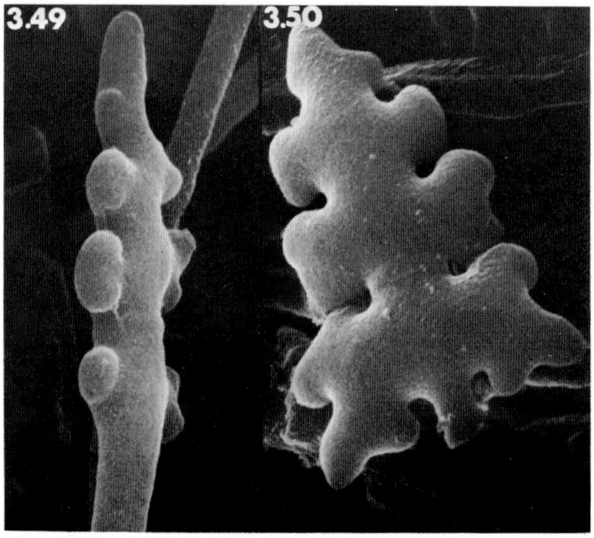

Figs. 3.49, 3.50, 3.52B *Dendrosporium lobatum.* Lobed, conidial initial (3.49) and mature, seceded conidium (3.50). Conidia produced by sympodially proliferated conidiogenous cells (3.52B). Arrow in 3.52B locates conidium initial. 3.49, ×5,865; 3.50, × 6,540.

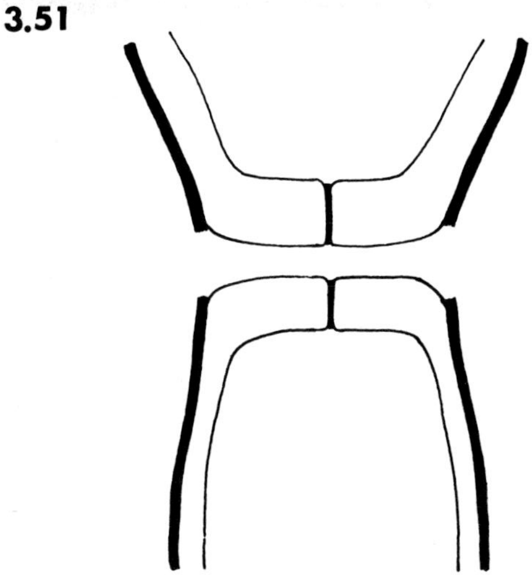

Fig. 3.51 Interpretation of wall relations during conidial secession in *Sympodiophora stereicola* (cf. 3.48).

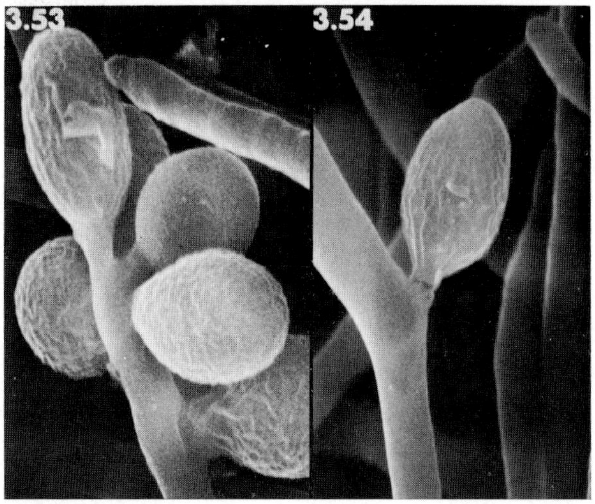

Figs. 3.53, 3.54 *Sporothrix schenckii.* Cluster of holoblastic conidia at fertile cell apex (3.53), and solitary, lateral conidium (3.54). ×7,062.

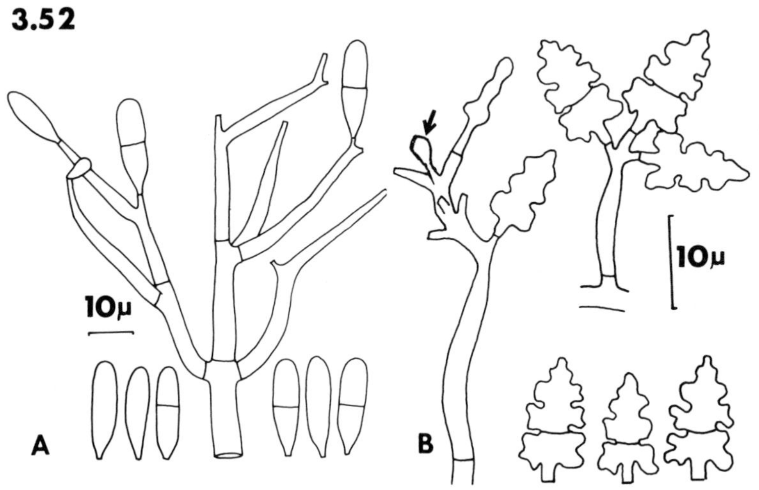

Fig. 3.52 For caption see page 38.

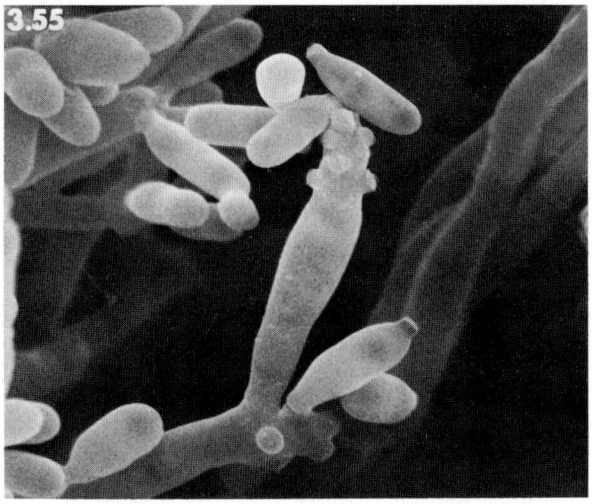

Fig. 3.55 *Fonsecaea pedrosoi.* Cluster of denticles at apex of conidiogenous cell, and seceded conidia. ×4,160.

elongates and becomes somewhat flattened and lobed, with 3 lobes developing on opposite sides of the cell (Fig. 3.49). These lateral outgrowths subsequently broaden and become secondarily lobed (Fig. 3.50).

Some Deuteromycetes produce denticulate conidiogenous cells which apparently undergo sympodial proliferation but the conidia are densely clustered. S. J. Hughes (1953) referred to these propagules as 'botryose terminus spores'.

Holoblastic (botryose) conidia

Proliferous conidiogenous cells

A cluster of conidia, each produced on short denticles, is shown at the apex of the fertile cells of *Sporothrix schenckii* Hektoen & Perkins in Fig. 3.53. The apex sometimes swells as a result of conidiogenesis. Occasionally, the conidiogenous cell resumes vegetative growth for a time before giving rise to another succession of conidia at its apex (arrows in Fig. 3.56A). Conidia may also arise directly from the hyphal surface (Fig. 3.54). Cole (1976a) has illustrated some of these features in a time-lapse analysis of *S. schenckii.* Similar features are seen in *Fonsecaea pedrosoi* (Brumpt) Negroni (Fig. 3.55) as well as other species of this genus (Schol-Schwarz, 1968). De Hoog (1974) illustrated developmental similarities between representatives of *Sporothrix* Hektoen & Perkins ex Nicot & Mariat, *Calcarisporiella* de Hoog, and *Calcarisporium* Preuss (Fig. 3.56C, D). However, a high degree of pleomorphism is evident in this group of hyphomycetous fungi, well demonstrated by *Sporothrix* and *Fonsecaea* Negroni even when grown under standard conditions in pure culture (Nicot and Mariat, 1973; de Hoog, 1974). For example, formation

3.56

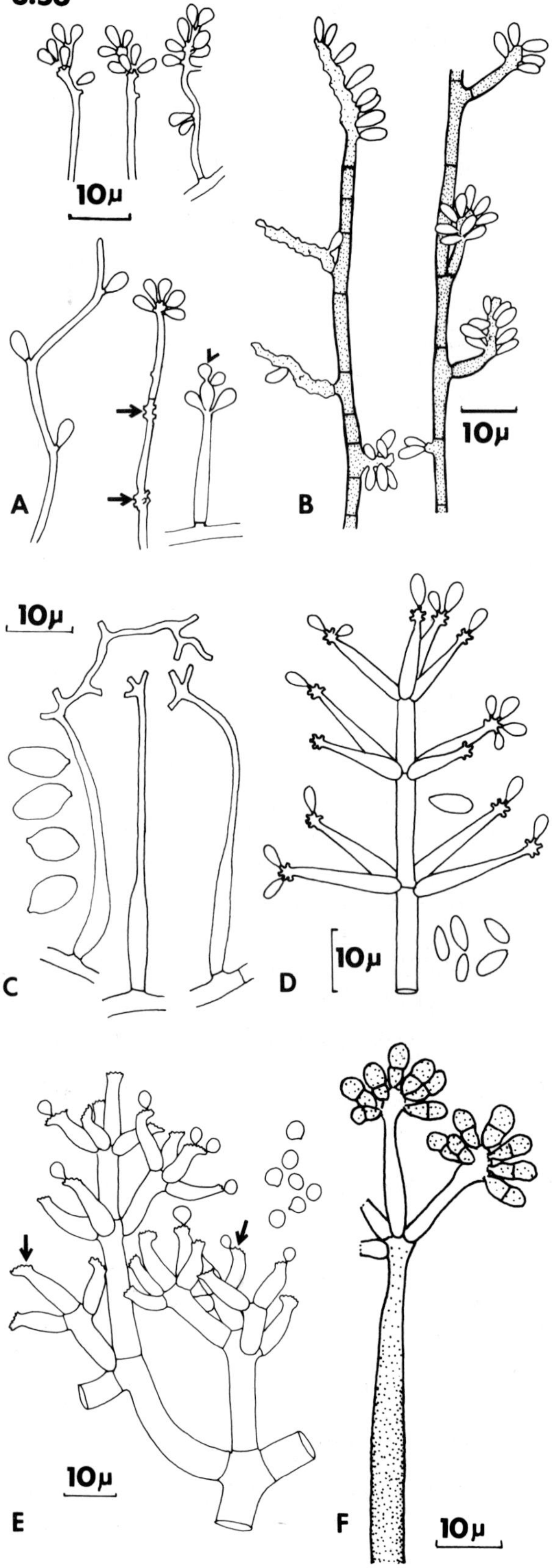

Fig. 3.56A *Sporothrix schenckii.* Variations in conidiogenous cell development showing botryose and solitary conidial arrangements. Arrows indicate clusters of denticles left after percurrent proliferation of the fertile cell. Arrowhead locates proliferated, secondary conidium.

Fig. 3.56B *Rhinocladiella atrovirens.*

Fig. 3.56C *Calcarisporiella thermophila* (after de Hoog, 1974).

Fig. 3.56D *Calcarisporium arbuscula.*

Fig. 3.56E *Costantinella micheneri.* Each conidiogenous cell demonstrates recurved, fertile apex or cervix (arrows).

Fig. 3.56F *Pseudobotrytis terrestris.*

of secondary and tertiary conidia by proliferation of primary conidia may occur in *S. schenckii* (arrowhead in Fig. 3.56A), although regular acropetal chains have not been observed. The solitary lateral conidia in this species (Fig. 3.54) often become pigmented and thick-walled and have been referred to as *chlamydospores* (Dodge, 1935), but develop as typical, holoblastic conidia. When isolates of *Sporothrix* are grown in agitated liquid culture at 37°C, yeastlike colonies are formed. Schol-Schwarz (1970) and Calley (1976) demonstrated this same phenomenon in several hyphomycetous human pathogens, including members of *Fonsecaea* and *Phialophora.*

In 1952 Nannfeldt and Ericksson proposed the name 'cervix' for the fertile region of conidiogenous cells of *Costantinella* Matruchot (Fig. 3.56E). Conidia form along one side of the swollen and ultimately recurved apex, which assumes the appearance of a tiny cockscomb. The denticles are very close together and only a minute proliferation of the conidiogenous cell occurs between successive conidia. Similar development occurs in *Pseudobotrytis terrestris* (Timonin) Subram. (Fig. 3.56F), in which conidia are formed on closely juxtaposed denticles. In these last two cases, the distinction between a proliferous and determinate conidiogenous cell is not clear.

Determinate conidiogenous cells

Barron (1968) included species which produce 'botryo-aleuriospores' (i.e., holoblastic, botryose conidia) in a separate *subseries* of his classification and interpreted their development as follows: 'A plurality of conidia is formed by the development of successive spores laterally and below the terminal spore to give an apical cluster of conidia . . .' and the conidiogenous cell is determinate (e.g., *Echinobotryum* Corda). Examination of only the mature conidiogenous cell can lead to misinterpretation of its development, as in *Nodulisporium hinnuleum* (Preuss) G. Smith (Fig. 3.57A–D). The conidia are produced on distinct denticles and arranged alternately at the slightly swollen apex of the conidiogenous cell (Fig. 3.57A), which has been interpreted as evidence that the latter proliferates sympodially during conidiogenesis (S. J. Hughes, 1953; Tubaki, 1958; Chesters

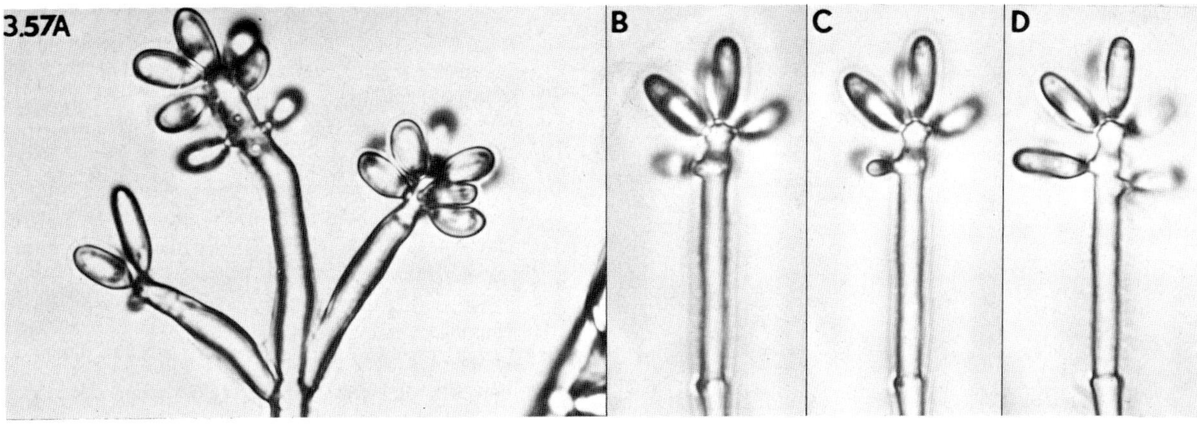

Fig. 3.57A–D *Nodulisporium hinnuleum.* Botryose arrangement of conidia produced by determinate conidiogenous cells (A). Time-lapse, photomicrographic sequence (B–D). T = 0, 1 and 12 hours, respectively. A, ×1,600; B–D, ×1,440.

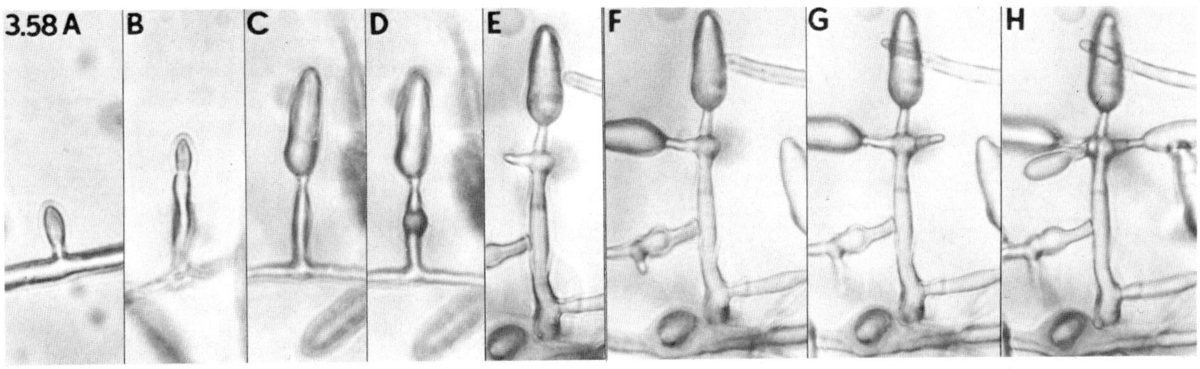

Fig. 3.58A–H *Asteromyces cruciatus.* Time-lapse sequences (C, D, E–H) of botryose, holoblastic conidial development from determinate conidiogenous cells. T = 0, $1\frac{1}{2}$ hours (C–D); 0, $5\frac{1}{2}$, $6\frac{3}{4}$, $23\frac{1}{4}$ hours (E–H, respectively). ×1,150.

Fig. 3.59 *Asteromyces cruciatus.* ×1,520

Fig. 3.60 *Wardomyces simplex.* Cluster of holoblastic conidia attached to apex of determinate conidiogenous cell. ×4,335.

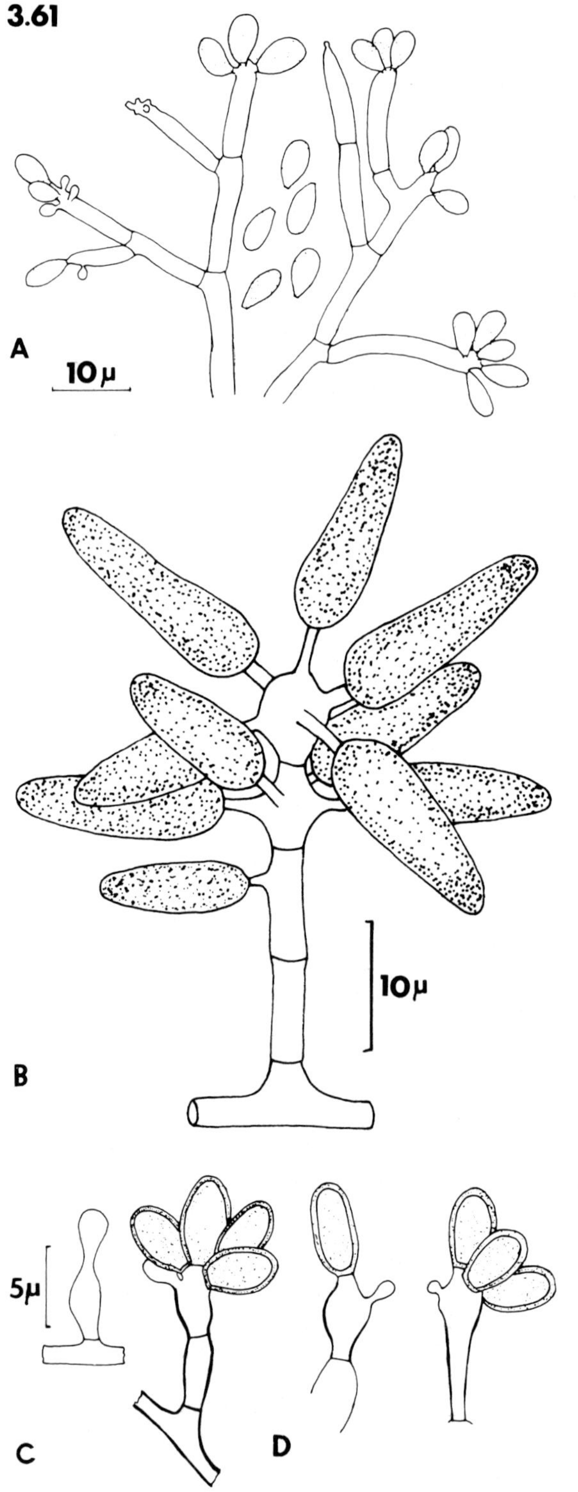

Fig. 3.61A *Nodulisporium griseo-brunneum.*

Fig. 3.61B *Asteromyces cruciatus.* Two clusters of conidia which have formed basipetally.

Fig. 3.61C, D *Wardomyces* spp. Stages of botryose conidial formation from determinate conidiogenous cells of *W. anomalus* (C) and *W. inflatus* (D).

and Greenhalgh, 1964; Rogers, 1966; Greenhalgh and Chesters, 1968; Barron, 1968). However, the time-lapse sequence (Fig. 3.57B–D) reveals that conidia are actually formed in basipetal succession (Cole, 1971). Taxonomically related Hyphomycetes, mainly conidial states of Xylariaceae, may undergo similar development and should be examined more carefully (e.g., *Nodulisporium griseo-brunneum* R. S. Mehrotra, Fig. 3.61A). In a 16 mm time-lapse sequence of *N. hinnuleum* (Cole and Kendrick, 1969d) it was demonstrated that more than one cluster of conidia may arise from different fertile regions of the same determinate conidiogenous cell. This also occurs in the marine Hyphomycete *Asteromyces cruciatus* F. & Mme. Moreau ex Hennebert (Hennebert, 1962). The time-lapse, photomicrographic sequences (Fig. 3.58A–H) show that the conidiogenous cell first appears as a lateral branch (Fig. 3.62A) which elongates and later produces a conidium at its apex (Fig. 3.58B, C). Subsequently, a succession of up to 6–8 conidia is formed below the first, producing a whorled arrangement (Fig. 3.58G, H). A second cluster of conidia may later develop below the first (Fig. 3.61B) and, as in *N. hinnuleum,* the length of the conidiogenous cell does not change. Successive conidia may also form in clusters from determinate conidiogenous cells of *Wardomyces simplex* Sugiyama *et al.* (Fig. 3.60), *W. anomalus* Brooks & Hansf. (Fig. 3.61C) and *W. inflatus* (Marchal) Hennebert (Fig. 3.61D). Careful studies will undoubtedly reveal that other genera of Deuteromycetes behave similarly.

Several species produce synchronous clusters of holoblastic conidia from fertile apical vesicles, or ampullae (e.g., *Botrytis* Pers. ex Fr., *Botryosporium* Corda, *Spiniger* Stalpers, *Gonatobotryum* Sacc.). These are discussed below.

Conidiogenous cells (proliferous or determinate) which form fertile ampullae

One rather uncertain member of this group is *Blastobotrys nivea* Klopotek (Figs. 3.62–3.64). Clusters of conidia form at successive apices of the sympodially proliferating conidiogenous cell. The vesicular structures at first appear to be ampullae producing conidia but the intact cluster commonly secedes as a unit (Figs. 3.62, 3.64), while individual, rough-walled cells are apparently detached only with difficulty from the vesicle (arrow in Fig. 3.62). De Hoog (1974) has suggested that the vesicles are actually primary conidia which in turn give rise to secondary, globose propagules. The number of rough-walled secondary conidia produced is variable (Fig. 3.64), and surface replicas (Fig. 3.63) show that the irregular ornamentations revealed by the SEM are ridges of wall material. Whether the ampulla-like structure actually represents a conidium requires further investigation.

In *Botrytis cinerea* Pers. ex Fr., the fertile branches of the conidiophore terminate as ampullae which produce holoblastic conidia on short denticles (Figs. 3.65, 3.66, 3.70A). The synchrony of conidial development on several different ampullae of the same conidiophore is suggested by Figs. 3.65 and 3.66. The conidia secede readily. In species of *Botryosporium*, each fertile branch produces several ampullae on which conidia develop synchronously (Fig. 3.70B).

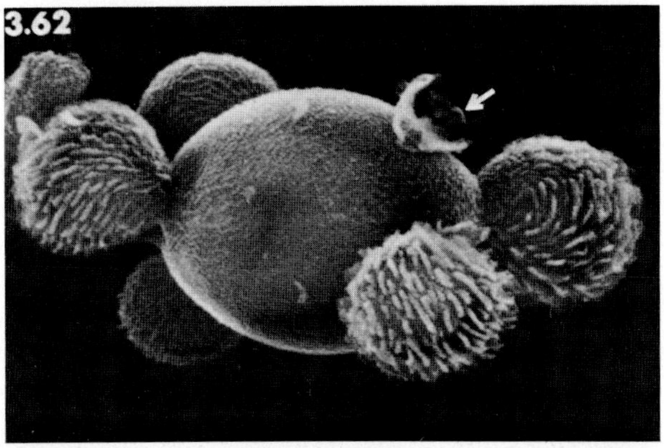

3.64

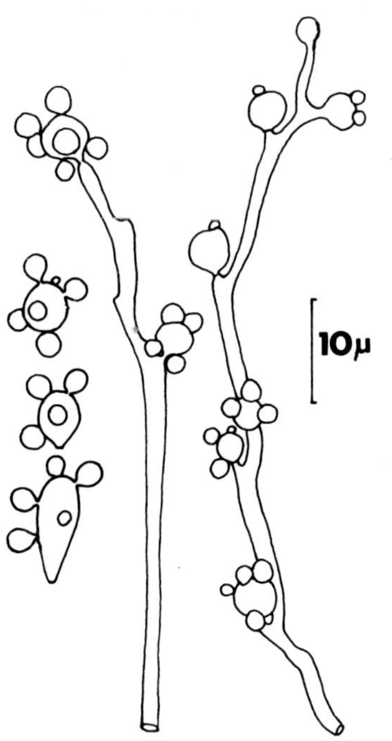

Figs. 3.62–3.64 *Blastobotrys nivea.* Large primary and rough-walled secondary conidia (3.62). Arrow locates abstriction scar of secondary conidium. Surface replica of secondary conidium (3.63). Sympodially proliferated conidiogenous cells showing attached and seceded conidia (3.64). 3.62, ×11,430; 3.63, ×46,500.

Ultrastructural aspects of conidiogenesis in the *Ostracoderma* state of *Peziza ostracoderma* Korf (Figs. 3.67, 3.68) were reported by G. C. Hughes and Bisalputra (1970). The unbranched, main axis of the conidiophore in this species produces up to 12 long, divergent fertile hyphae at its apex, each forming conidia synchronously over its entire surface (Fig. 3.70C), like the ampullae described above. Early stages of conidiogenesis (Figs. 3.67 and 3.68) are clearly holoblastic. Similar development occurs in species of *Oedocephalum* Preuss (Fig. 3.74D) and an ultrastructural examination of conidiogenesis in *Oedocephalum roseum* Cook was reported by Cook (1972, 1974). This genus includes conidial states of Pezizales (Stalpers, 1974a) and when a new *oedocephaloid* imperfect state with basidiomycetous affinities was discovered, it was necessary to accommodate it in the new genus, *Spiniger* Stalpers (Stalpers, 1974b). The terminal ampullae of *S. meineckellus* (Olson) Stalpers give rise synchronously to conidia on distinct cone-shaped denticles (Fig. 3.69).

As one could predict, deviations from this pattern of botryose conidium formation occurs. In *Riessia semiophora* Fresenius (Fig. 3.70E), another basidiomycetous imperfect state, microconidia are produced synchronously in groups of four at the apex of slightly swollen fertile hyphae, while other conidiogenous cells form single, holoblastic stauroconidia at their apex (Goos and Tubaki, 1973). The conidiophores of the imperfect state of *Peziza quelepidotia* Korf & O'Donnell do not form swollen, fertile branches but instead produce widespaced conidia synchronously from apparently undifferentiated branches (O'Donnell *et al.*, 1976). The authors misinterpreted the conidiogenous cells as retrogressive (the fertile cells are determinate).

Gonatobotryum apiculatum (Peck) Hughes produces aerial conidiogenous cells with conspicuous, terminal ampullae, each forming a synchronous cluster of primary conidia (Fig. 3.71A–D). Subsequently these cells asynchronously produce acropetal chains of secondary conidia (Fig. 3.71E–G). After conidiogenesis has ceased, the fertile cell may proliferate percurrently, producing a new fertile ampulla, and the process of conidial development is repeated (Fig. 3.71H, 3.72). Thin-sections have demonstrated the wall relations associated with conidial development (Cole, 1973b) and the arrangement of septa in denticles separating the primary conidia and conidiogenous cell (Fig. 3.75), and between adjacent secondary conidia in a chain (Fig. 3.73D). It is logical that the ampulla serves as a cytoplasmic reservoir for the aerial chains of cells. The necessity to prevent interruption of this food supply, even though one or more chains may be damaged during the developmental process, has probably led to evolution of the elaborate kind of septal ontogeny which provides a measure of *safety* against loss of cytoplasm from this vulnerable biological system. A

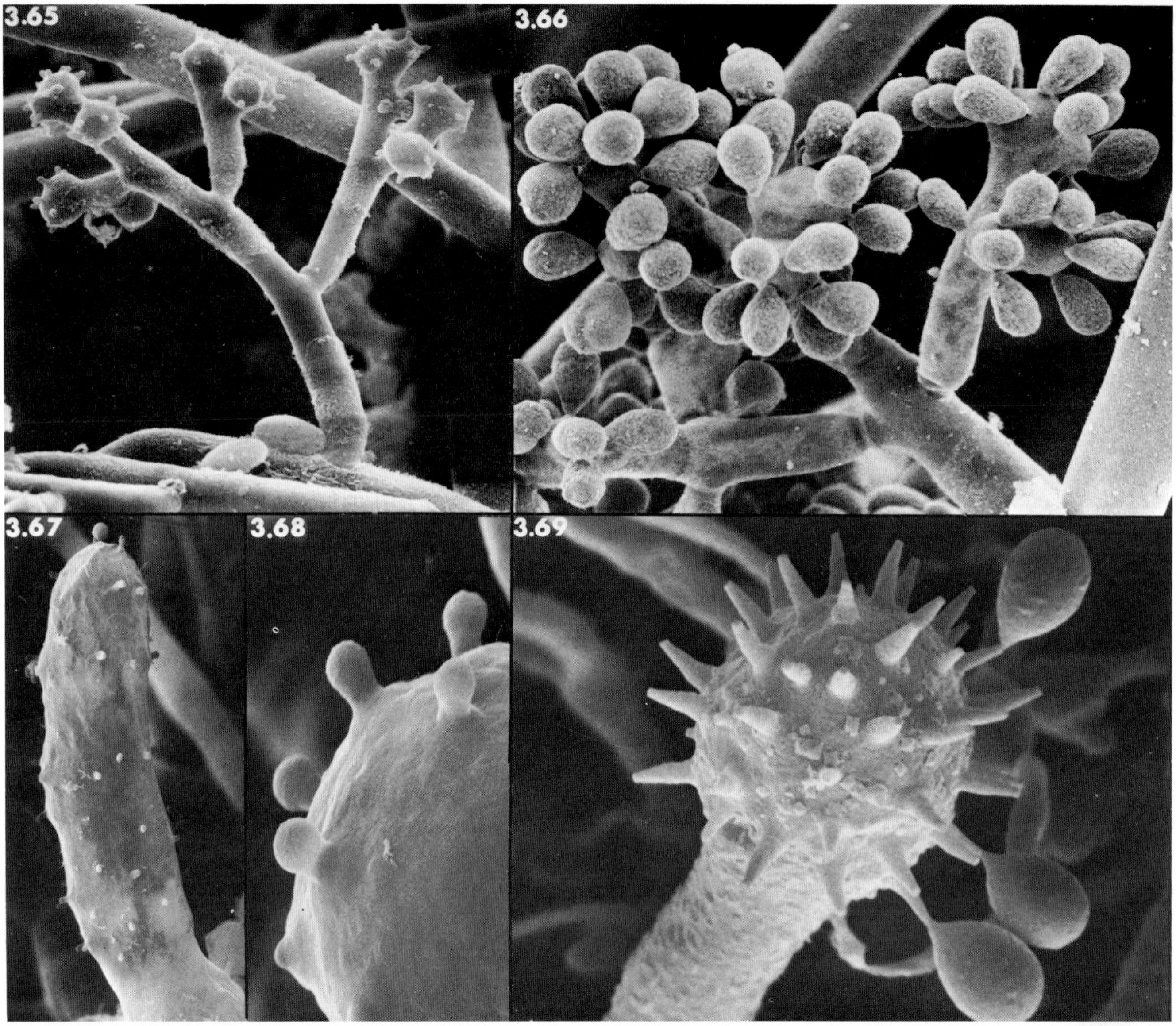

Figs. 3.65, 3.66 *Botrytis cinerea.* Synchronized development of holoblastic conidial initials from fertile branches of conidiophore (3.65). Later stage of conidial formation (3.66). ×905 and 5,000, respectively.

Figs. 3.67, 3.68 *Ostracoderma* state of *Peziza ostracoderma.* Holoblastic, conidial initials in stages of synchronized development. ×1,335 and 6,090, respectively.

Fig. 3.69 *Spiniger meineckellus.* Fertile ampulla of conidiogenous cell showing conical denticles and three attached conidia. ×8,600 (courtesy J. A. Stalpers).

diagrammatic interpretation of wall differentiation in *G. apiculatum* is presented in Fig. 3.73A–D. Karyological studies of this species (Kendrick and Chang, 1971) have demonstrated that nuclei migrate through the septal pores after the conidial initials have partially differentiated. However, when the conidium secedes, or damage occurs, the pores are rapidly closed by Woronin bodies (Fig. 3.73D). Conidial secession is rhexolytic and results in loss of some cytoplasmic material from the separating cells (arrow in Fig. 3.75). Similar developmental and ultrastructural features are demonstrated by species of *Nematogonium* Desm. (Fig. 3.74A), in which acropetal chains of conidia are also produced, and by *Gonatobotrys simplex* Corda (Fig. 3.74B), whose ampullae produce synchronous clusters of only primary conidia; the fertile cell proliferates percurrently, and the same process of conidiogenesis is repeated at the new apex. Rhexolytic secession also occurs in *Peziza quelepidotia* (O'Donnell *et al.*, 1976). Presumably, once the fertile cells have contributed their cytoplasmic material to development of the conidia, their function is complete and they degenerate (Fig. 3.76). Similar developmental and ultrastructural features may be revealed in *Botrytis, Botryosporium* and *Ostracoderma* Fr.

The earlier discussion of secondary conidium development in *G. apiculatum* (Fig. 3.75E–G) introduces the next group of deuteromycetous fungi.

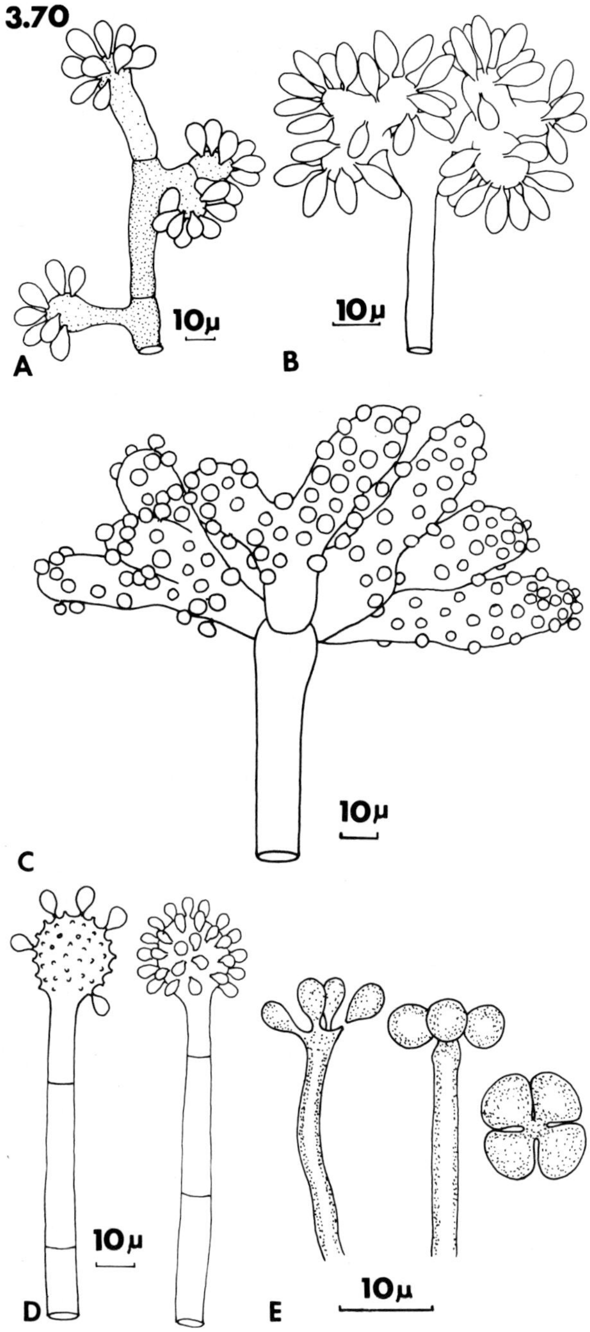

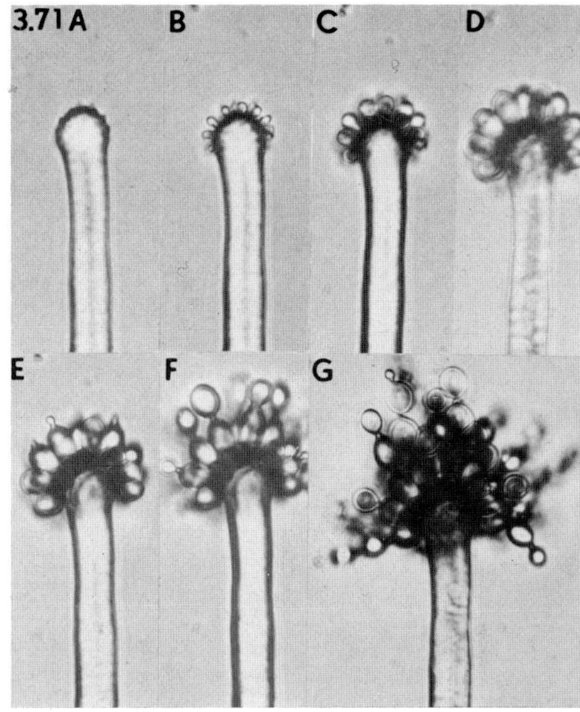

Fig. 3.71A–H *Gonatobotryum apiculatum.* Time-lapse sequence of synchronous, holoblastic conidial formation (A–D), asynchronous development of acropetal chains (E–G), and percurrent proliferation of the conidiogenous cell (H). T = 0, $\frac{1}{2}$, $\frac{3}{4}$, $1\frac{3}{4}$, $2\frac{3}{4}$, $3\frac{3}{4}$, $5\frac{3}{4}$ and 10 hours, respectively. ×780.

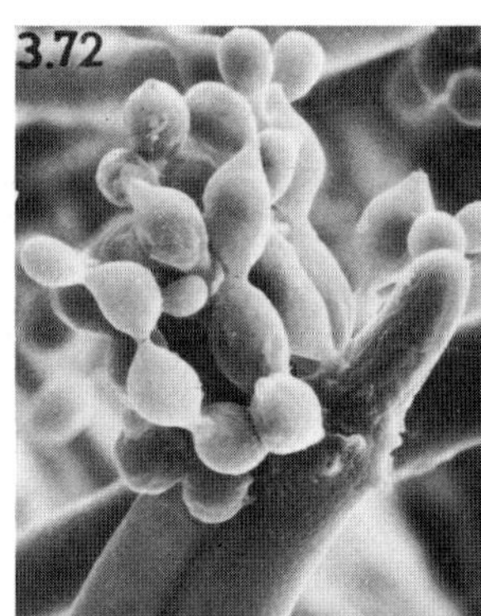

Fig. 3.70A *Botrytis cinerea.*

Fig. 3.70B *Botryosporium longibrachiatum* (Oudemans) Maire.

Fig. 3.70C *Ostracoderma* state of *Peziza ostracoderma.*

Fig. 3.70D *Oedocephalum beticola* Oudemans.

Fig. 3.70E *Riessia semiophora.*

Fig. 3.72 *Gonatobotryum apiculatum.* Early stage of percurrent proliferation. Denticles and conidial chains associated with previous fertile apex also visible. ×1,491.

3.73

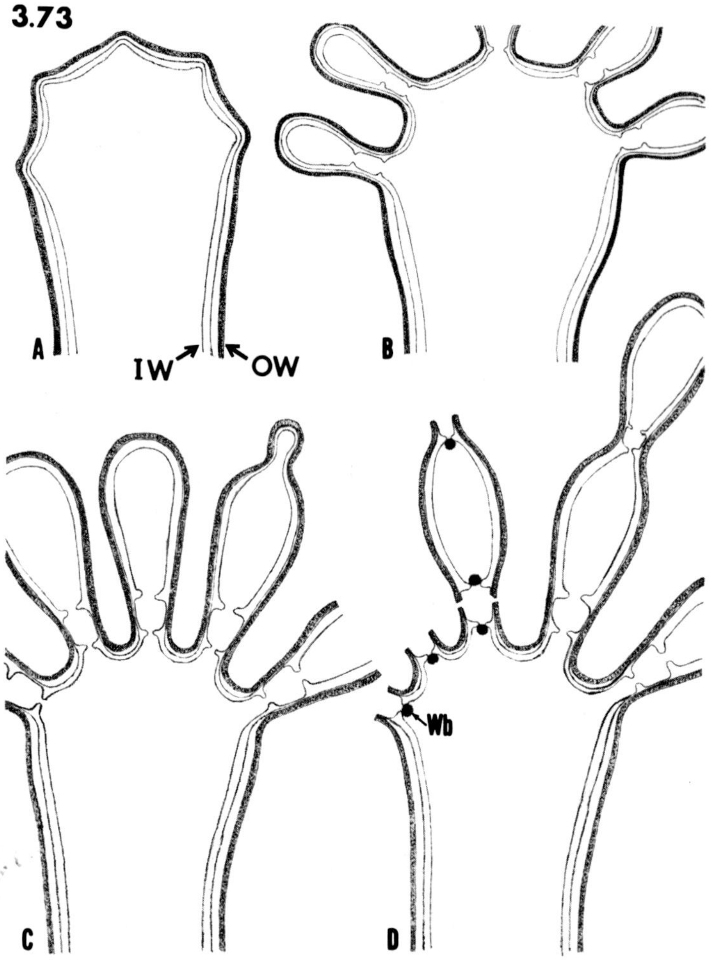

Fig. 3.73A–D Interpretation of wall differentiation during synchronous, primary and asynchronous, secondary conidial formation in *Gonatobotryum apiculatum.* IW, inner wall layer; OW, outer wall layer.

3.74

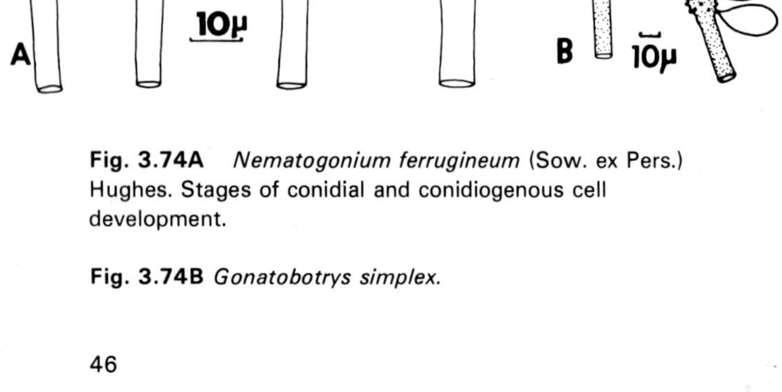

Fig. 3.74A *Nematogonium ferrugineum* (Sow. ex Pers.) Hughes. Stages of conidial and conidiogenous cell development.

Fig. 3.74B *Gonatobotrys simplex.*

3.75

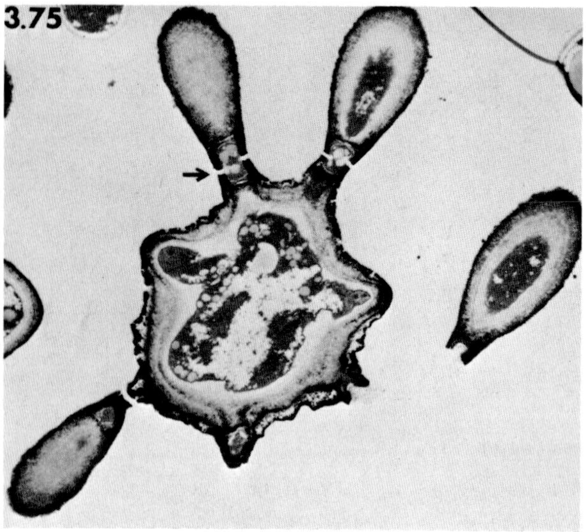

Fig. 3.75 *Gonatobotryum apiculatum.* Thin-section of ampulla and holoblastic conidia. Arrow points to separating cell, which is sacrificed as a result of rhexolytic secession. ×8,040.

3.76

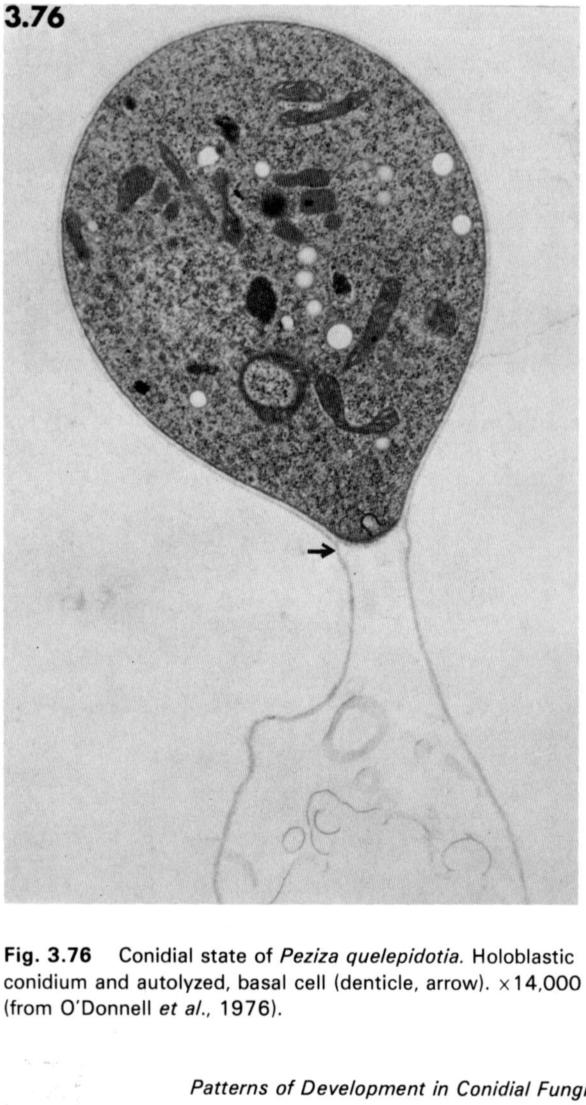

Fig. 3.76 Conidial state of *Peziza quelepidotia.* Holoblastic conidium and autolyzed, basal cell (denticle, arrow). ×14,000 (from O'Donnell *et al.*, 1976).

Holoblastic (catenulate) conidia

Formation of acropetal chains of conidia is illustrated by the time-lapse sequence of *Cladosporium sphaerospermum* Penz. in Fig. 3.77A–H. The first two conidia at the apex of the fertile hypha are morphologically distinct from the terminal conidium. Later, these same cells may branch, initiating development of new acropetal chains (Fig. 3.80A). Because of the typical ontogeny of these basal cells, they have appropriately been referred to as *ramoconidia* (S. J. Hughes, 1953; M. B. Ellis, 1971b). Similar development occurs in *C. herbarum* (Pers.) Link ex S. F. Gray (Fig. 3.78) and has been illustrated by Hashmi *et al.* (1973). The apex of the fertile hypha of *C. herbarum* usually becomes swollen as a result of conidiogenesis (Figs. 3.80B, 3.83), and distinct abstriction scars remain on its surface (Fig. 3.83). Both sympodially and percurrently proliferating conidiogenous cells occur in species of *Cladosporium* Link ex Fr. (Fig. 3.80C and D, respectively), while other species produce well-differentiated conidiophores (e.g., *C. resinae* (Lindau) de Vries, Fig. 3.80E, and *C. chlorocephalum* (Fresen.) Mason & M. B. Ellis, Fig. 3.80F).

An unusual mechanism of acropetal chain development is demonstrated by *Torula herbarum* (Pers.) Link ex S. F. Gray (Fig. 3.81), which has been examined both by time-lapse, light microscopy (Hashmi *et al.*, 1973) and by electron microscopy (D. H. Ellis and Griffiths, 1975a, b). A lightly pigmented cell, first arising laterally or terminally from the vegetative hypha, forms the base of the conidial chain (arrows in Fig. 3.81A, C). Ellis and Griffiths (1975a) referred to this as a *subtending cell* rather than a conidiogenous cell as suggested by Hashmi *et al.* (1973). Instead, the former authors stated that the subtending cell gives rise to the conidiogenous cell whose wall is differentiated into a thickened basal region and thinner walled, apical proliferating zones (Fig. 3.82A, B; arrowheads in Fig. 3.81A–C). They concluded that conidium initiation occurs 'by protoplasmic pressure following the structural modification of the conidiogenous cell; the deposition of melanin solely within lateral walls results in the formation of an inextensible barrier which canalizes any mechanical pressure towards the apex'. The conidia of *T. herbarum* are usually 4–5 celled (Fig. 3.81B, D) and the same conidiogenous cell may produce more than one propagule (Fig. 3.81A–C). In an ultrastructural examination of *Alysidium resinae* (Fr.) M. B. Ellis (= *Torula ramosa* Peck), D. H. Ellis and Griffiths (1977) have shown that conidia form enteroblastically.

One final example of acropetal chain formation is demonstrated by species of *Periconia* Tode ex Fr. The unusual feature of this fungus is that the holoblastic conidia

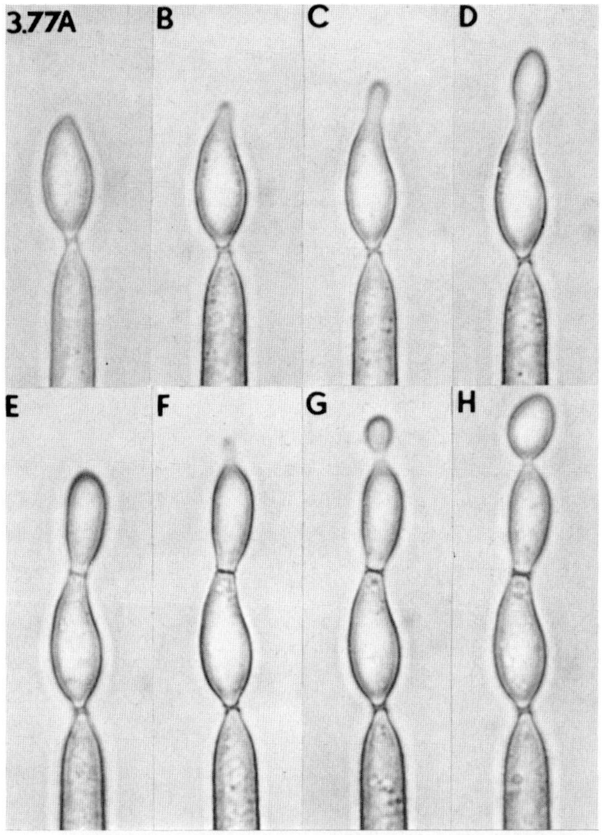

Fig. 3.77A–H *Cladosporium sphaerospermum.* Time-lapse sequence of acropetal chain formation. T = 0, 15, 25, 30, 75, 115, 125 and 140 minutes, respectively. ×2,835.

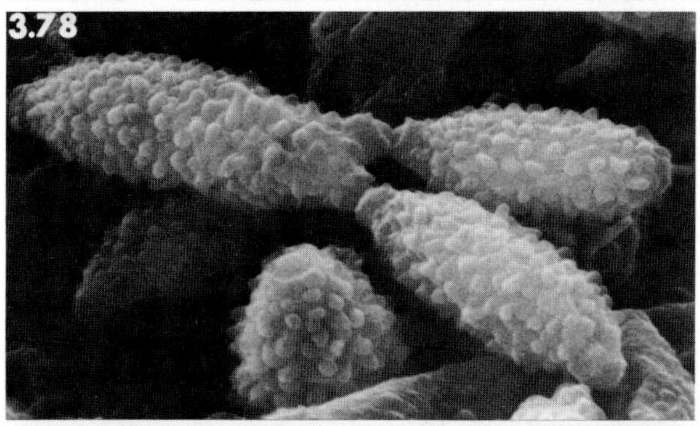

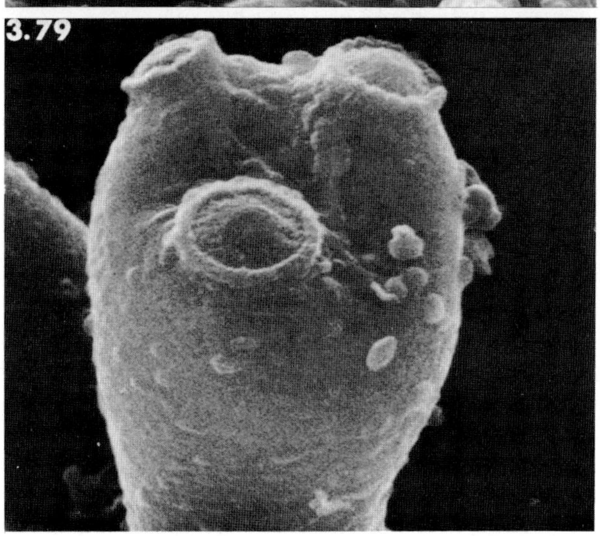

Figs. 3.78, 3.79 *Cladosporium herbarum.* Branched ramoconidium (3.78) and apex of fertile hypha showing abstriction scars (3.79). ×6,825 and 11,000 respectively.

3.80

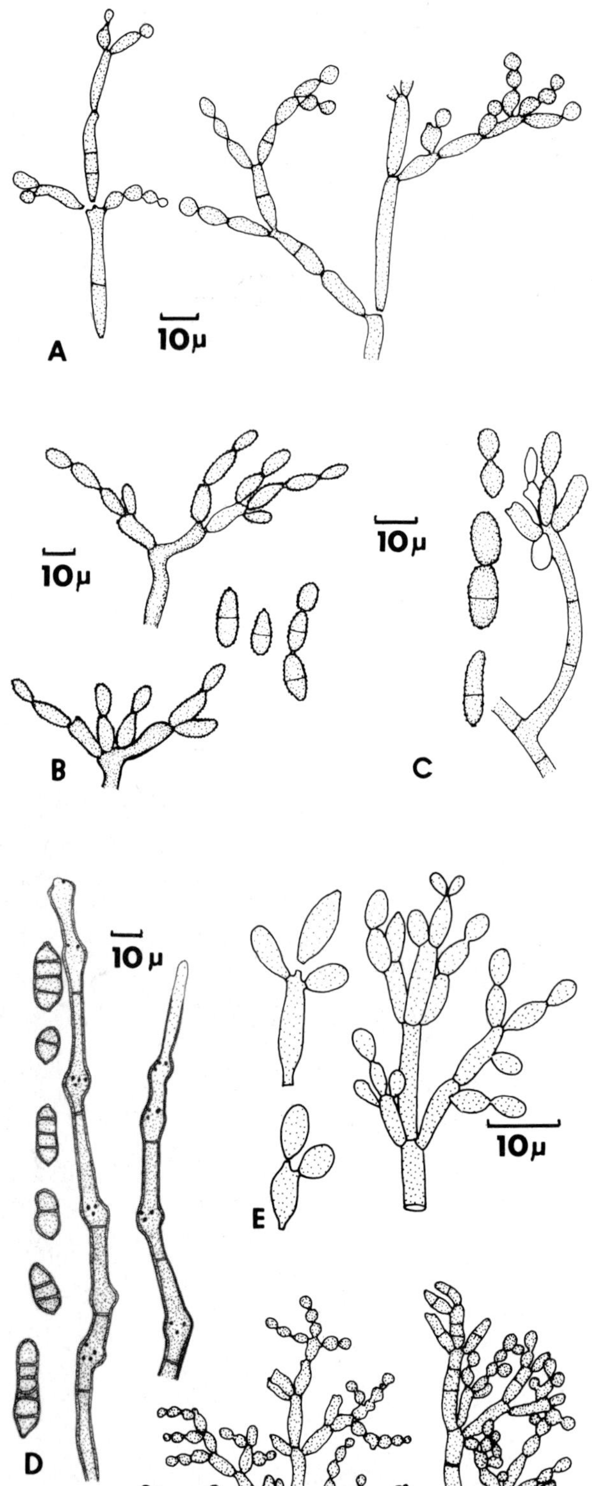

Fig. 3.80A–F *Cladosporium* spp. *C. sphaerospermum* (A), *C. herbarum* (B), *C. variabile* (C), *C. colocasiae* Sawada (D), *C. resinae* (E), *C. chlorocephalum* (F) (A–D and F after M. B. Ellis, 1971b).

3.81

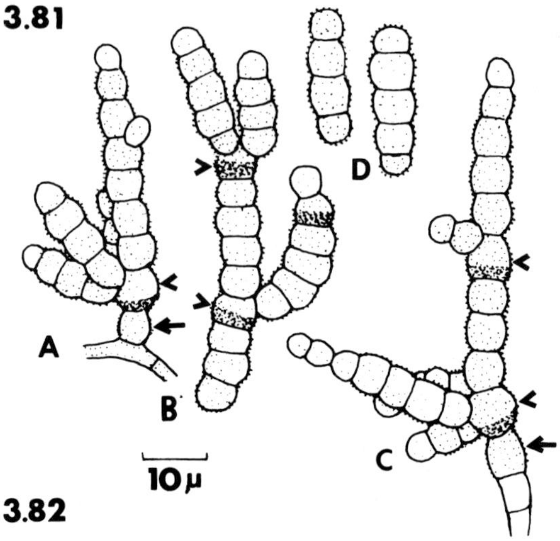

3.82

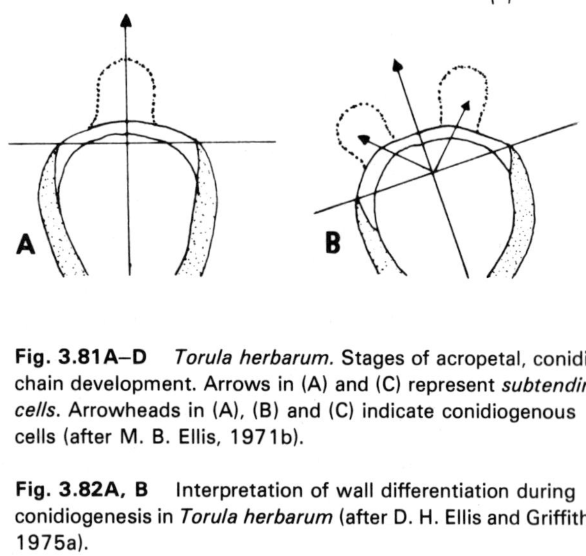

Fig. 3.81A–D *Torula herbarum.* Stages of acropetal, conidial chain development. Arrows in (A) and (C) represent *subtending cells.* Arrowheads in (A), (B) and (C) indicate conidiogenous cells (after M. B. Ellis, 1971b).

Fig. 3.82A, B Interpretation of wall differentiation during conidiogenesis in *Torula herbarum* (after D. H. Ellis and Griffiths, 1975a).

Figs. 3.83–3.88 *Periconia cookei.* Stages of conidial development. Synchronous development of holoblastic primary conidia (3.83) which bud apically (3.84) and form acropetal chains of secondary conidia (3.85, 3.86). The latter mature basipetally (3.87, 3.88). ×4,000, 2,600, 2,720, 3,200, 1,197 and 4,770, respectively.

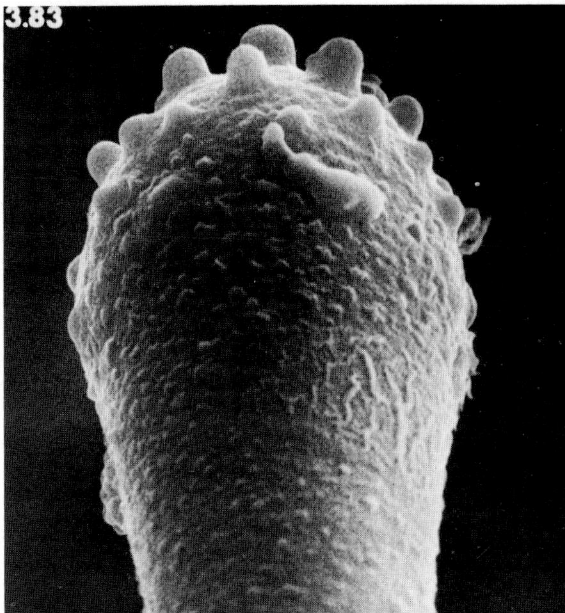
3.83

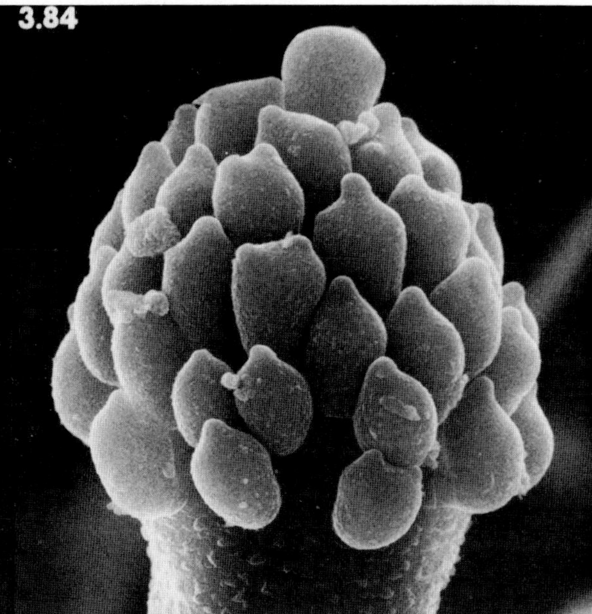
3.84

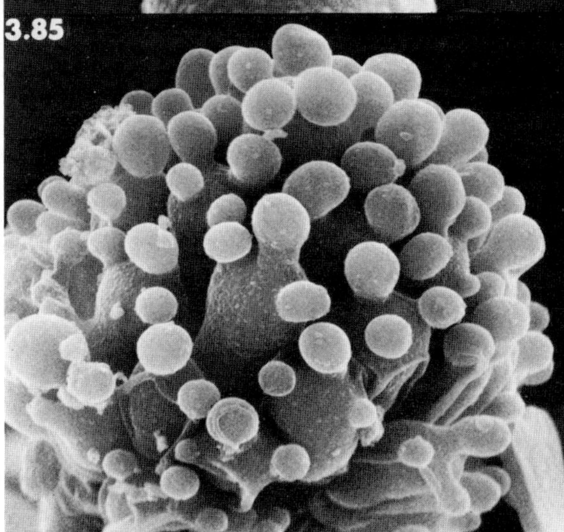
3.85

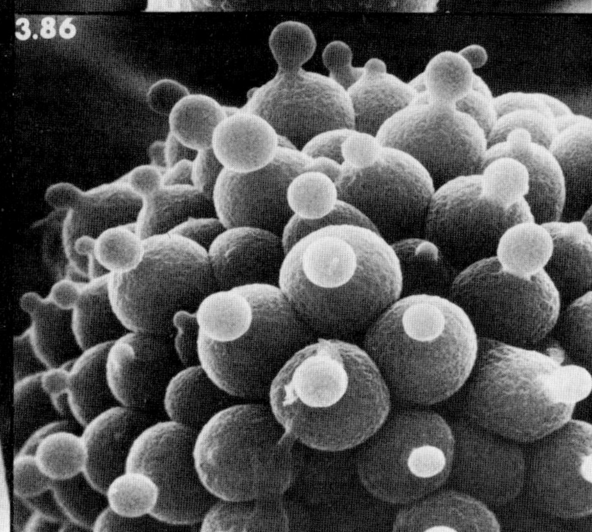
3.86

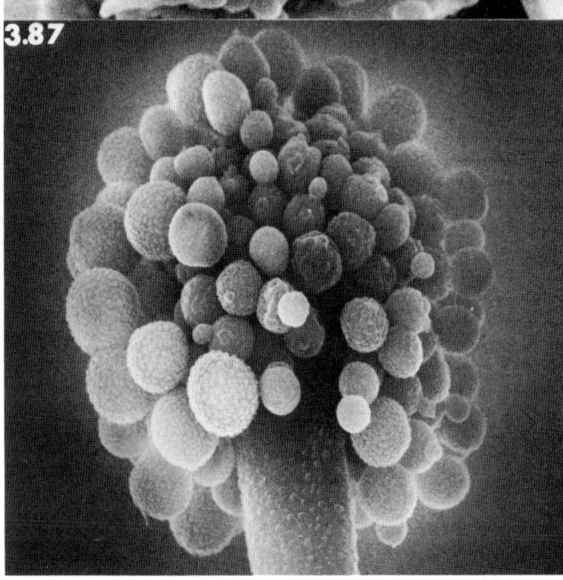
3.87

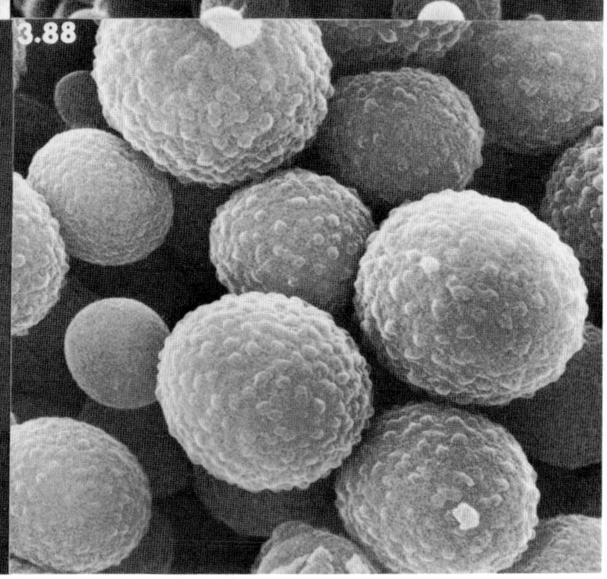
3.88

arise in acropetal succession, but mature basipetally (M. B. Ellis, 1971b). Various stages of conidiogenesis in *P. cookei* Mason & M. B. Ellis are illustrated in Figs. 3.83–3.88. The more or less synchronously produced primary conidial initials at the swollen conidiogenous cell apex, or ampulla in Fig. 3.83, later enlarge forming 1–3 protuberances at their apices (Fig. 3.84), which is indicative of an early stage in the acropetal development of secondary conidia (Figs. 3.85, 3.86). Conidial maturation is recognized by development of large, spherical, verrucose conidia which first appear at the apex of the chains (Fig. 3.87) and then differentiate basipetally (Fig. 3.88). A similar mode of development probably occurs in other conidial fungi, such as species of *Haplobasidion* Eriksson and *Lacellinopsis* Subram., and comparative studies of these fungi are proposed.

As stated at the beginning of this chapter, blastic ontogeny has been separated into two processes, holoblastic and enteroblastic development, based on distinguishing features of conidial wall differentiation. This dichotomy originally emerged from interpretation of data provided by light-microscopic investigations (Kendrick, 1971a). However, ultrastructural studies have not entirely supported these earlier developmental concepts, as pointed out in the following discussion.

Poroconidia

S. J. Hughes (1953) determined that the main character of poroconidial ontogeny is the formation of a minute pore in the wall of the conidiophore. He stated that the fertile cell 'is nearly always rounded, and the outer and inner walls come to an end abruptly, thus delimiting a more or less cylindrical pore'. M. B. Ellis (1971a) suggested that 'enteroblastic conidiogenous cells are *tretic* when they produce conidia, usually solitary but sometimes in acropetal chains, by protrusion of the inner wall through one or more channels in the outer wall'. This concept of the formation of *tretoconidia* was based on light-microscopic observations of *Helminthosporium velutinum* Link ex Ficinus & Schubert (M. B. Ellis, 1971b) as well as thin-sections of *Alternaria brassicicola* (Berk.) Sacc. presented by Campbell (1968, 1969). Campbell (1968) proposed that the channel through the conidiogenous cell wall was formed by localized mural enzyme activity and a diagrammatic interpretation of his micrographs is presented in Fig. 3.89A. The channel, or pore, provides a pathway for translocation of materials between the conidiogenous cell and conidium and it seems likely that it plays a significant role in conidiogenesis. Cole (1973a) examined the ultrastructure of *Drechslera sorokiniana* (Sacc.) Subram. & Jain and supported Ellis' (1971a, b) concept of the tretoconidium. Cole suggested that the innermost wall layer of the fertile cells is continuous with the developing wall of the conidial initial, while the outer wall layers break down (enzymatically?), as illustrated in Fig. 3.89B. Carroll (1972), and Carroll and Carroll (1974a) presented thin-section micrographs of *Stemphylium botryosum* Wallroth and *Ulocladium atrum* Preuss, respectively, which indicated that the fertile wall remains intact during conidiogenesis, suggesting that the conidia are holoblastic (Fig. 3.89C). These authors challenged earlier interpretations of thin-sections presented by Campbell (1968, 1969) and Cole (1973a), and proposed that *A. brassicicola* and *D. sorokiniana* also produce holoblastic conidia. Brotzman *et al.* (1975) supported Carroll's concept of poroconidial development with thin-section examinations of *Helminthosporium maydis* Nisikado and Miyake. Unlike *Ulocladium atrum* (Carroll and Carroll, 1974a), however, the conidiogenous cell wall of *H. maydis* is differentiated into at least three distinct layers which Brotzman *et al.* (1975) concluded are continuous with the wall of the conidial initial. Our interpretation of their micrographs is presented in Fig. 3.89D.

Although this interpretative problem of wall differentiation has not been completely resolved, we have opted to consider the poroconidium as a special kind of holoblastic propagule after careful examination of current ultrastructural data. It is evident that aspects of conidial and conidiogenous cell wall differentiation during conidiogenesis cannot be easily incorporated into a classification of the Deuteromycetes (Harvey 1974b; D. H. Ellis and Griffiths, 1977). However, Hughes's (1953) Section VI in his experimental classification,

Fig. 3.89A–D Various interpretations of wall differentiation during poroconidial development based on thin-section examinations (A, Campbell, 1968, 1969; B, Cole, 1973a; C, Carroll, 1972; Carroll and Carroll, 1974a; D, Brotzman *et al.*, 1975).

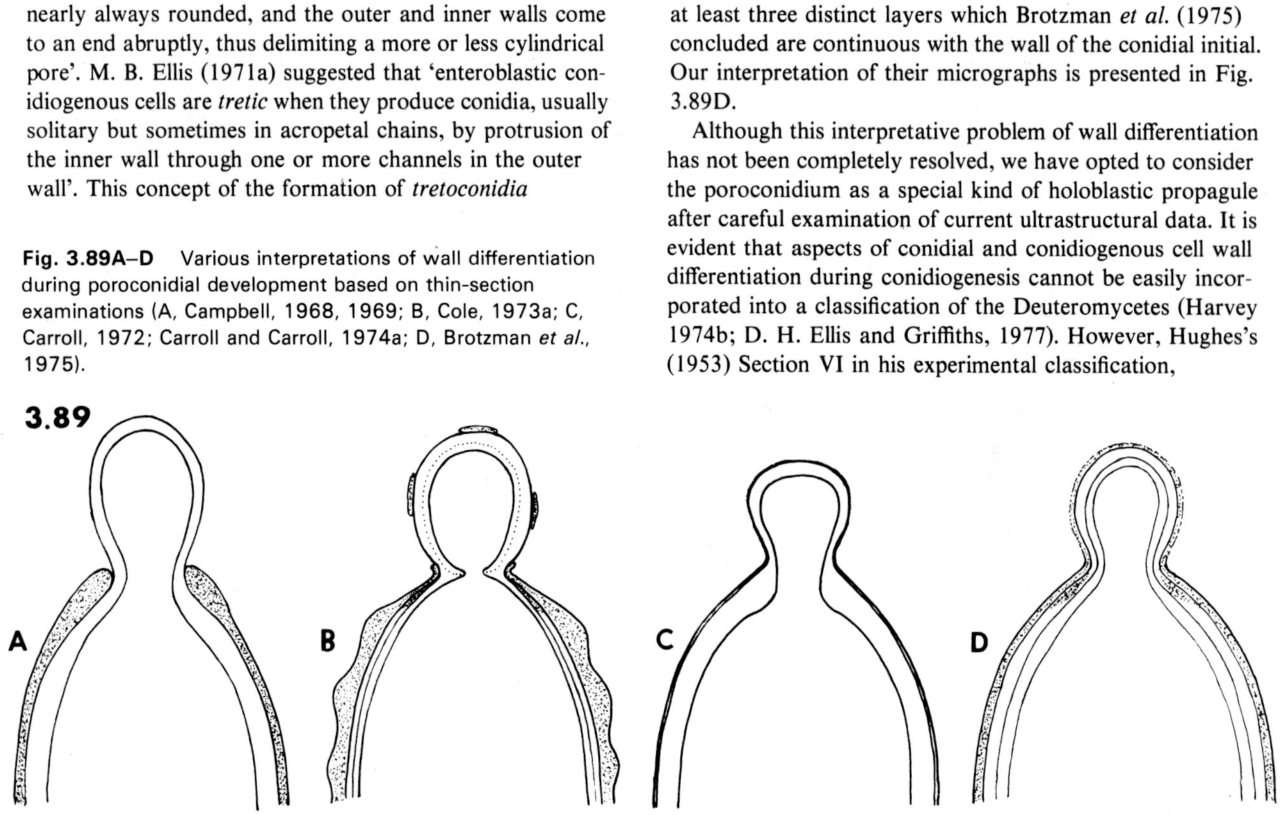

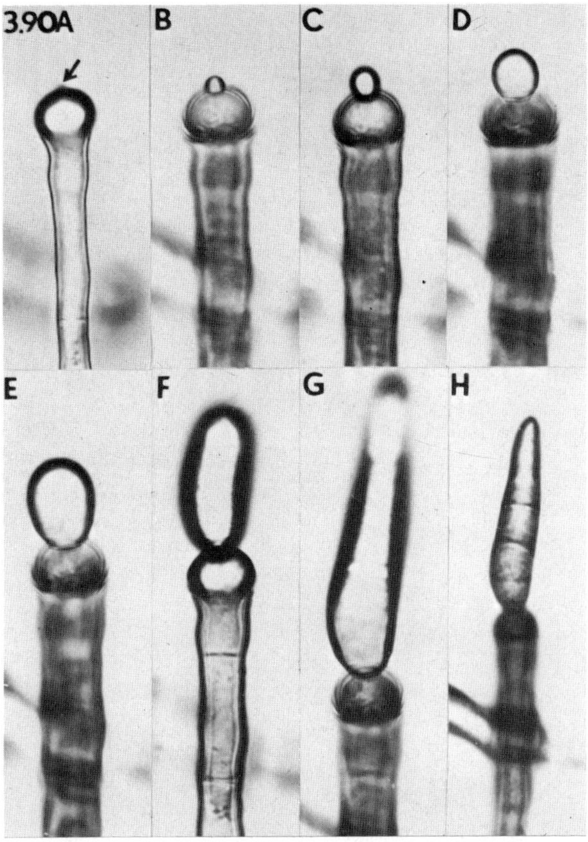

Fig. 3.90A–H *Deightoniella torulosa.* Time-lapse sequence of poroconidial formation (A–G). Arrow in (A) locates protuberance which becomes conidial initial (B). T = 0, 15, 30, 50, 100, 175 and 385 minutes, respectively. A–G, ×800; H, ×512.

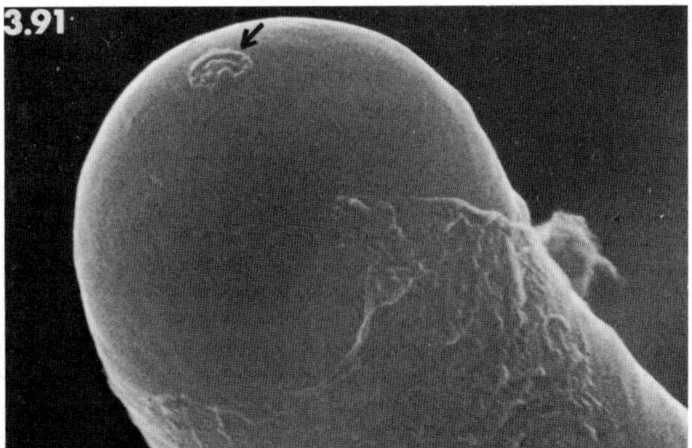

Fig. 3.91 *Helminthosporium solani.* Occluded pore at apex of conidiogenous cell. Arrow indicates thickened ring or annulus surrounding pore. ×9,350.

accommodating species which he distinguished by the 'development of solitary conidia at minute, single, or numerous pores in the wall of the conidiophore', is still a logical assemblage of conidial fungi. Carroll and Carroll (1974a) described the ultrastructure of the pore complex in *Ulocladium atrum* and pointed out that these structures have been seen only in species which were included in Hughes's Section VI. The authors also suggested that the pores 'should perhaps be considered as micromorphological indicators of a natural taxon of Hyphomycetes'.

Differentiation of a poroconidium is illustrated by time-lapse photomicrography in *Deightoniella torulosa* (Syd.) M. B. Ellis (Fig. 3.90A–H). A tiny protuberance arises from a restricted region of the conidiogenous cell apex (Fig. 3.90A, B) and rapidly enlarges (Fig. 3.90C–F) to become the mature, septate conidium (Fig. 3.90H). Similar developmental stages occur in *Curvularia inaequalis* (Shear) Boedijn (Kendrick and Cole, 1968). After conidial secession, a tiny occluded pore is revealed at the apex of the conidiogenous cell, typically surrounded by a thickened ring of wall material (e.g., *Helminthosporium solani* Dur. & Mont., Fig. 3.91). Similar, occluded pores are present at the base of mature seceded conidia. If a conidium is prematurely removed from the apex of the conidiogenous cell, as often happens during preparation for the SEM, a prominent, open pore is visible (Cole, 1973a). The conidial initial of *Ulocladium atrum* in Fig. 3.92 appears to be holoblastic. At a later stage of development, a circumscissile rupture of the outer wall layer occurs at the base of the conidium (Fig. 3.93) and secession reveals the occluded pore at the conidiogenous cell apex (arrow in Fig. 3.94). The channel through the conidiogenous cell wall can be detected in many Deuteromycetes by light microscopy.

Fertile cells which produce poroconidia may be determinate or proliferous. The slightly collapsed, lateral swelling of the conidiogenous cell of *Ulocladium atrum* in Fig. 3.94 is the new proliferating apex which would have elongated and given rise to a new conidium (Figs. 3.95 and 3.96). Percurrent proliferation is demonstrated by *Deightoniella torulosa* (Cole, 1971, Fig. 3.97A). The poroconidia of *Dendryphiella vinosa* (Berk & Curt.) Reisinger (1968) may be arranged in clusters at the fertile apex or proliferate, giving rise to simple and branched, acropetal chains of secondary conidia (Fig. 3.97B). In *Helminthosporium velutinum* Link ex Ficinus & Schubert, poroconidia are formed acro-pleurogenously from the long, erect conidiophores (Fig. 3.97C). *Dichotomophthora portulacae* Mehrlich & Fitzpatrick ex M. B. Ellis (Figs. 3.98, 3.99A) produces botryose poroconidia which arise synchronously from the fertile branches of the conidiogenous cell (Cole, 1977). Typical of *Alternaria alternata* (Fr.) Keissler is the ability of poroconidia to develop in simple and branched, acropetal chains (Fig. 3.99B). In summary, the arrangement of poroconidia may be solitary, botryose or catenulate, and the associated *porogenous*, fertile cells may be either determinate or proliferous.

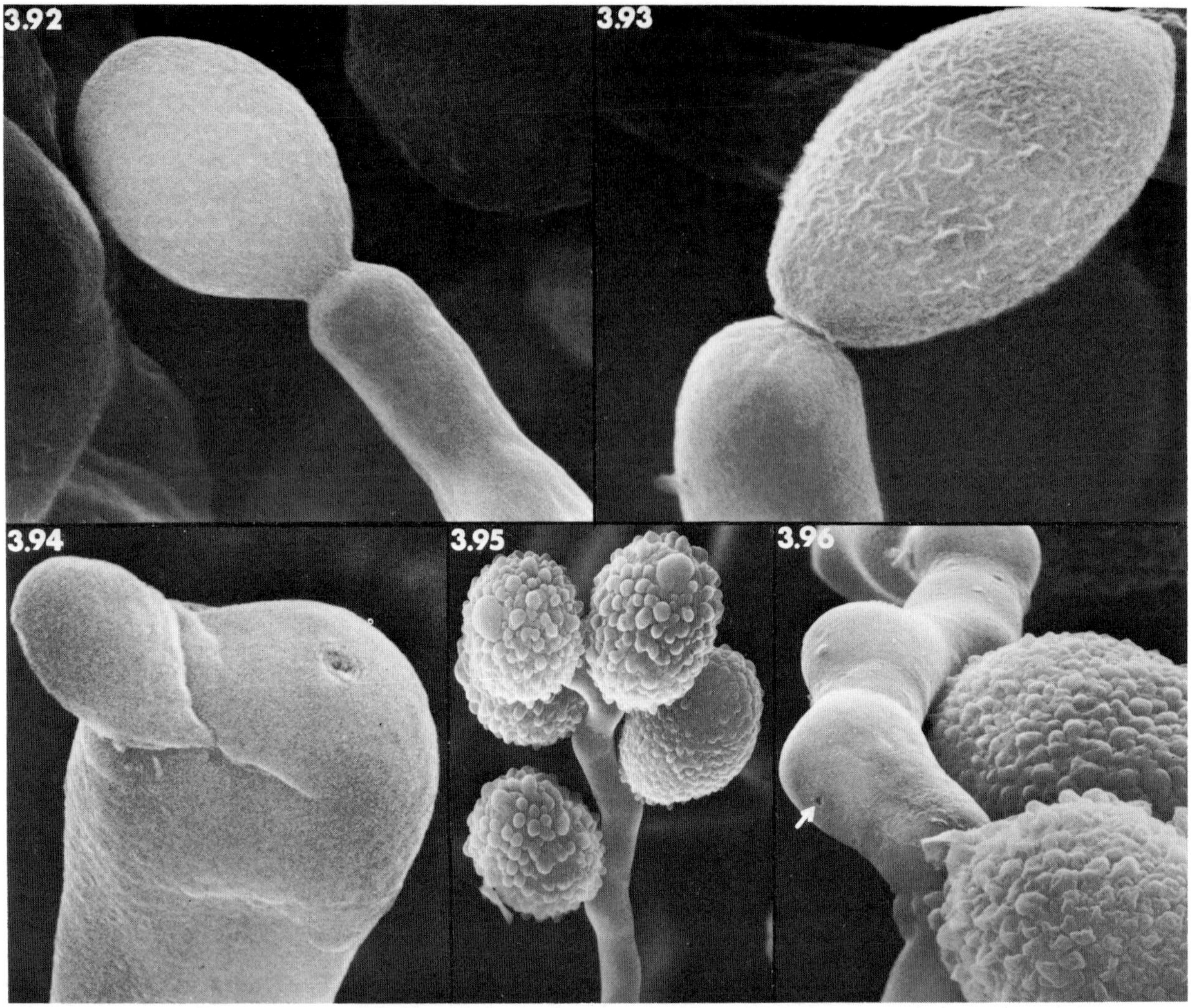

Figs. 3.92–3.96 *Ulocladium atrum.* Stages of poroconidial formation and sympodial proliferation of the conidiogenous cell. Arrow in 3.96 locates pore in wall of fertile cell. ×6,615, 8,980, 14,220, 1,680 and 4,400, respectively.

Enteroblastic conidia

The original concept of enteroblastic development, as a process 'in which the outer layer(s) of the wall of the conidiogenous cell is (are) not involved in the formation of the conidium wall' (Kendrick, 1971b), applies to the formation of conidia from phialides and annellides. An illustrated discussion of enteroblastic conidiogenesis and developmental aspects of these special kinds of conidiogenous cells follows in Chapter 4.

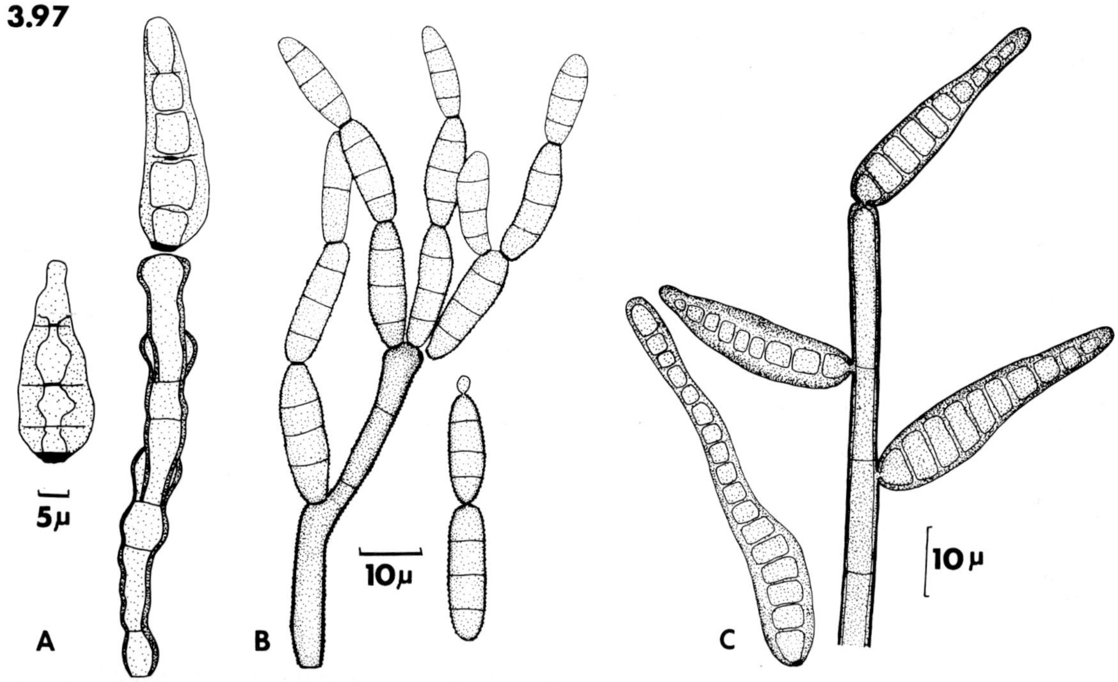

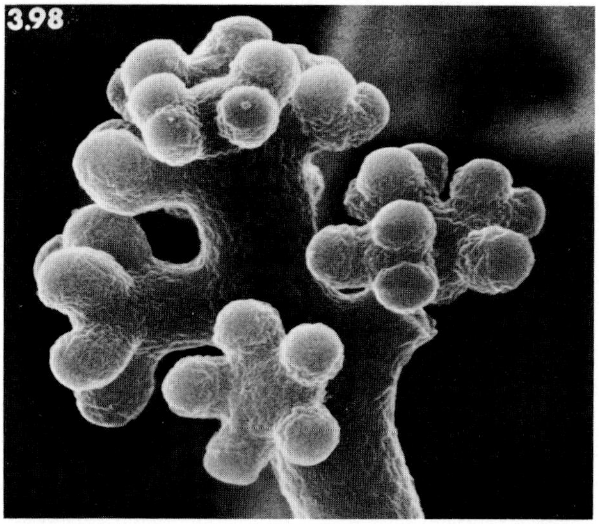

Fig. 3.97A *Deightoniella torulosa.* Percurrently proliferated conidiogenous cell (after M. B. Ellis, 1971b).

Fig. 3.97B *Dendryphiella vinosa.* Sympodially proliferated conidiogenous cell which has produced both botryose and catenulate poroconidia (after Reisinger, 1968).

Fig. 3.97C *Helminthosporium velutinum.*

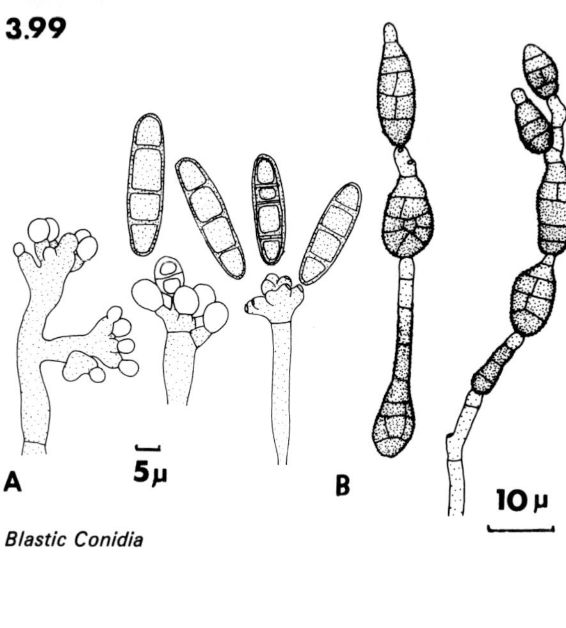

Figs. 3.98, 3.99A *Dichotomophthora portulacae.* Botryose arrangement of poroconidia at apices of fertile branches. 3.98, ×1.965.

Fig. 3.99B *Alternaria alternata.* Catenulate arrangement of poroconidia.

4 Phialidic and Annellidic Development

Many Deuteromycetes produce a plurality of blastic conidia in basipetal succession from the fertile apex, or apices, of specialized conidiogenous cells called phialides or annellides.

Earlier concepts of phialidic development

The term 'phialide' was coined by Vuillemin (1910b) in the description of his division 'Conidiosporés'. He stated: 'Cet élément, isolé à sa base par une cloison, a la forme d'une bouteille avec un ventre plus ou moins renflé et un col plus ou moins effilé ... J'ai designé l'élément différencie en forme de bouteille qui porte les conidies, sous le nom de phialide'. Mason (1933) amended the original definition to read 'fusiform truncate, fusiform beaked, or acuminate terminal portion of a hypha, from the apex of which, or within which, thin-walled conidia are abstricted'. In 1937 Mason elaborated further suggesting that the phialide apex could be considered 'a practically open growing point', which could produce a single 'terminus phialospore' or a succession of 'meristem phialospores'. The former was a single conidium terminating the growth of the conidiogenous cell, and is synonymous with the solitary, holoblastic conidium terminating the growth of each apex of a sympodially proliferating conidiogenous cell (e.g., *Tritirachium oryzae* (Vincens) de Hoog, Chapter 3, Figs. 3.38A–J). Mason's (1937) description of 'meristem phialospore' formation, however, agrees with current ideas of phialoconidiogenesis. Nevertheless, 16 years elapsed before S. J. Hughes (1953) revived the discussion. In defining Section IV of his classification, he restricted the term *phialide* to 'those unicellular structures which are usually terminal, but sometimes intercalary as well, on simple or branched conidiophores; they are oval to subcylindrical to flask-shaped or subulate often with a well-differentiated basal swelling and a narrower distal neck with or without a terminal collarette; from the apex of each phialide develops a basipetal succession of phialospores without an increase in the length of the phialide itself'. Later definitions of phialidic development incorporated these same ideas (Subramanian, 1962a, b, 1965; Tubaki, 1958, 1963; Barron, 1968). However, Cole and Kendrick (1969a) added that despite morphological variation, 'the existence of a localized endogenous meristem producing a basipetal succession of phialospores through a rupture of the primary wall ...' indicated 'that the overriding criterion used in defining the phialide must be the method of conidium formation, not the morphology of the sporogenous cell'. A modified definition, based mainly on light-microscopic observations follows: A phialide is a fertile cell which produces conidia (phialoconidia) 'in a basipetal succession from a fixed meristem', or conidiogenous locus, 'whose position may vary in different fungi from the apex of the cell to deep within the body of the cell'. The first-formed phialoconidium usually differentiates within the unruptured, outer wall of the phialide apex. Initiation of the second conidium often causes the rupture of that wall and subsequent conidia are extruded through the resulting open end of the cell. In some species e.g., *Chalara* (Corda) Rabenh., a chain of several conidia may develop within the phialide before its outer wall ruptures. 'Each conidium is clad in a *secondary wall* especially laid down during its differentiation. The sporogenous cell wall remaining between the point of rupture and the endogenous meristem is called the collarette'. The length of the collarette is the distance between the final position of the meristem and the exact position of the circumscissile rupture of the outer wall. The collarette 'varies from a scarcely visible ring of wall material to a long cylindrical structure enclosing several conidia'.

Developmental and ultrastructural aspects

These features of phialide and phialoconidial formation are illustrated by *Phialophora lagerbergii* (Melin & Nannf.)

Fig. 4.1A–D *Phialophora lagerbergii.* Time-lapse sequence showing emergence of primary conidium from phialide apex (B) and initiation of endogenous, secondary conidial development at base of collarette (C, D). col, collarette. T = 0, 7, 13, 50 minutes respectively. ×2,760.

Fig. 4.2A–D *Penicillium corylophilum.* Time-lapse sequence of phialoconidial development. Arrows in (A) and (B) locate apex of collarette. Note that conidial development occurs above the collarette (cf. 4.1A–D). The conidia are held together in chains by *connectives* between adjacent cells (arrow in D). T = 0, 60, 100, 170 minutes, respectively. ×4,284.

Fig. 4.3A–D *Chalara paradoxa.* Time-lapse sequence of phialoconidial development (A–C). Arrowheads in A–C locate a common reference point. The outer conidiogenous cell wall is eventually ruptured by basipetally developing chain of endogenous conidia but point of rupture is not clear under light microscope. Conidia often adhere in chains at phialide apex (D). (A–C), T = 0, 65, 105 minutes, respectively. A–C, ×1,480; D, ×432.

Conant (Fig. 4.1A–D), *Penicillium corylophilum* Dierckx (Fig. 4.2A–D) and *Chalara paradoxa* (de Seynes) Sacc. (Fig. 4.3A–D). In spite of the variation in phialide morphology among these three species, their mechanisms of conidiogenesis are developmentally similar.

Conidiogenesis in Phialophora

In *Phialophora lagerbergii* each new conidial initial first appears at the constricted neck of the phialide which, therefore, represents the location of the fixed conidiogenous locus (Cole and Kendrick, 1969a). Successive conidia separate from each other upon release from the collarette, but often accumulate in slimy droplets at the phialide apex.

Several deuteromycetous fungi produce distinctive collarettes, which have been used as important taxonomic characters (Cole and Kendrick, 1973). The striking,

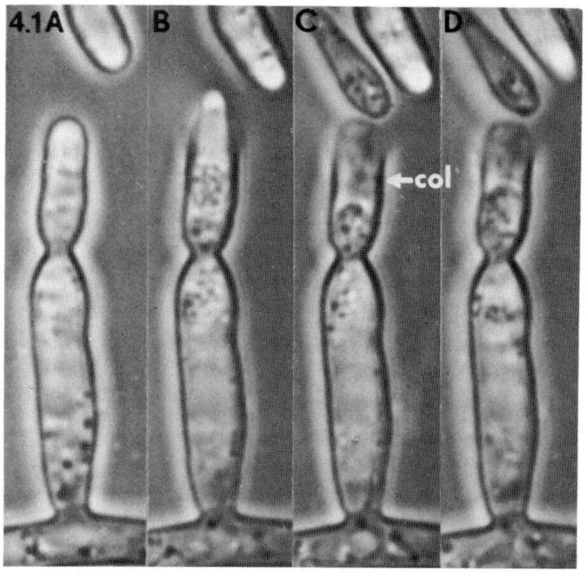

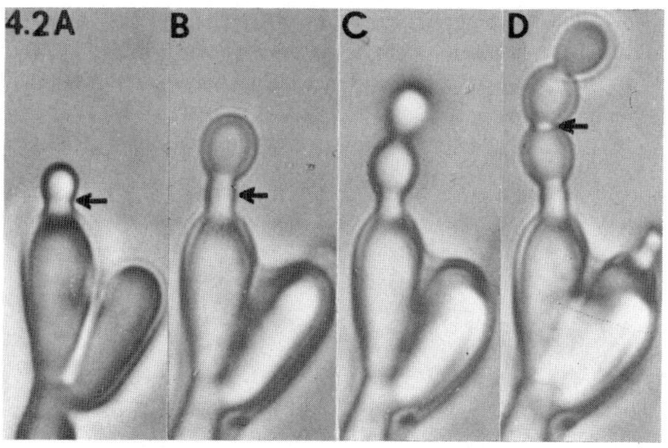

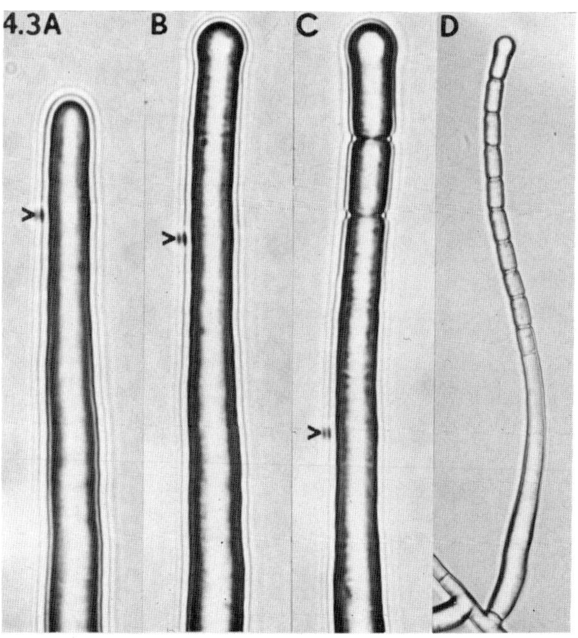

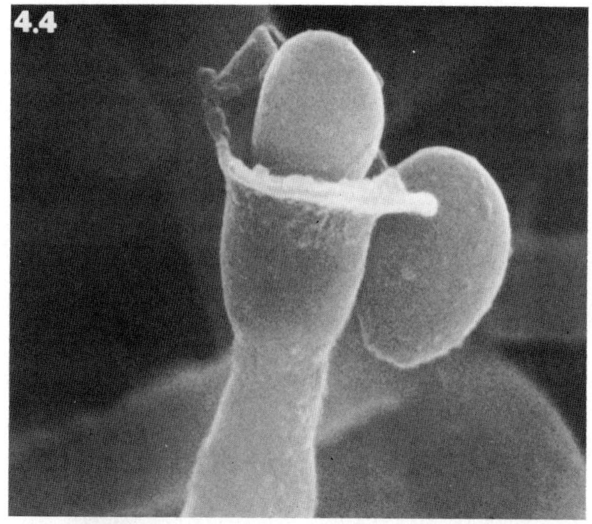

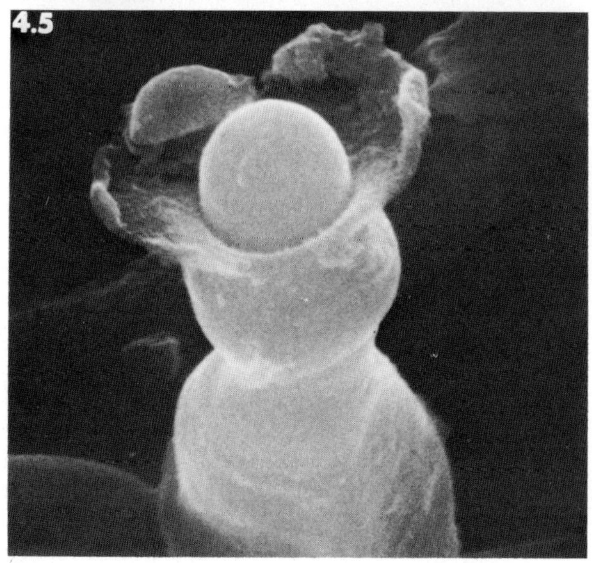

Figs. 4.4, 4.5 *Phialophora verrucosa.* Ruptured apices of phialides showing characteristic vase-shaped collarettes and partially emerged conidia. ×12,600 and 12,635, respectively.

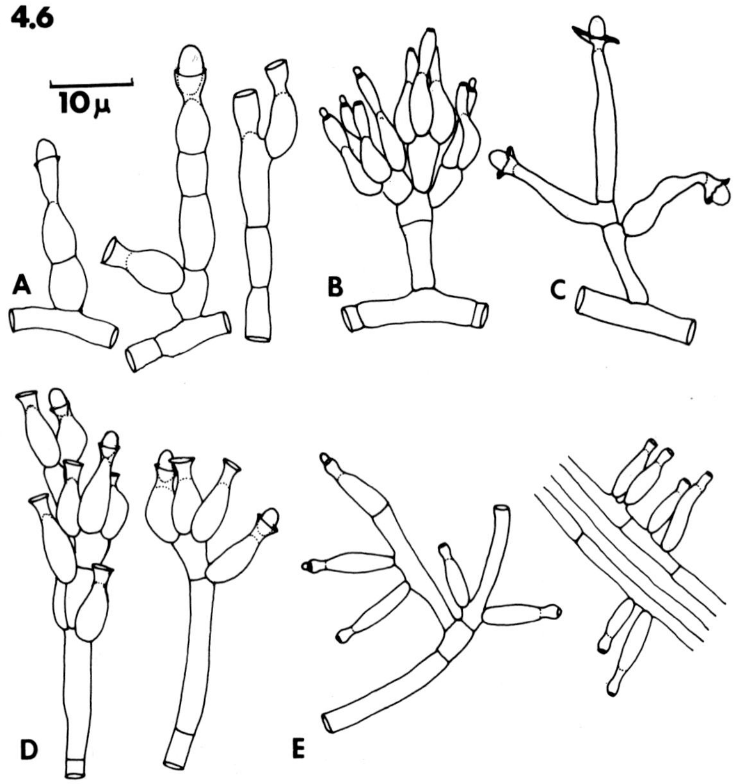

Fig. 4.6A–E *Phialophora* spp. Variation in collarette morphology. A, *P. verrucosa*; B, *P. botulispora* Cole & Kendrick; C, *P. richardsiae* (Nannf.) Conant; D, *P. fastigiata* (Lagerb. & Melin) Conant; E, *P. melinii* (Nannf.) Conant (after Cole and Kendrick, 1973).

Fig. 4.7A–K Variation in phialide morphology and arrangement. Phialides are produced directly from vegetative hyphae (A, *Acremonium strictum* W. Gams; B, *Monocillium dimorphosporum* W. Gams) or are arranged in various patterns on erect conidiophores. A verticillate arrangement is shown in *Verticillium tenerum* (Nees ex Pers.) Link (C, after Domsch and Gams, 1972), and similar whorls of phialides arise from the main axis and lateral branches of conidiophores of *Stachylidium bicolor* Link ex S. F. Gray (D). Irregular clusters of divergent phialides develop from a single conidiophore (E, *Phialophora verrucosa*). Phialides are also produced directly from the conidiophore apex (F, *Stachybotrys cylindrospora* Jensen), or on metulae which arise as branches of the conidiophore apex (G, *Penicillium expansum* Link ex S. F. Gray). A series of metulae is sometimes borne at the conidiophore apex which supports the phialides (H, *Phialocephala dimorphospora* Kendrick). The fertile branches of *Phialocladus zsoltii* Kreisel (I) produce dense clusters of randomly arranged phialides. Regular arrangements of phialides are produced by species of *Aspergillus* in which the conidiogenous cells are formed either from the apex of short metulae attached to the swollen conidiophore apex or ampulla [J, *Aspergillus nidulans* (Eidam) Wint.] or directly from the surface of the ampulla (K, *Aspergillus fumigatus* Fres.).

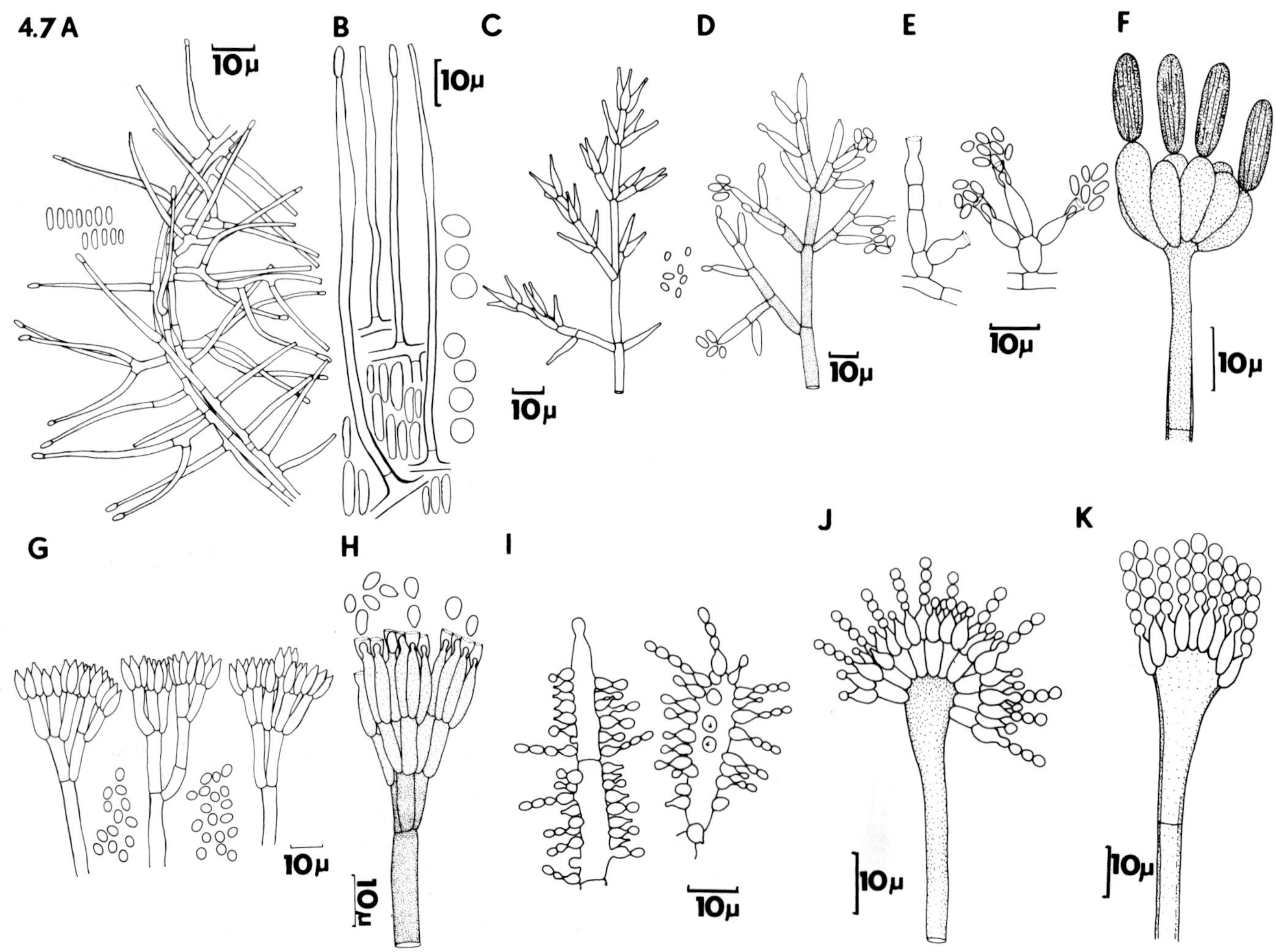

campanulate collarettes of *Phialophora verrucosa* Medlar, type species of the genus, are illustrated in Figs. 4.4, 4.5 and 4.6A. Other wood-inhabiting members of *Phialophora* Medlar also possess characteristic collarettes (Fig. 4.6A–E).

Phialides are found on structures and in arrangements of varying size and complexity, as seen in Figs. 4.7–4.11. Although phialide arrangement has also been used extensively as a taxonomic character, the high degree of variation of this feature displayed even within the same genus limits its value in a classification. The synchrony of early phialidic development in species of *Aspergillus* Mich. ex Fr. is demonstrated by the scanning electron microscope (SEM) in Figs. 4.12 and 4.13. Formation of primary phialoconidia also occurs more or less synchronously, followed by production of chains of secondary conidia through inconspicuous collarettes (Figs. 4.14–4.17). The examples of phialide arrangement in Figs. 4.7–4.11 and 4.15–4.17 take into account only a few of the different patterns demonstrated by hyphomycetous fungi.

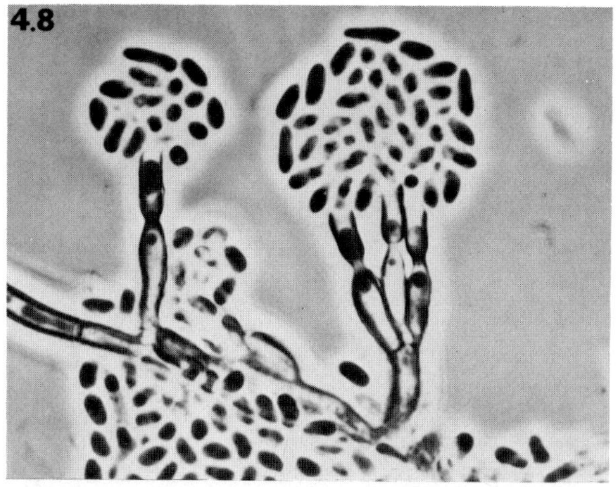

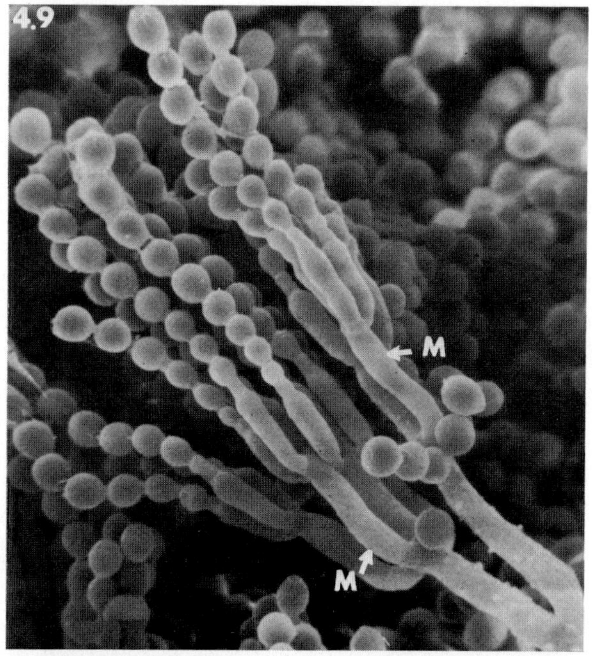

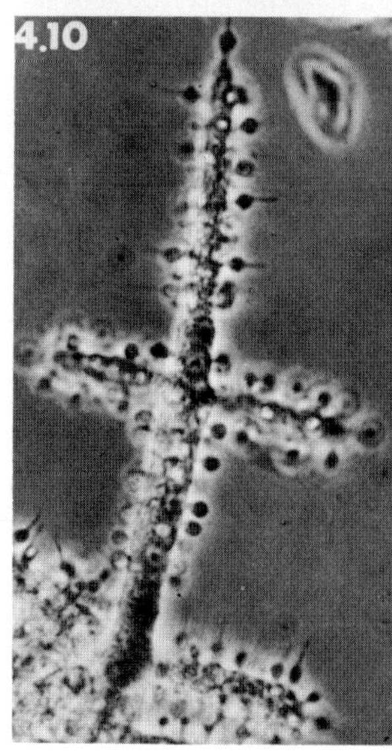

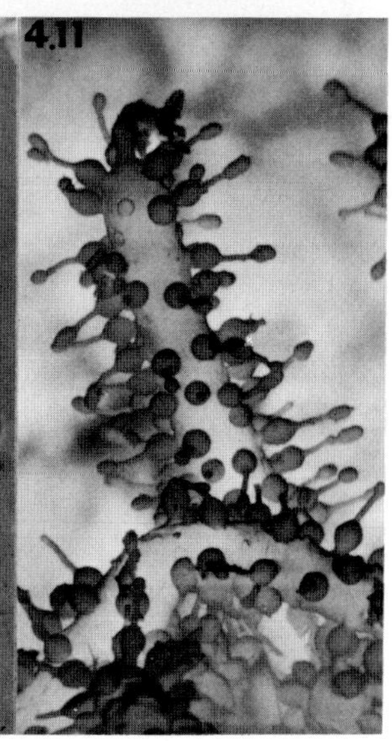

Fig. 4.8 *Phialophora lagerbergii.* Phialides arranged in cluster at apex of short branch and phialide borne directly on vegetative hypha. Note common droplet of conidia at apices of three phialides at right. ×840.

Fig. 4.9 *Penicillium roqueforti* Thom. Phialides arranged in parallel clusters at apices of metulae. Chains of conidia have been preserved. M, metula. ×7,800

Figs. 4.10, 4.11 *Phialocladus zsoltii.* Dense arrangement of phialides covering surface of conidiophore branches. ×990 and 1,600, respectively.

4.12
4.13
4.14
col
4.15
cn
4.16
4.17

Figs. 4.12–4.17 *Aspergillus* spp. Stages of phialidic and phialoconidial development showing synchrony of conidiogenous cell initiation (4.12, *A. fumigatus;* 4.13, *A. flavus* Link ex Fr.). Emergence of conidia through ruptured phialide apex (4.14, 4.15) and formation of conidial chains (4.16) are also demonstrated by *A. flavus.* Typical, dense clusters of phialides and conidia are shown in 4.17 (*A. fumigatus*). col, collarette, cn, conidial initial. ×2,970, 3,960, 6,600, 1,900, 2,160 and 823, respectively.

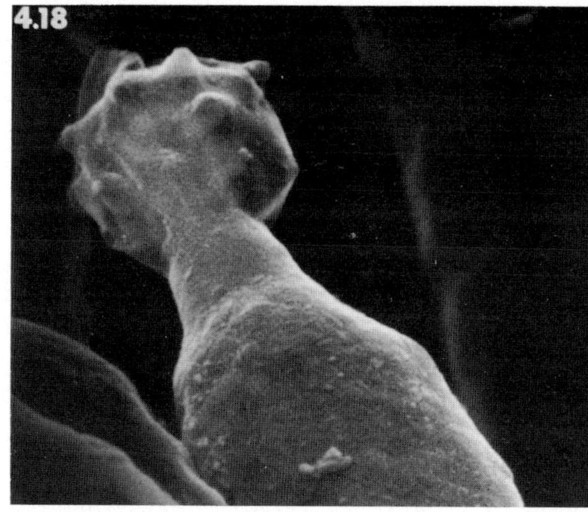

Figs. 4.18–4.20 *Penicillium nigricans.* Stages in development of conidial chains. The absence of a collarette in 4.18 suggests that this is the first-formed holoblastic conidium of the phialide. Collarettes, formed by rupture of fertile cell apex, are shown in 4.19 (arrows) and connectives between adjacent conidia in the chain are illustrated in 4.20. Arrows in 4.20 locate transverse septa through connectives. Collarettes at apices of phialides in 4.20 (arrowheads) also visible. cn, conidial initial. ×16,600, 8,000 and 9,000, respectively.

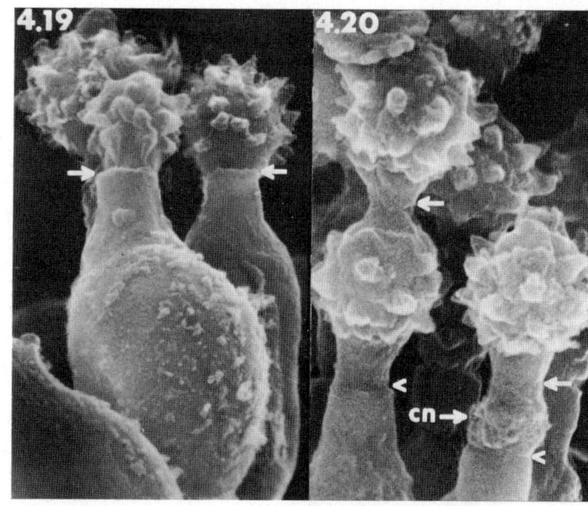

Figs. 4.21, 4.22 *Paecilomyces marquandii.* Tapered apex of phialide and liberated conidium (4.21). Chain of conidia held together by connectives (arrows in 4.22). ×12,750 and 8,625.

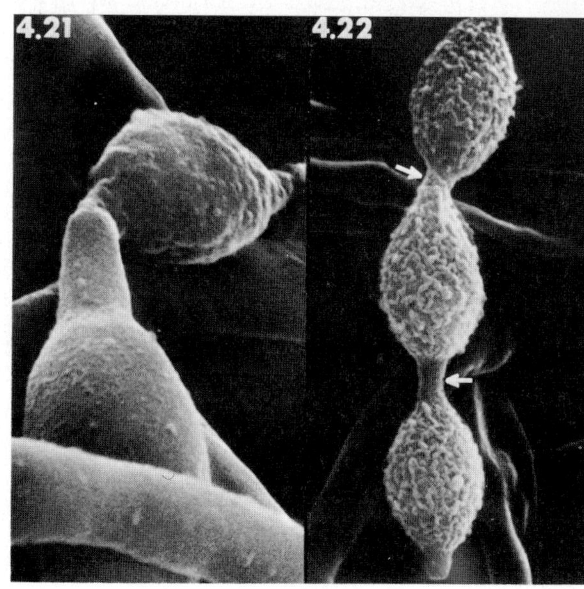

Conidiogenesis in Penicillium

The time-lapse developmental sequences of *Penicillium corylophilum* (Fig. 4.2A–D) and *P. expansum* Link ex S. F. Gray (Laane, 1969) demonstrate that the conidiogenous locus is located close to the phialide apex, in contrast to *Phialophora lagerbergii* (Fig. 4.1A–D) in which it is fixed at the base of the collarette. This apical location of the fertile zone results in an inconspicuous collarette in species of *Penicillium* as well as several other phialidic Hyphomycetes (e.g., species of *Aspergillus, Metarrhizium* Sorok., *Myrothecium* Tode ex Fr. and *Torulomyces* Delitsch.). The SEM has demonstrated that the first-formed conidium of *Penicillium nigricans* Bain. ex Thom develops holoblastically. Figure 4.18 shows the absence of a collarette. Initiation of the second conidium results in a circumscissile rupture of this outer wall at the base of the mature conidium, leaving a collarette (arrows in Fig. 4.19). Conidia are held together in chains by connectives (Fig. 4.20). Adjacent cells, however, eventually separate schizolytically at transverse septa (arrows in Fig. 4.20). Conidial chains possessing intercellular connectives are also produced by *Paecilomyces marquandii* (Massee) Hughes (Figs. 4.21, 4.22). Distinct connectives form between the conidia of *Phialotubus microsporus* Roy & Leelavathy (Fig. 4.23A). The large, verrucose phialoconidia of *Phialomyces macrosporus* Edward (Fig. 4.23B) are similarly linked in chains. In *Metarrhizium anisopliae* (Metsch.) Sorok. (Fig. 4.23C), the fragile chains are held together by slight thickenings of wall material at the apex of each conidial initial which is bound to the lamellate, basal septum of the previously formed conidium (Hammill, 1972a). The conidial chains of *Mariannaea elegans* (Corda) Samson (Fig. 4.23D) are not formed by intercellular connectives but instead the cells are arranged in an imbricate pattern, due to lateral growth of successive conidia at the phialide apex, and

4.23

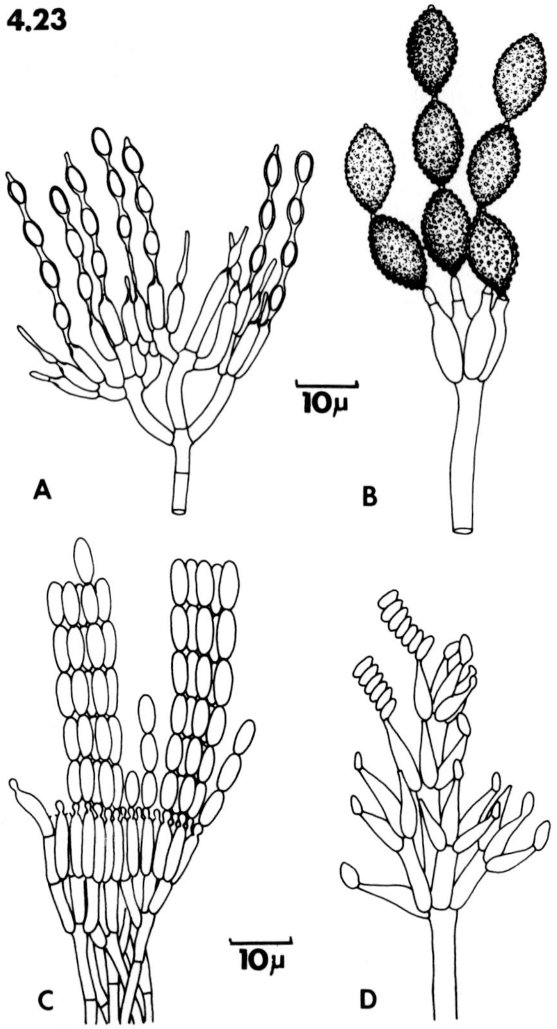

4.24

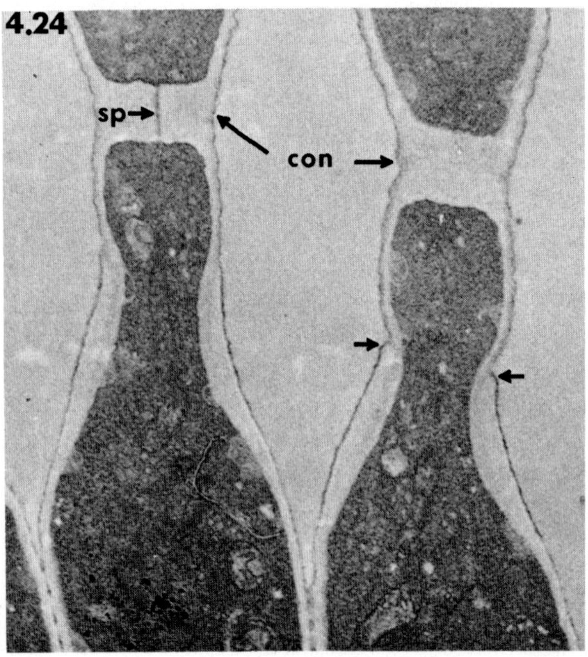

Fig. 4.23A–D Examples of chains of phialoconidia. A, *Phialotubus microsporus*; B, *Phialomyces macrosporus*; C, *Metarrhizium anisopliae*; D, *Mariannaea elegans.*

Fig. 4.24 *Aspergillus clavatus.* Thin-section of phialides and conidial chains showing connectives (con), narrow septal pore (sp) and collarette (arrows) (from Hanlin, 1976). ×14,926.

are held together by a slimy exudate (Samson, 1974). Thin-section examinations of phialoconidial formation in *Penicillium* (Zachariah and Fitz-James, 1967; Fletcher, 1971) and *Aspergillus* (Tanaka and Yanagita, 1963; Trinci *et al.*, 1968; Tsukahara, 1970; Oliver, 1972; Ghiorse and Edwards, 1973; Tokunaga *et al.*, 1973; Bojović-Cvetić and Vuyicić, 1972; Hanlin, 1976) have also demonstrated ultrastructural features of the connectives between adjacent conidia. In *Aspergillus clavatus* Desm. (Fig. 4.24), the connective has formed as a result of thickening of the septal wall between adjacent cells. An earlier stage of septation in this same species is shown in Fig. 4.33. A channel is visible through the thickened septum in Fig. 4.24 (sp) but is too narrow to provide for significant cytoplasmic movement between adjacent cells.

Conidiogenesis in Chalara

Chains of phialoconidia are also produced by *Chalara paradoxa* (Fig. 4.3A–D). The time-lapse sequence of conidiogenesis indicates that early phialide development is difficult to distinguish from vegetative hyphal growth (Fig. 4.3A). However, the appearance of an apical swelling signals emergence of a phialoconidium through the ruptured, outer phialide wall (Fig. 4.3B). The conidia continue to be produced in basipetal succession through this open end of the fertile cell forming a long chain (Fig. 4.3C, D). Cole and Kendrick (1969a) suggested that during early stages of phialide differentiation in *C. paradoxa*, growth of the outer wall of the conidiogenous cell and conversion of its protoplast into endogenous conidia proceed simultaneously. Several conidia usually differentiate within the conidiogenous cell while its outer wall remains intact. Huang and Patrick (1974) showed that the order of septation within the intact phialides of *Chalara elegans* Nag Raj & Kendrick (= *Thielaviopsis basicola* (Berk. & Br.) Ferr.) is initially basipetal and the protoplasm differentiates into a chain of endogenous conidia. Apical growth then ceases while a new growth zone develops at the base of the conidial chain giving rise to a new conidial initial and causing the terminal cell to rupture the outer wall at the phialide apex. This zone of protoplasmic differentiation (conidiogenous locus) apparently remains fixed in position during formation of successive conidia. However, Hawes and Beckett (1977b) have presented evidence that phialoconidia of *Chalara* (Corda) Rabenh. 'are delimited from the conidiogenous cell by septation at a variable point' within the phialide wall. The collarette, which is the cylindrical portion of the phialide wall above the conidiogenous locus, is commonly quite long in *Chalara* and related genera (Nag Raj and Kendrick, 1975).

The phialide apex of *C. paradoxa* in Fig. 4.25 is located by

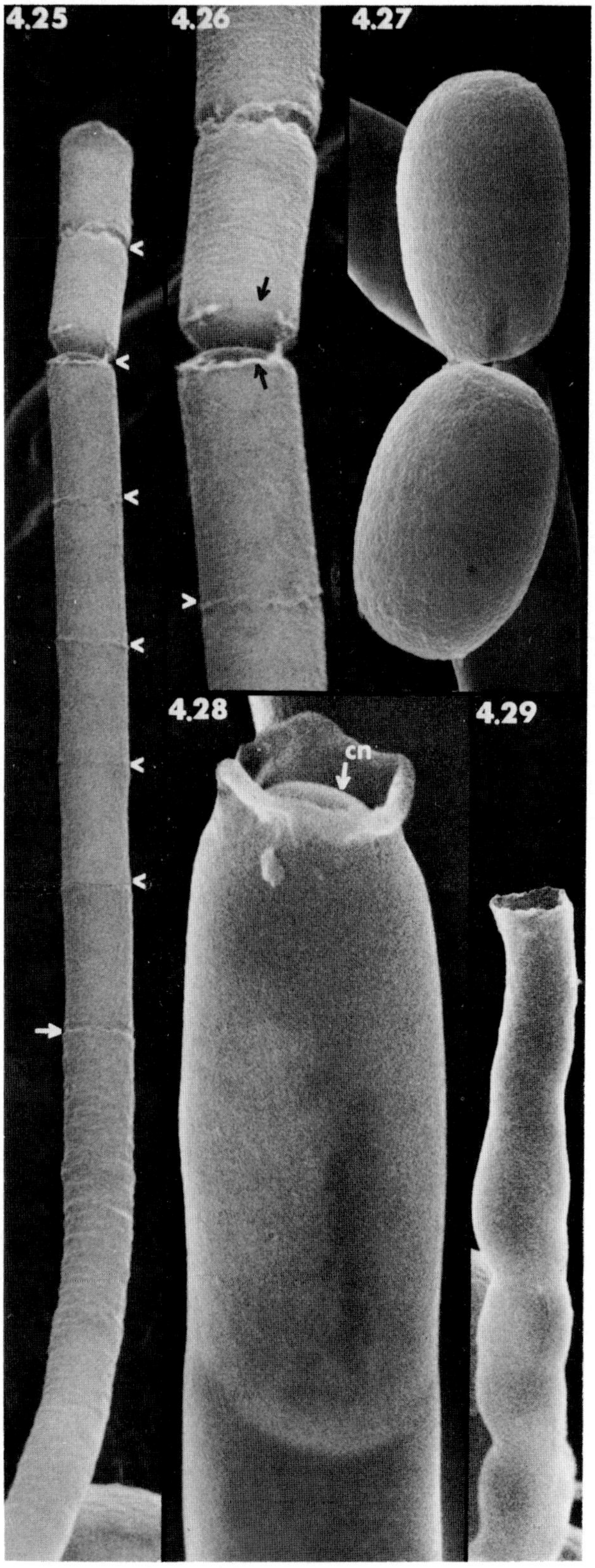

Figs. 4.25–4.29 *Chalara* spp. Chain of conidia attached to phialide apex of *C. paradoxa* (4.25). Arrow in 4.25 locates collarette and arrowheads indicate septa between adjacent conidia. At high magnification (4.26), outer wall shown to be continuous (arrowhead) between adjacent conidia. Arrows indicate rings of torn wall material remaining after conidial secession. Similar wall relations shown between adjacent phialoconidia of *Chalara* state of *Ceratocystis adiposa* (4.27). Endogenous conidium (cn) contained by collarette of phialide of *Chalara paradoxa* (4.28). Empty phialide of *Ceratocystis adiposa* (4.29). ×1,580, 5,500, 4,100, 9,526 and 2,640, respectively.

the arrow, above which is a chain of seven conidia. The septa of adjacent conidia near the upper end of the chain have undergone partial schizolytic separation (Fig. 4.26). Each septum is surrounded by a ring of torn wall material (arrows in Fig. 4.26) resulting from rupture of the outer conidial wall. The lower pair of conidia in Fig. 4.26 have not separated and the outer wall layer is continuous between them (arrowhead). The location of the transverse septum is recognized by a slight folding of the outer wall layer (arrowheads in Fig. 4.25). Similar wall relations are demonstrated by two adjacent conidia of the *Chalara* state of *Ceratocystis adiposa* (Butler) C. Moreau in Fig. 4.27. It is interesting to note that both the mechanism of secession and arrangement of wall layers in the chains of phialoconidia of *Chalara* are very similar to features of holoarthric development in *Geotrichum candidum* Link ex Pers. (cf. Chapter 7, Fig. 7.9). The mechanisms of conidiogenesis in species of *Chalara, Chaetochalara* Sutton & Pirozynski, *Sporoschisma* Berkeley & Broome apud Berkeley, *Sporendocladia* Arnaud ex Nag Raj & Kendrick, *Ascoconidium* Seaver, and *Bloxamia* Berk. & Br. are probably similar.

Hawes and Becket (1977a, b) examined stages of phialidic development in the *Chalara* state of *Ceratocystis adiposa* and determined that a *sheath* exists around the persistent chain of conidia. Khan and Aldrich (1973) have examined ultrastructural aspects of conidiogenesis in *Termitaria snyderi* Thaxter, the phialides of which also produce chains of several endogenous conidia before the outer wall is ruptured apically and the cells are liberated. In contrast to *Chalara,* the conidiogenous locus becomes fixed at the constricted region of the phialide during development of the first conidium (Fig. 4.30A–E). Their electron micrographs demonstrate that wall continuity at first exists between the conidial initial and the adjacent, differentiated conidium but this wall layer is soon ruptured as the double-layered septum differentiates and separates centripetally (Fig. 4.30C), leaving an inconspicuous ring of wall material at the conidiogenous locus and base of each conidium (arrows in Fig. 4.30C, E).

In *Chalara paradoxa* two processes of wall differentiation may occur during phialoconidiogenesis (Fig. 4.31A–C). The conidia may remain together in long chains (e.g., Fig. 4.25) with a continuous outer wall layer between adjacent cells due to the delayed schizolytic separation of the double septa at their base (Fig. 4.31A, B). Delvecchio *et al.* (1969) presented thin-sections of *C. elegans* showing electron-translucent material between the conidial and phialide wall which may represent a lubricant for the conidia as they move upward

4.30

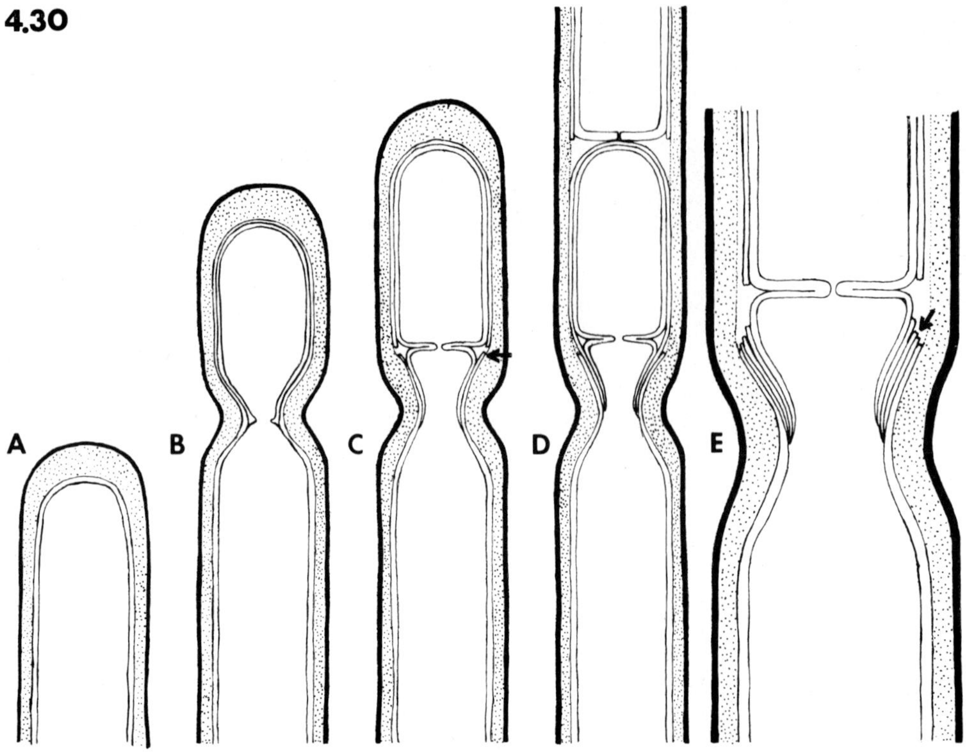

4.31

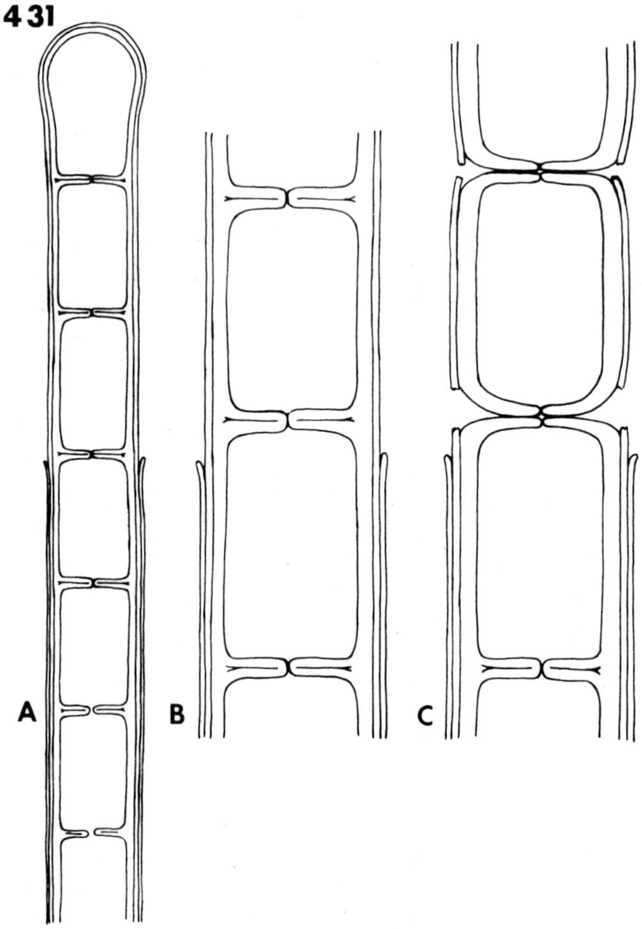

4.32

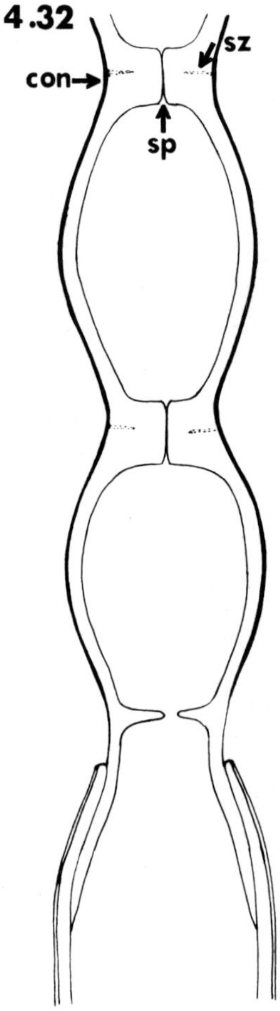

Fig. 4.30A–E Interpretation of wall differentiation during various stages of phialoconidial formation in *Termitaria snyderi.* Arrows in C and E indicate rings of wall material which appear as successive conidia are formed within the collarette of the phialide.

Fig. 4.31A–C Interpretation of wall relations during phialoconidial development in *Chalara*, demonstrating chains of cells held together by a continuous, outer wall layer (A, B), and conidia which secede as they emerge from the collarette (C).

Fig. 4.32 Diagrammatic interpretation of wall differentiation during phialoconidial chain development in *Aspergillus.* con, connective: sp, septal pore; sz, schizolytic zone.

through the collarette (Cole and Kendrick, 1969a). Species of *Penicillium* and *Aspergillus* also possess a continuous outer wall layer between adjacent conidia (Fig. 4.32). As the interconidial septa or connectives thicken, this outer wall layer presumably stretches and remains intact.

On the other hand, phialoconidia of *Chalara paradoxa* are not always produced in chains, but may secede as they emerge from the collarette causing the outer conidial wall layer to fracture (Fig. 4.31C). In Fig. 4.28, a young phialoconidium of *C. paradoxa* is still within the cylindrical collarette. This image is provided by 2° electrons which are emitted from the conidial wall layer after the 1° electrons have penetrated the outer wall of the phialide. A mature, empty phialide of the *Chalara* state of *Ceratocystis adiposa* is shown in Fig. 4.29.

Revised concept of phialidic development

Examinations of ultrastructural aspects of phialides have provided the basis for a revised concept of phialidic development.

Formation of primary phialoconidia

Hanlin (1976) showed that the primary phialoconidium of *Aspergillus clavatus* is holoblastic (Fig. 4.33), a phenomenon also demonstrated for several other phialidic species (see earlier references on transmission electron microscope studies of *Penicillium* and *Aspergillus*: Boerema, 1965; Brewer and Boerema, 1965; Wilsenach and Kessel, 1965b; Lowry *et al.*, 1967; Oulevey-Matikian and Turian, 1968; Buckley *et al.*, 1969; Willets and Calonge, 1969; Hammill, 1972a, 1974a; Campbell, 1972; Ola'h, 1972; Roquebert and Abadie, 1973; Oláh and Reisinger, 1974; Carroll and Carroll, 1974b; R. Marchant, 1975). The immature septum delimiting this conidium in Fig. 4.33 is uniperforate and its wall is continuous with the inner phialide wall (arrows). The latter extends down the neck of the phialide and then tapers to become indistinguishable from the primary wall of the conidiogenous cell. Just below the conidial septum, the outer phialide wall has ruptured (arrowheads in Fig. 4.33), probably as a result of becoming inelastic, or *set* (Da Riva Ricci and Kendrick, 1972), while new extension growth of the inner phialide wall has occurred. The point of rupture of this outer wall layer in other phialidic fungi is probably associated with its degree of elasticity which, in turn, results in a variation of collarette shape among species (cf. Fig. 4.6).

Formation of secondary phialoconidia

Extension growth of the inner phialide wall of *Aspergillus clavatus*, which probably occurs by secondary intussusception (Hunsley and Burnett, 1968), is combined with swelling and differentiation of the second phialoconidium and thickening of the basal septum of the first.

Conidia remain in chains due to the intact septa between adjacent cells and the continuous outer conidial wall layer (Fig. 4.32). Similar wall differentiation occurs in *Chalara paradoxa* (Fig. 4.31A, B).

Formation of secondary phialoconidia in *Metarrhizium anisopliae* (Hammill, 1972a) and *Trichoderma saturnisporum* Hammill (Hammill, 1974a) involves an alternative mechanism of wall differentiation suggested in Figs. 4.30 and 4.31C. A thin-section of *M. anisopliae* (Fig. 4.34) reveals that the inner phialide wall and conidial wall are continuous (arrows) between the two layers of the basal septum. The narrow septal pore is too small to permit translocation of organelles and is later occluded by deposition of electron-opaque material. Woronin bodies, which function as septal plugs in many uniperforate septa, are apparently absent from the fertile apices of phialides (Carroll and Carroll, 1974b). However, Hawes and Beckett (1977b) have demonstrated an exception to this generality in their ultrastructural study of the *Chalara* state of *Ceratocystis adiposa.* Initiation of secondary conidia in *M. anisopliae* results in rupture of the conidial and inner phialide wall, leaving a ring of torn wall material at the open end of the phialide (arrows in Fig. 4.35). A new wall layer has also differentiated (arrowheads) which will encompass the next conidium. This sequence of events is repeated many times at the phialide apex and results in accumulation of rings of torn wall material inside the collarette, as demonstrated by *Trichoderma saturnisporum* in Fig. 4.36. In longitudinal section, these wall layers appear as lamellations. Under the light microscope, this fertile region of the phialide is a thickened, pigmented zone shown in camera lucida drawings of several Hyphomycetes in Fig. 4.37A–D. Although this apical thickening of the fertile cell is a useful character for identification of phialidic development, many conidiogenous cells suspected to be phialides are too minute to reveal such features under the light microscope.

The fertile regions of the phialides in Figs. 4.36 and 4.37 probably become plugged with wall material after several conidia have developed and seceded. One could speculate that the number of conidia produced by a phialide from one conidiogenous locus is related to the width of the fertile cell at the region where wall lamellae accumulate as well as to the thickness of these residual wall layers. Gams (1973) suggested that phialides of certain species (e.g., *Aphanocladium* W. Gams, *Sibirina* Arnold) occasionally produce solitary conidia, while others formed by the same mycelium yield many. Gams thought that the narrow openings at the apex of the former became plugged during the development of the primary conidium.

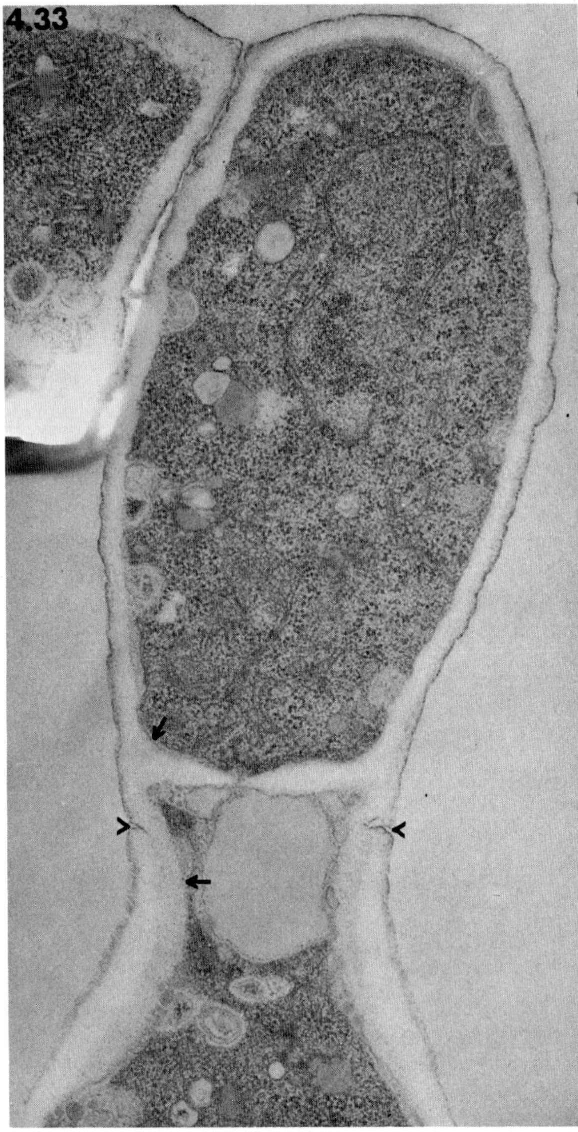

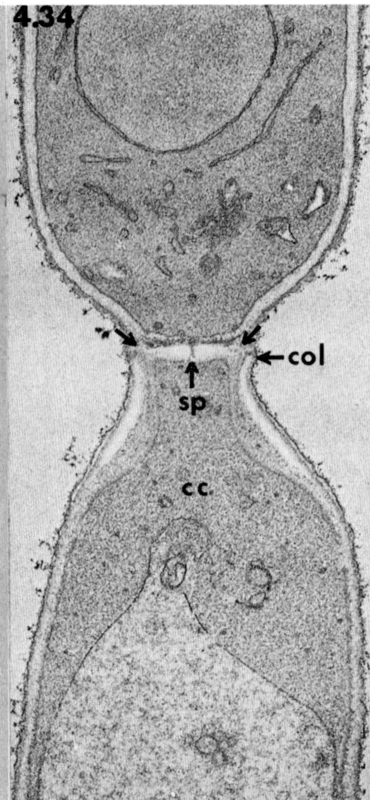

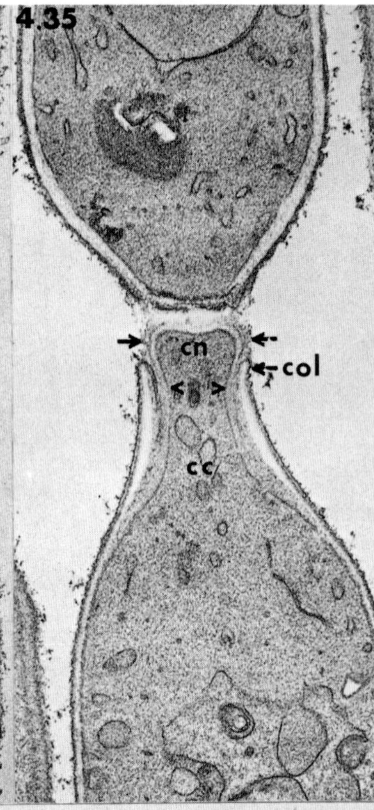

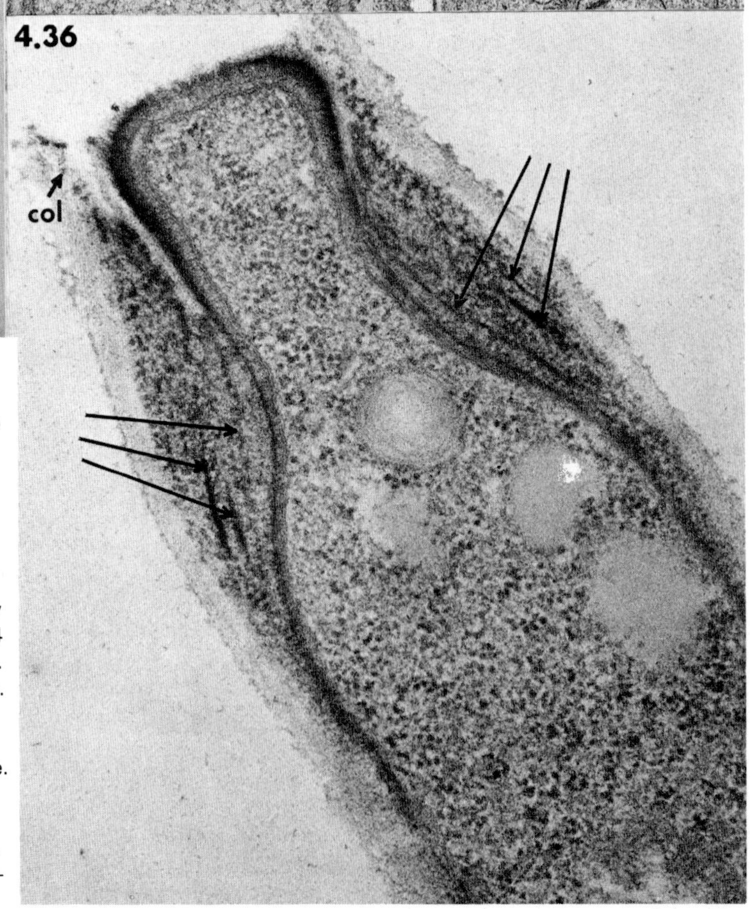

Fig. 4.33 *Aspergillus clavatus.* First-formed conidium attached to phialide apex. Arrowheads point to ruptured outer wall layer which becomes collarette of phialide. Arrows indicate continuity of septal wall and inner wall of phialide and phialoconidium. First-formed phialoconidium is holoblastic. ×26,780 (from Hanlin, 1976).

Figs. 4.34, 4.35 *Metarrhizium anisopliae.* Thin sections through basal septa of differentiated conidia and apices of phialides, prior to initiation of new conidium (4.34) and at early stage of secondary conidial development (4.35). Arrows in 4.34 indicate continuity of inner phialide wall and outer conidial wall. Arrows in 4.35 locate ring of torn wall material inside collarette. Arrowheads in 4.35 indicate newly differentiated inner wall layer which will encompass next conidium. cc, conidiogenous cell (phialide); cn, conidial initial; col, collarette; sp, septal pore. ×21,600 and 18,200, respectively (from Hammill, 1972a).

Fig. 4.36 *Trichoderma saturnisporum.* Thin section through mature phialide apex showing rings of wall material (arrows) inside collarette (col). ×47,000 (from Hammill, 1974a).

4.38

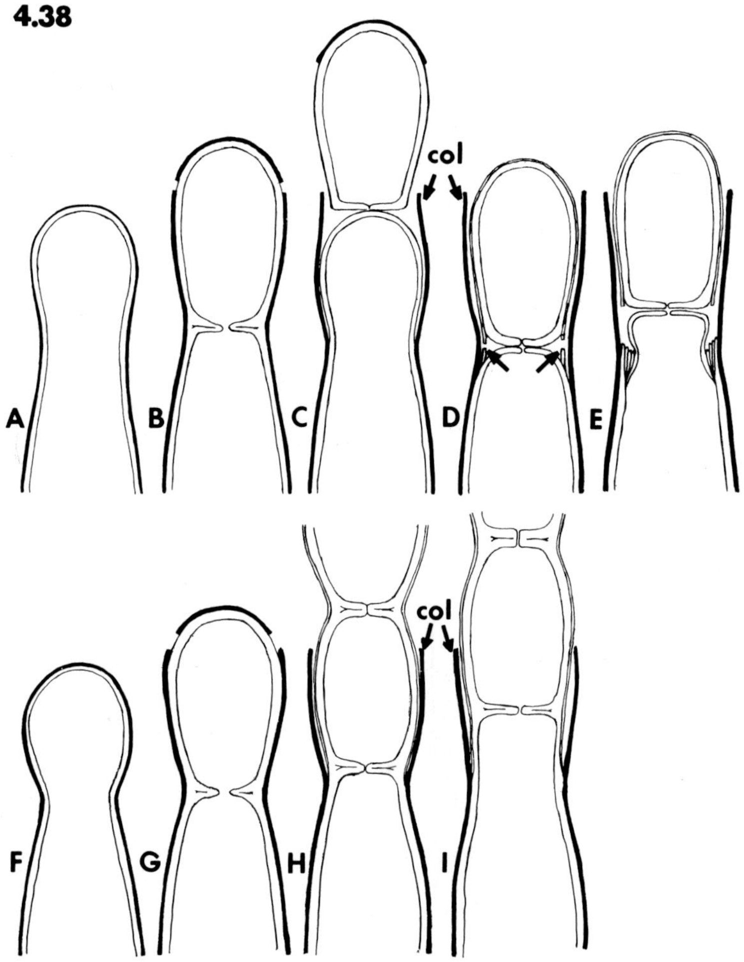

4.37A

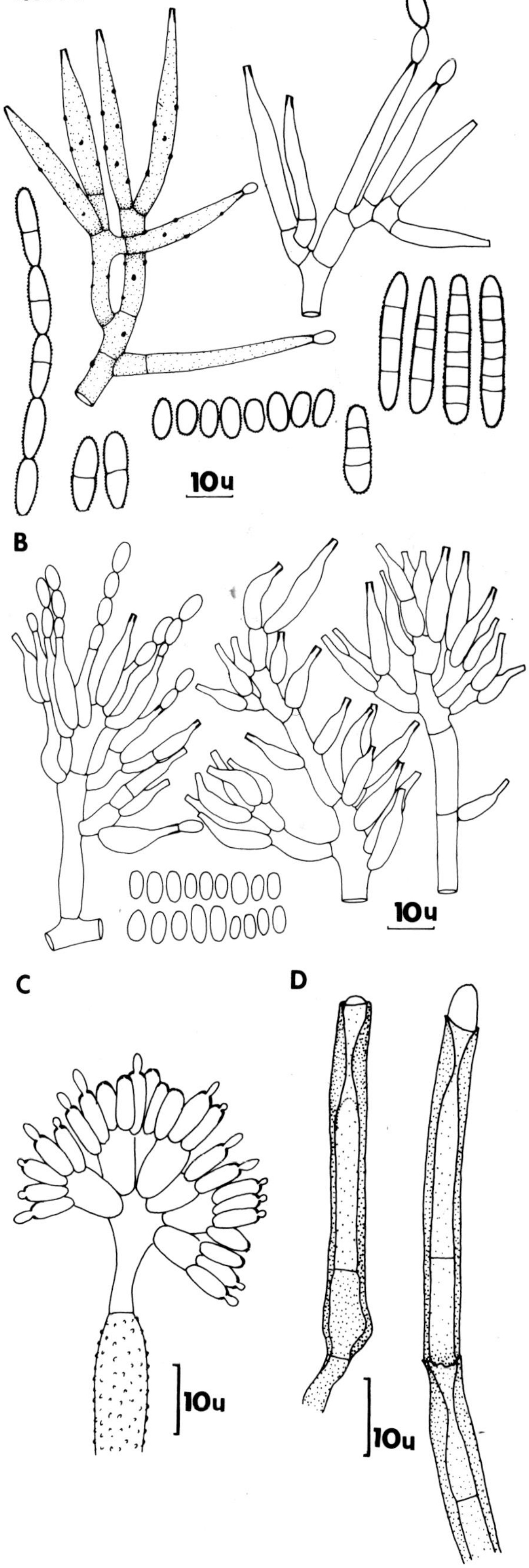

Fig. 4.37A–D Phialides which demonstrate thickened apices associated with successive conidial development (cf. 4.36, 4.38A–E). A, *Septofusidium berolinense* (Oláh and Ackermann) Samson; B, *Paecilomyces crustaceus* Apinis and Chester; C, *Gibellula pulchra* (Sacc.) Cavara; D, *Fusichalara minuta* Holubova-Jechova.

Fig. 4.38A–I Interpretations of wall differentiation associated with two basic mechanisms of phialoconidial formation. A–E. Conidia readily secede from the phialide apex, form droplets or adhere in 'false' chains (cf. 4.23D). F, G. Conidia remain in a chain at the phialide apex because of thick-walled connectives between adjacent cells and/or a thin, continuous, outer wall layer encompassing the conidial chain. Note that the first-formed conidium in each case is holoblastic (B, G) and initiation of the second conidium causes the outer phialide wall to rupture forming the collarette (col). Arrows in (D) locate the ruptured outer wall of the second conidium which resulted from initiation of third conidial initial.

Summary of wall differentiation during phialidic development
A diagrammatic interpretation of the two recognized processes of wall differentiation associated with phialidic development is presented in Fig. 4.38A–I.

Variations in phialidic development

As one would expect, variation of these two basic modes of phialoconidial and phialide ontogeny occurs, and a few selected examples are presented below.

Formation of secondary phialides in Chalara
In addition to the development of cylindrical to ellipsoidal smooth-walled, hyaline conidia from elongated phialides of the *Chalara* state of *Ceratocystis adiposa* (Figs. 4.27, 4.29), thick-walled, pigmented conidia are also produced from shorter phialides on the same mycelium (Fig. 4.39A–H) two days after inoculation. Nag Raj and Kendrick (1975) considered age to be an important factor in determining the principal mode of conidiogenesis displayed by this fungus, suggesting that morphogenetic variations may be reactions to limiting nutrient supply and/or accumulation of staling substances (Brown, 1923). A culture of *C. adiposa* of approximately the same age was also examined with the SEM (Figs. 4.42–4.44). A conidium has partially emerged through the collarette in Fig. 4.42. The circumscissile scar of the outer conidial wall (arrow) and septal plug (arrowhead) were left after secession of the previous conidium. Adjacent conidia are normally held together by the continuous outer wall (Fig. 4.43) or *sheath* (Hawes and Beckett, 1977a, b). The mature conidia (approx. 5-day-old cultures) are ovoid to globose and their outer wall layer is distinctly verrucose (Figs. 4.41, 4.44). *Chalara paradoxa* also produces ellipsoidal conidia from short phialides (Fig. 4.45). The conidia swell after liberation from the fertile cell (Fig. 4.46), becoming thick-walled and darkly pigmented. Nag Raj and Kendrick (1975) concluded that irrespective of the shape of the phialoconidia produced by these species, they 'all become dark brown and thick-walled, and function as chlamydospores, as do some of the phialides and hyphal elements'. This latter phenomenon is demonstrated by *C. paradoxa* in Fig. 4.47, in which the phialide apex has become swollen and thick-walled, terminating conidiogenesis. Additional developmental variation in *Chalara* is demonstrated by *C. ovoidea* Nag Raj & Kendrick in which solitary, holoblastic, conidia are often formed at successive apices of a sympodially proliferating fertile hypha (Nag Raj and Kendrick, 1975). Formation of arthric conidia in mature cultures of *C. elegans* (Chapter 7, Fig. 7.37H) represents yet another developmental variation in this genus.

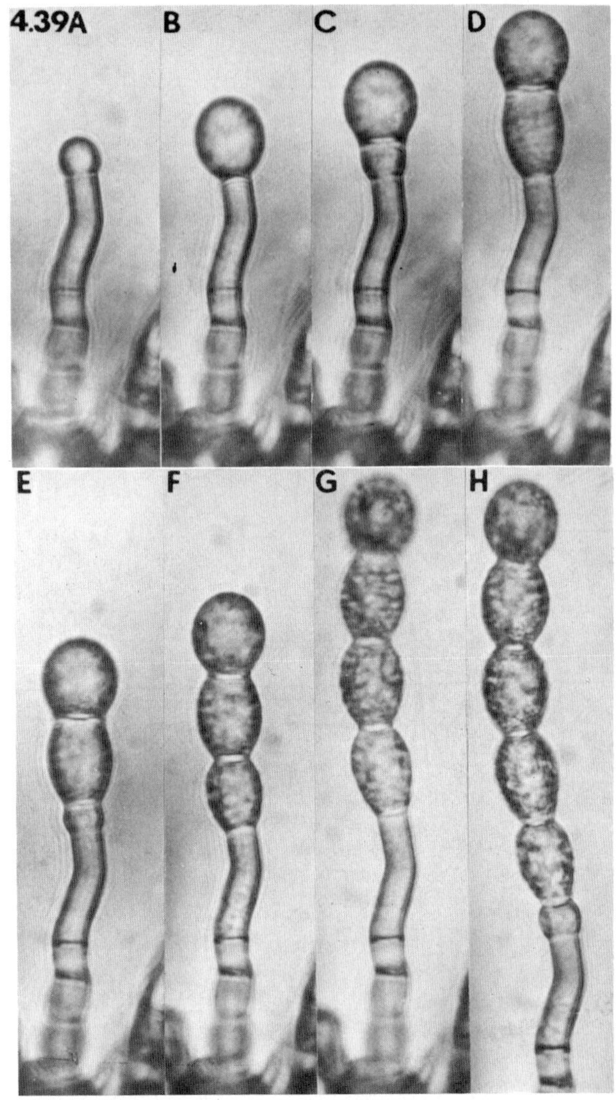

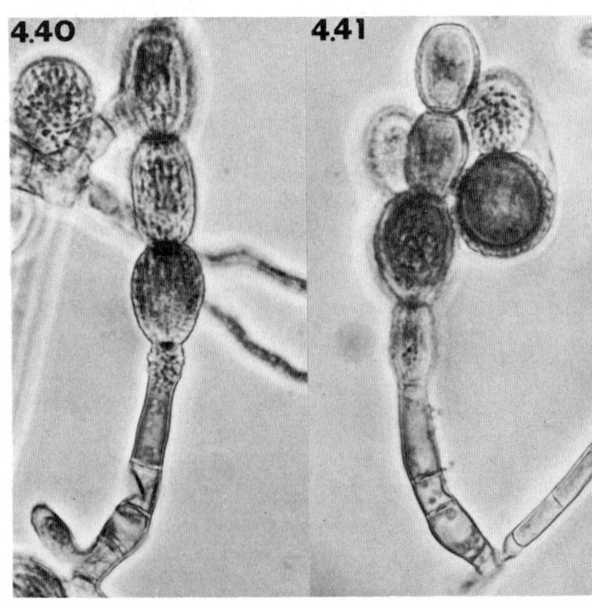

Figs. 4.39–4.41 *Chalara* state of *Ceratocystis adiposa*. 4.39A–H. Time-lapse sequence of phialoconidial development. Conidia remain together in chains. After conidia have emerged from phialide, they become verrucose, thick-walled and darkly pigmented (H). 4.40, 4.41. Mature phialoconidia. 4.39A–H; T = 0, 65, 100, 160, 190, 235, 320, 495, respectively. All × 1,260.

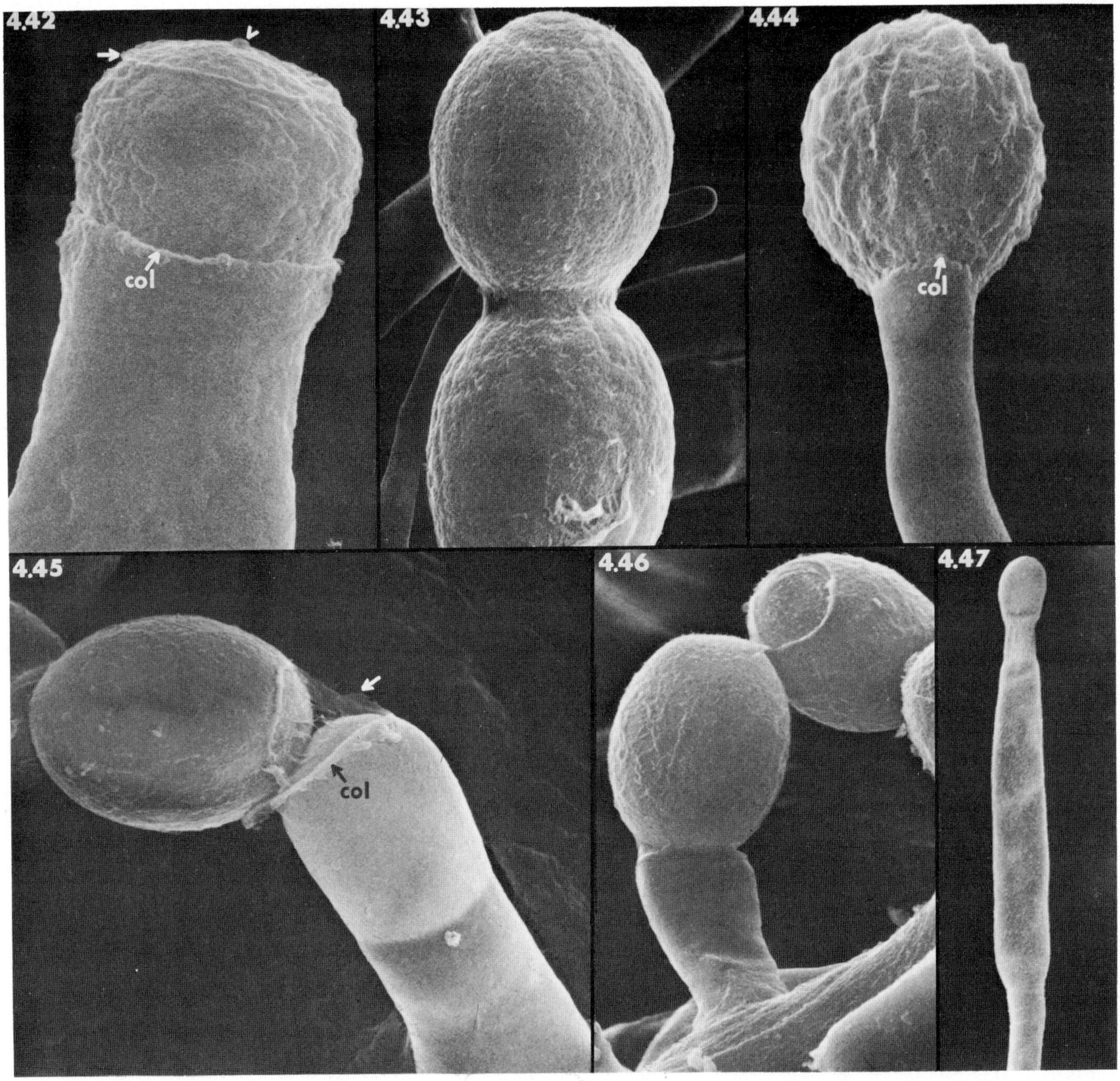

Figs. 4.42–4.44 *Chalara* state of *Ceratocystis adiposa.* 4.42. Early stage of emergence of conidium through cylindrical collarette. Note circumscissle scar (arrow) and septal plug (arrowhead) at apex of conidium. 4.43. Adjacent conidia united by continuous wall layer. 4.44. Mature, verrucose conidium at phialide apex. col, collarette. ×9,500, 5,100 and 3,500, respectively.

Figs. 4.45–4.47 *Chalara paradoxa.* 4.45, 4.46. Phialoconidia at stages of emergence from short, cylindrical phialides. Note that the thin outer conidial wall layer (arrow in 4.45), which normally binds conidia together in a chain, is distinct from the ruptured, outer phialide wall (i.e., collarette, col). 4.47. Phialide and young apical conidium have become thick-walled and function as a chlamydospore. ×3,740, 3,250 and 1,840, respectively.

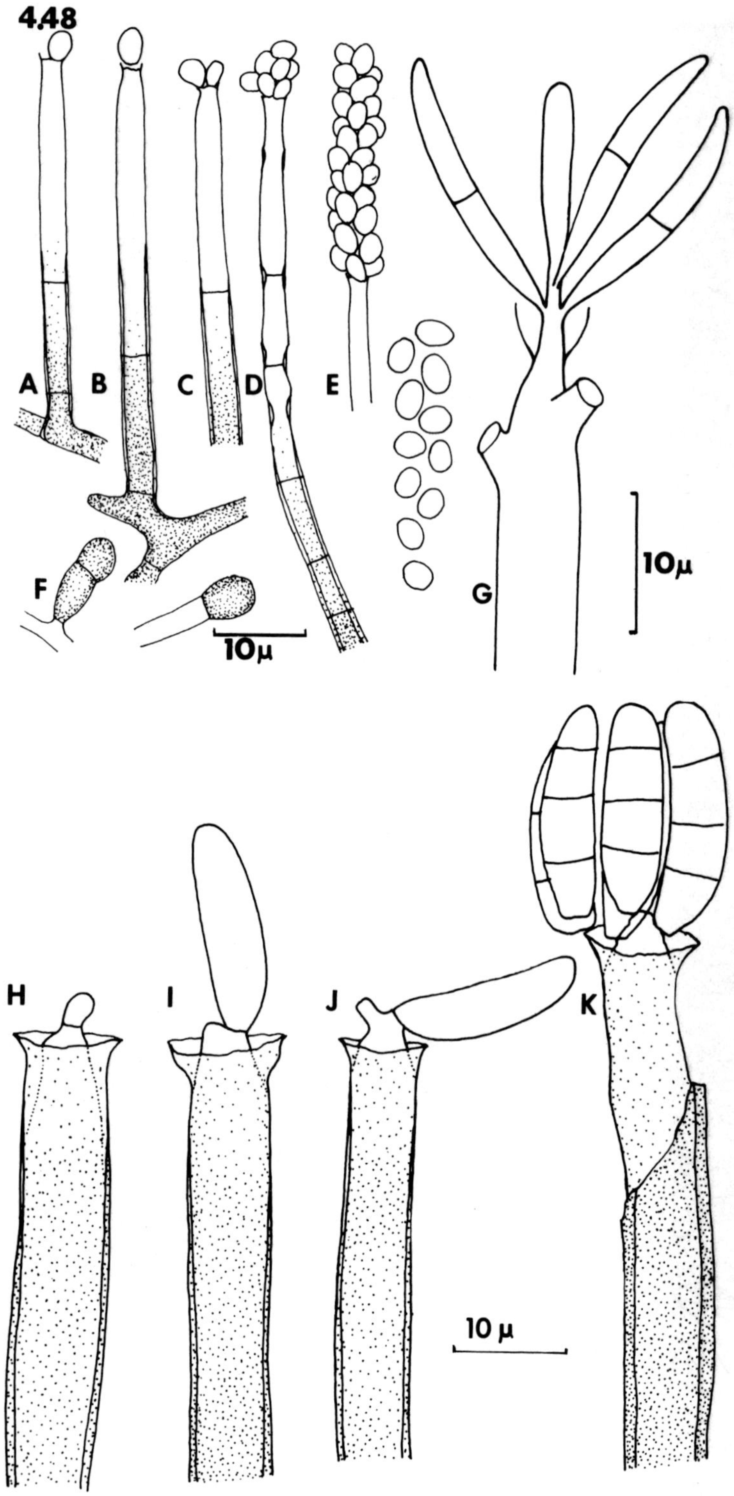

Conidiogenesis in Chloridium

Species of *Chloridium* Link ex Fr. are generally considered to produce phialoconidia (M. B. Ellis, 1971b), but the minute nature of the fertile cells has made it impossible to define this developmental process fully with the light microscope alone. The ellipsoidal to subspherical propagules are arranged in slimy masses and often in long columns at the apex of conidiogenous cells (Fig. 4.48E), which often proliferate percurrently (Fig. 4.48D). However, conidiogenesis in *Chloridium* spp. is unlike typical phialoconidium formation, a fact which was first suspected by Mangenot (1952) in his study of *C. chlamydosporis* (van Beyma) Hughes, and later confirmed by Hammill (1972c). Figures 4.49–4.52 show several stages of conidial development in the type species, *C. virescens* (Pers. ex Pers.) W. Gams & Hol.-Jech. (= *C. viride* Link). Successive conidia are formed from alternate sides of the slightly proliferated conidiogenous cell apex, above the collarette. The fertile cell of *C. virescens* does not proliferate after each conidium, like the holoblastic forms discussed in Chapter 3 (cf. Figs. 4.50, 4.51). As the conidia secede, they accumulate in droplets (Fig. 4.52) or form a cirrus at the apex of the fertile cell (Fig. 4.48E). Hammill (1972c) made electron microscope observations of wall differentiation during conidiogenesis in *C. chlamydosporis* (Figs. 4.53, 4.54). In Fig. 4.53, the collarette surrounds the base of a young conidial initial which has begun to grow from below and to one side of the previously formed conidium. A double-layered septum has formed at the base of the mature conidium. Later (Fig. 4.54), three conidia have formed at the fertile apex, and one has seceded, rupturing a thin, outer wall layer (arrows). The outer wall layer enclosing the opposite conidium (arrowheads) is still intact. As successive conidia abstrict, rings of ruptured wall layers accumulate within the collarette in a manner like that described for the phialides of *Trichoderma saturnisporum* (Figs. 4.36, 4.38). A diagrammatic interpretation of conidiogenesis in *Chloridium* spp. is presented in Fig. 4.55A–D which is generally in agreement with those presented by Harvey (1974b) and Beckett *et al.* (1974).

Goos (1969) described a similar mechanism in *Cacumisporium capitulatum* (Corda) Hughes (Fig. 4.48H–K). S. J. Hughes and Kendrick (1968) illustrated

Fig. 4.48A–F *Chloridium virescens.* Stages of conidial development (A–C). Columns of ellipsoidal conidia attached to phialide apex (D, E). Percurrently proliferated conidiogenous cells (D) and chlamydospores (F) also demonstrated by this species.

Fig. 4.48G *Codinaea setosa.* Unusual phialide demonstrating alternate arrangement of conidia attached to proliferated, fertile apex which has ruptured the outer wall layer leaving a distinct collarette. Collarettes also visible subapically indicating occurrence of sympodial proliferation of the conidiogenous cell (after S. J. Hughes and Kendrick, 1968).

Fig. 4.48H–K *Cacumisporium capitulatum.* Stages of conidiogenesis at phialide apex (H–J) and percurrently proliferated conidiogenous cell (K).

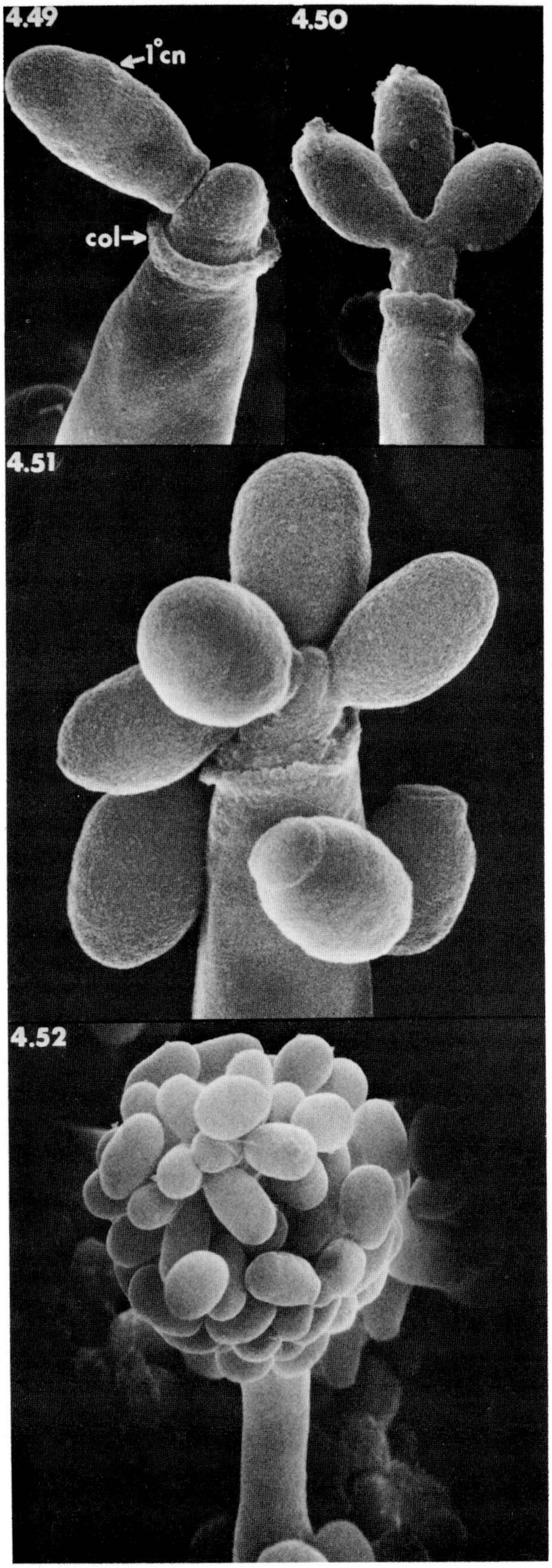

another example of this mode of conidiogenesis in *Codinaea setosa* Hughes and Kendrick (Fig. 4.48G). Species of *Chloridium, Cacumisporium* Preuss and *Codinaea* Maire seem to represent a special, aberrant type of phialidic development, since they have collarettes and accumulations of ruptured wall layers within these collarettes (Fig. 4.55D). Hammill (1972c) referred to their cells as *sympodulo-phialides.*

Polyphialides

S. J. Hughes (1951) proposed the term *polyphialide* for the conidiogenous cells of *Lasiosphaeria hirsuta* (Fries) Ces. & de Not., which he described as having 'more than one open end from each of which a basipetal succession of phialospores is produced'. This definition was extended by Hughes and Kendrick (1968), in their monograph of *Codinaea*, to read, 'the proliferation is sympodial, that is, the proliferation develops, not through the collarette, but from a new growing point just behind and to one side of the first formed collarette; a succession of such proliferations often results in a geniculate polyphialide which bears laterally the remains of displaced collarettes . . . as the polyphialide elongates, successive septa may be laid down within it'. Several hyphomycetous fungi produce polyphialides, sometimes as the typical condition (e.g., *Codinaea simplex* Hughes & Kendrick, Fig. 4.56A; *Phialophora geniculata* van Emden, Figs. 4.56B, 4.57, 4.58), and sometimes as rare variations of normal phialidic development. Cole and Kendrick (1969a) suggested that in any functional polyphialide, only the most recently formed conidiogenous locus, that is, the apical one, is fertile. We now recognize, however, that although this is true of most polyphialides, there are exceptions in which two or more conidiogenous loci of the same fertile cell are active simultaneously. *Polypaecilum insolitum* G. Smith, whose development has been misinterpreted as annellidic (G. Smith, 1961; M. B. Ellis, 1971b; Yamashita and Yamashita, 1972), produces several simultaneously functioning phialidic apertures (Cole, 1976c) on a single fertile cell, or polyphialide (Figs. 4.56C, 4.59, 4.60). The definition of the polyphialide, therefore, should be amended as follows: polyphialides may rarely have two or more concurrently acting conidiogenous loci (e.g., *Polypaecilum insolitum).*

Figs. 4.49–4.52 *Chloridium virescens.* Successive stages of conidial development from fertile apex of conidiogenous cell. The first-formed conidium (1°cn) ruptures the outer wall of the latter leaving a collarette (col, 4.49). Subsequent conidia arise above the collarette from alternate sides of the slightly proliferated, fertile apex of the conidiogenous cell (4.50, 4.51). A droplet of conidia is shown at the conidiogenous cell apex in 4.52. ×8,250, 6,900, 10,000 and 3,900, respectively.

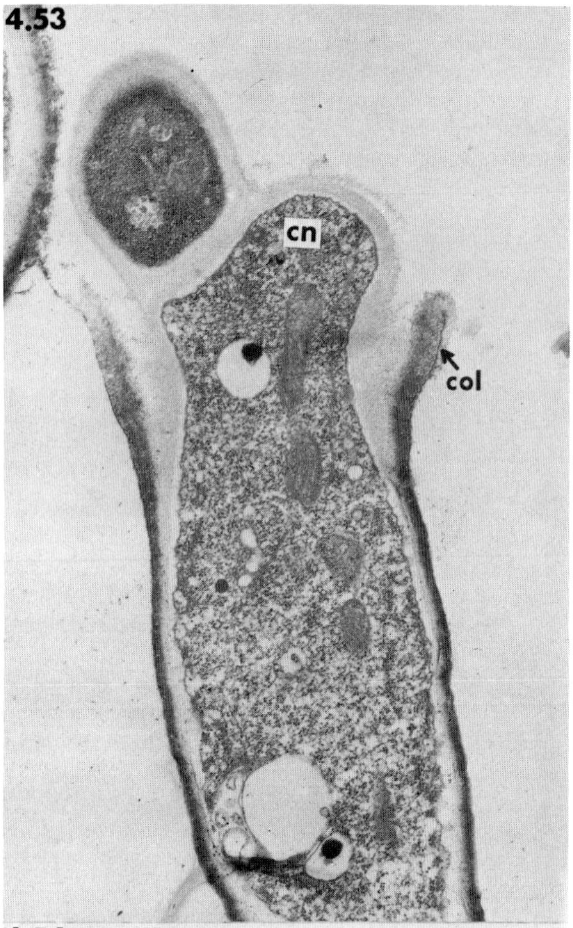

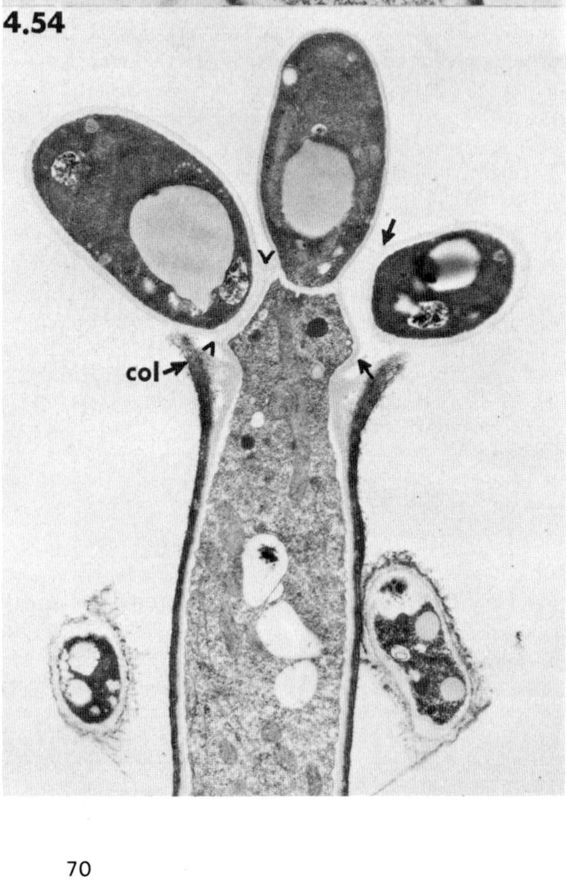

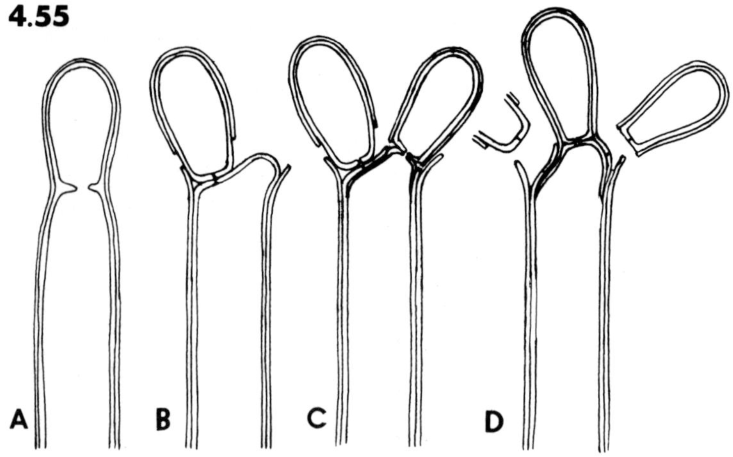

Figs. 4.53–4.54 *Chloridium chlamydosporis.* Thin sections of phialide apex showing conidial initial at base of mature conidium (4.53) and three conidia attached above the collarette (4.54). Arrowheads in 4.54 indicate continuous wall between conidium and fertile cell apex, while arrows locate wall rupture. cn, conidial initial; col, collarette. ×21,000 and 13,150, respectively (from Hammill, 1972c).

Fig. 4.55A–D Interpretation of wall differentiation during formation of successive conidia in *Chloridium chlamydosporis.*

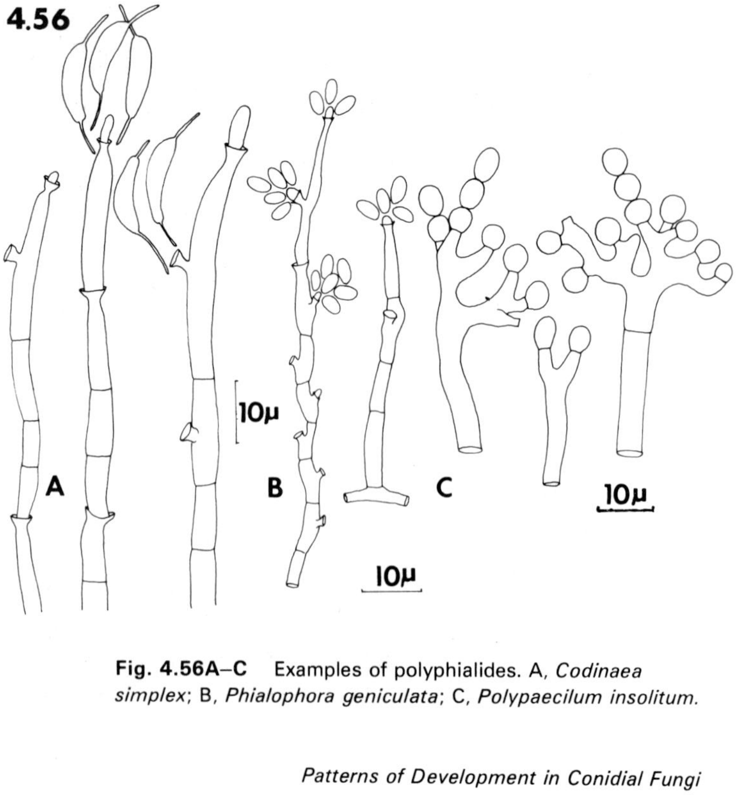

Fig. 4.56A–C Examples of polyphialides. A, *Codinaea simplex*; B, *Phialophora geniculata*; C, *Polypaecilum insolitum.*

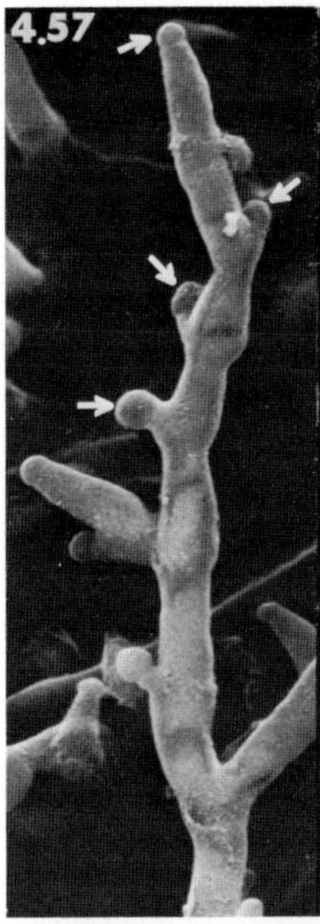

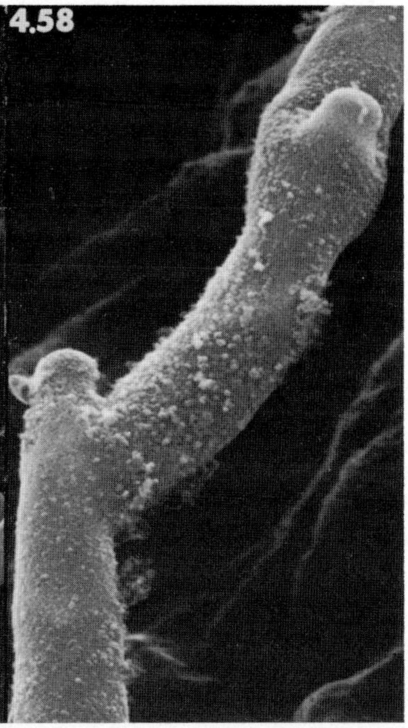

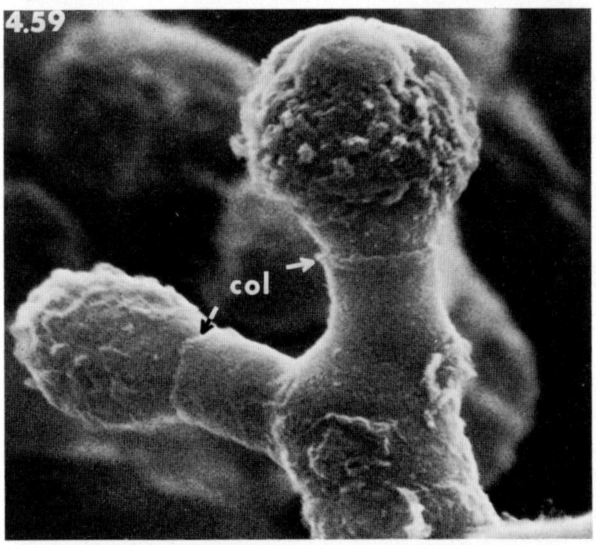

Figs. 4.57, 4.58 *Phialophora geniculata.* Alternate or geniculate arrangement of fertile apices (arrows in 4.57). ×2,140 and 7,875, respectively.

Figs. 4.59, 4.60 *Polypaecilum insolitum.* Pair of fertile apices showing collarettes (col) and solitary conidium attached to each (4.59). Cluster of fertile apices, each showing conidia in various stages of development (4.60). ×10,300 and 4,270, respectively.

Reduced phialides

Reduced phialides are found in *Phialophora heteromorpha* (Nannf.) Wang, in which sessile collarettes are formed directly from sclerotial bodies (Fig. 4.61A). Phialides may also be reduced to collarettes on the surface of vegetative hyphae (e.g., *Phialophora lignicola* (Nannf. apud Melin & Nannf.) Goid., Fig. 4.61B). Some species demonstrate considerable variation in the morphology of phialides produced by the same mycelium (e.g., *Phialophora parasitica* Ajello, Georg and Wang; Figs. 4.62–4.66). In Fig. 4.62, a typical phialide is shown with a septum at its base (arrow) and slightly tapering apex which terminates in a minute collarette. The phialide in Fig. 4.63 is more vase-shaped with a narrowed apex, whereas in Fig. 4.64, the conidiogenous cell is reduced to a short, morphologically undifferentiated fertile branch arising from the aerial hypha.

4.61

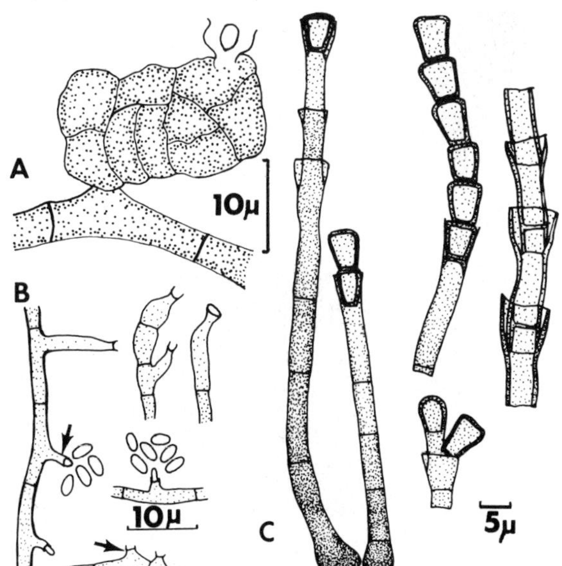

Fig. 4.61A *Phialophora heteromorpha.* Sessile collarette on the surface of sclerotial body (after Wang, 1964).

Fig. 4.61B *Phialophora lignicola.* Variation in phialide morphology. Note reduced phialides (arrows).

Fig. 4.61C *Catenularia* state of *Chaetosphaeria cupulifera* (Berk. & Br.) Sacc. Percurrently proliferated phialides (after M. B. Ellis, 1971b).

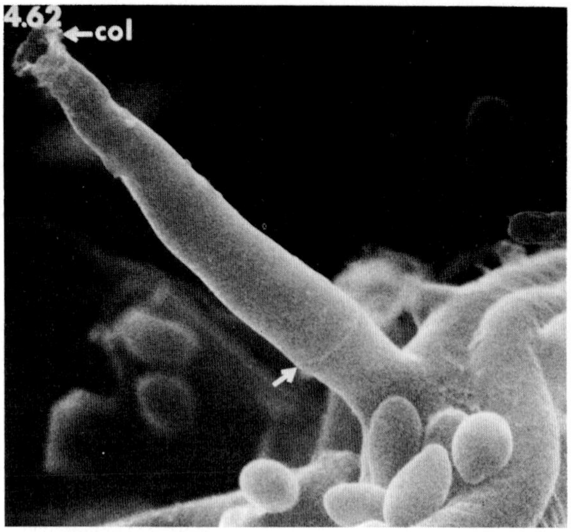

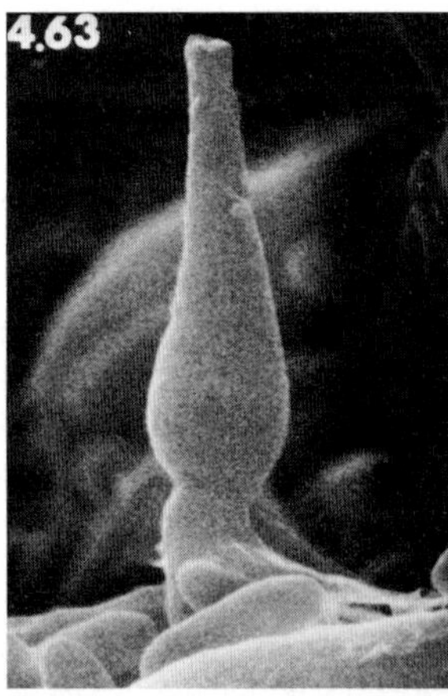

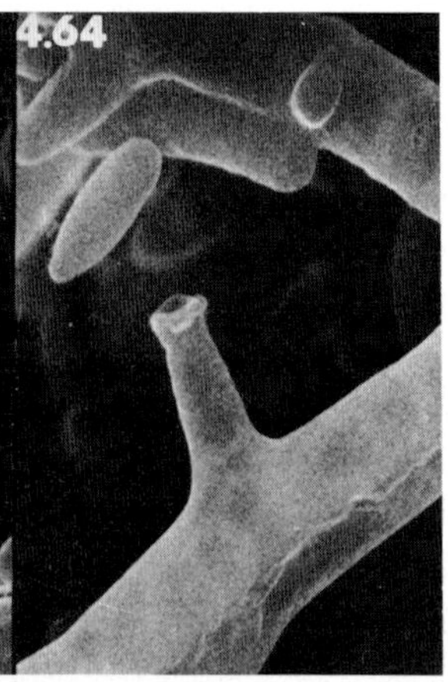

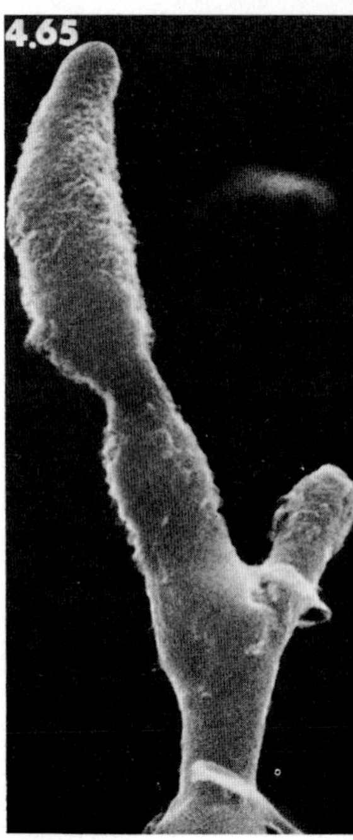

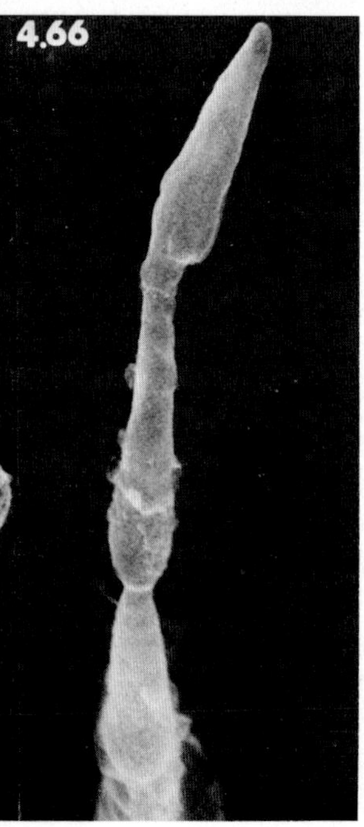

Figs. 4.62–4.66 *Phialophora parasitica.* Variation in phialide morphology. Typical phialide (4.62) showing basal septum (arrow), tapered apex and minute collarette (col). Vase-shaped (4.63), reduced (4.64) and percurrently proliferated phialides (4.65, 4.66) are also produced. ×5,800, 5,340, 4,950, 5,880 and 3,630, respectively.

Percurrently proliferated phialides

During percurrent proliferation of phialides, one conidiogenous cell gives rise to another by growing on through its erstwhile fertile apex, then ceases elongation and produces another basipetal succession of phialoconidia (e.g., *Catenularia* spp., Fig. 4.61C). This kind of phialidic development resembles annellidic ontogeny (Kendrick, 1971a), except that the latter involves formation of one rather than many conidia between proliferations. The problems of clearly distinguishing between these two processes have been discussed by Morgan-Jones *et al.* (1972) who questioned whether there 'are any fundamental differences between a phialide and an annellide or is there in fact a continuum.

Earlier concepts of annellidic development

S. J. Hughes (1951) described the conidiogenous cells of his new genus *Annellophora* Hughes as 'elongating by numerous successive conidial scars', which results in an annellated apex of the fertile cell (Fig. 4.67). In 1953, he referred to these annular scars as 'annellations' (arrowheads in Fig. 4.67) and introduced the term 'annellophore' for the conidiogenous cell which bears them. Hughes also recognized that the distance between successive annellations in certain species may be much smaller than observed in *Annellophora*, and the walls may be thinner, rendering annelations as such more difficult to resolve and identify. This problem, which will be considered later, provided for ambiguity at the outset in formulation of a clear definition of annellidic development.

Cole and Kendrick (1969b) presented a revised concept of annellidic development and the following definition appeared in Kendrick (1971b): 'a specialized form of holoblastic conidium ontogeny in which the growth of successive, very short, percurrent, vegetative proliferations of the conidiogenous cell is terminated by the development of a conidiogenous locus which gives rise to a single conidium. The conidiogenous cell usually becomes longer as it produces a basipetal sequence of conidia. What remains of each vegetative proliferation after the resulting conidium has seceded may be termed an annellation, and the conidiogenous cell itself may be described as annellated, or termed an annellide'.

Developmental and ultrastructural aspects

This developmental process is illustrated by the time-lapse sequence of *Scopulariopsis brevicaulis* (Sacc.) Bain. in Fig. 4.68A–H. The primary conidium forms holoblastically at the conidiogenous cell apex and is delimited at its base by a septum (Fig. 4.68A, B). Over a period of $54\frac{1}{4}$ h (Fig. 4.68B–H),

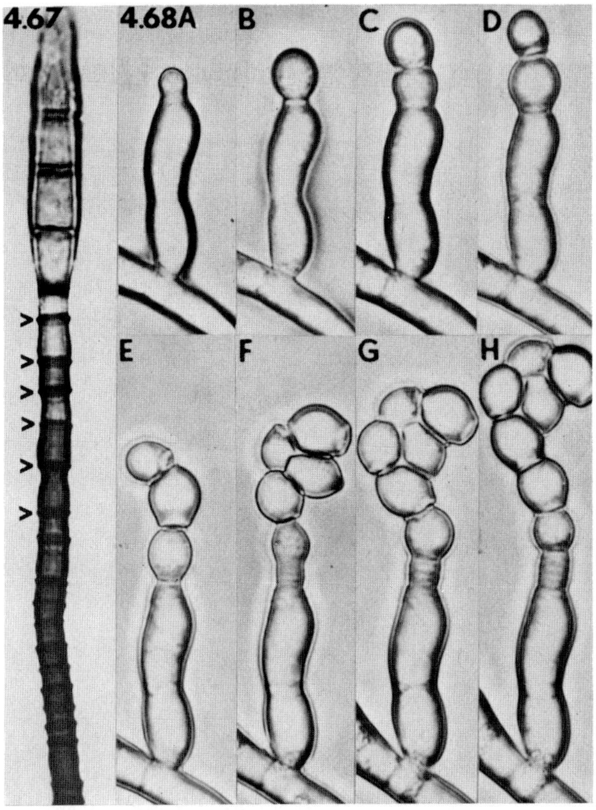

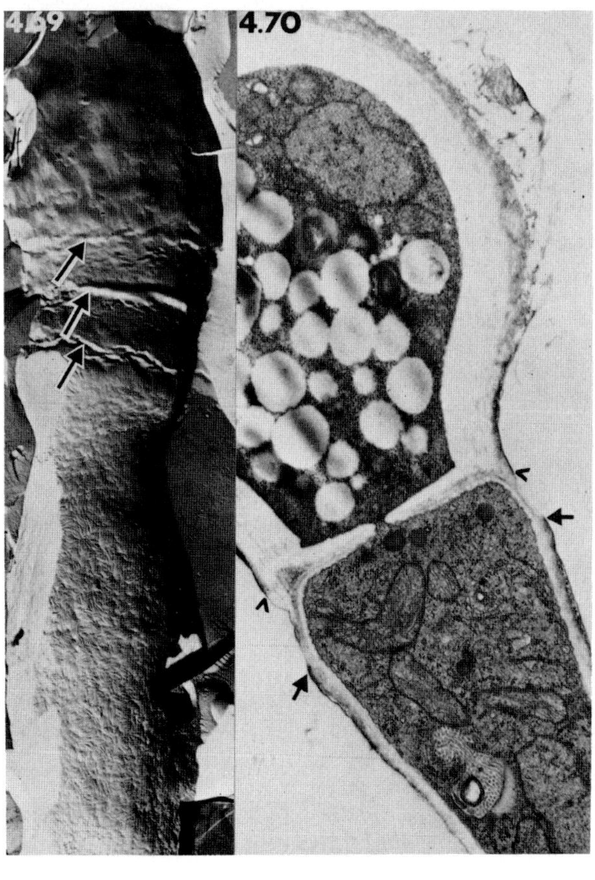

Fig. 4.67 *Annellophora africana* Hughes. Annellated conidiogenous cell showing terminal conidium. Arrowheads locate annellations at apex of fertile cell. ×1,180.

Fig. 4.68A–H *Scopulariopsis brevicaulis.* Time-lapse sequence showing successive conidium ontogeny and percurrent proliferation of the conidiogenous cell. Circumscissile scars (annellations) of previously formed conidia are visible on the proliferated apex of annellide in (H). T = 0, $3\frac{3}{4}$, 6, $8\frac{1}{2}$, 19, 31, $50\frac{1}{4}$, 58 hours, respectively. ×1,830.

Figs. 4.69, 4.70 *Scopulariopsis brevicaulis.* Freeze-fracture (4.69) and thin-section (4.70) of annellide apex. Arrows in 4.69 and 4.70 indicate annellations. Arrowheads in 4.70 locate circumscissile scar at conidial base. ×7,820 and 14,400, respectively.

six more conidia are produced in basipetal succession and the fertile cell, or annellide, forms an elongated neck which possesses recognizable annellations. Ultrastructural investigations of this fungus (Cole and Aldrich, 1971a; Hammill, 1971) have illustrated additional aspects of annellidic development. Annellations can be identified in both freeze-etch fractures (Fig. 4.69) and thin-sections (Fig. 4.70). The latter shows an early stage in percurrent proliferation in which the previously formed conidium has not yet seceded. The basal septum is uniperforate and several Woronin bodies are visible near the septal pore. Although these organelles have not been found at the fertile apices of phialides, they apparently function in annellides as septal plugs at the base of seceding conidia. The outer conidial wall layer, originally continuous with the wall of the annellide, has ruptured, leaving a circumscissile scar at the conidial base (arrowheads in Fig. 4.70) and an annellation at the apex of the conidiogenous cell (arrows). The rupture has been caused by the percurrent proliferation of the annellide. An even earlier stage of proliferation is illustrated in Fig. 4.71, showing the differentiated double-layered septum at the base of the terminal conidium. The thick, upper layer has become the basal wall of the conidium while the thin, lower layer is the proliferating apex of the conidiogenous cell. The medial, electron-transparent zone is presumably the site of schizolytic activity. A circumscissile rupture of the outer wall layer is visible between the conidium and conidiogenous cell (arrows in Fig. 4.71). Growth above the annellation at first results in cylindrical elongation (Fig. 4.72) and then the apex swells as the next conidium is initiated (Figs. 4.73, 4.74). The outer wall layer of the new conidium expands and appears to be degenerating (Fig. 4.73) while an inner wall layer begins to differentiate (arrowheads in Fig. 4.74). The cell membrane encompassing the initial becomes convoluted (Fig. 4.73), perhaps due to synthesis of excess plasma membrane prior to the marked increase in volume of the cell, since these convolutions later disappear (Fig. 4.74). The thin, secondary wall of the young conidium in Fig. 4.74 is clearly visible in Fig. 4.75 as well as the beginning of a basal septum. Cisternal rings at the apex of the conidiogenous cell early in conidium differentiation (Fig. 4.73) probably function as secretory organelles (dictyosome equivalents?), producing

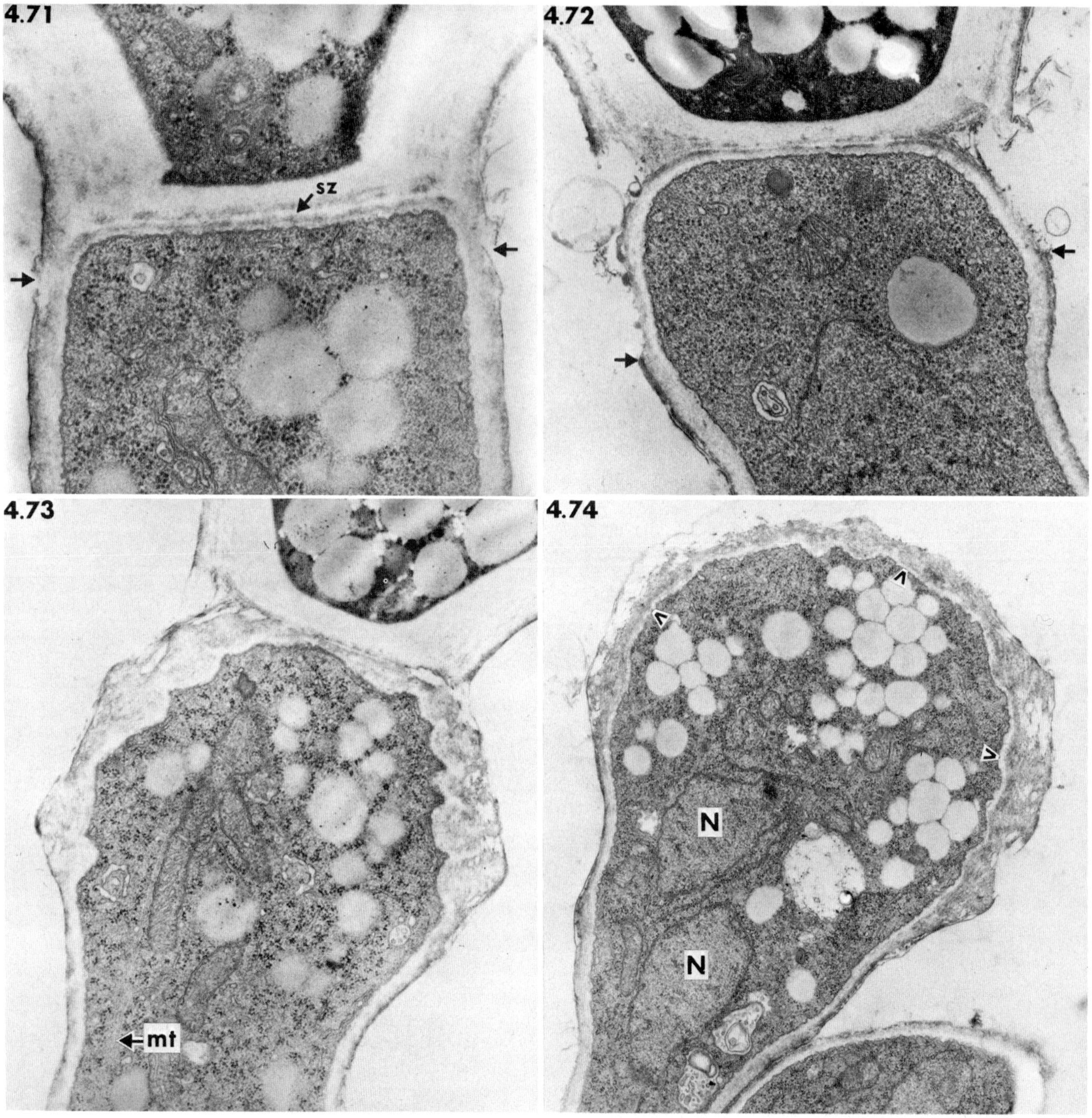

Figs. 4.71–4.74 *Scopulariopsis brevicaulis.* Thin-sections of early stages of conidiogenous cell proliferation (4.71, 4.72) and conidial development (4.73, 4.74). Arrows in 4.71 indicate ruptured outer wall layer which later becomes the annellation (arrows in 4.72). Arrowheads in 4.74 locate inner wall layer of conidial initial. mt, microtubules; N, nucleus; sz, schizolytic zone. ×29,900, 23,200, 18,000 and 13,500, respectively.

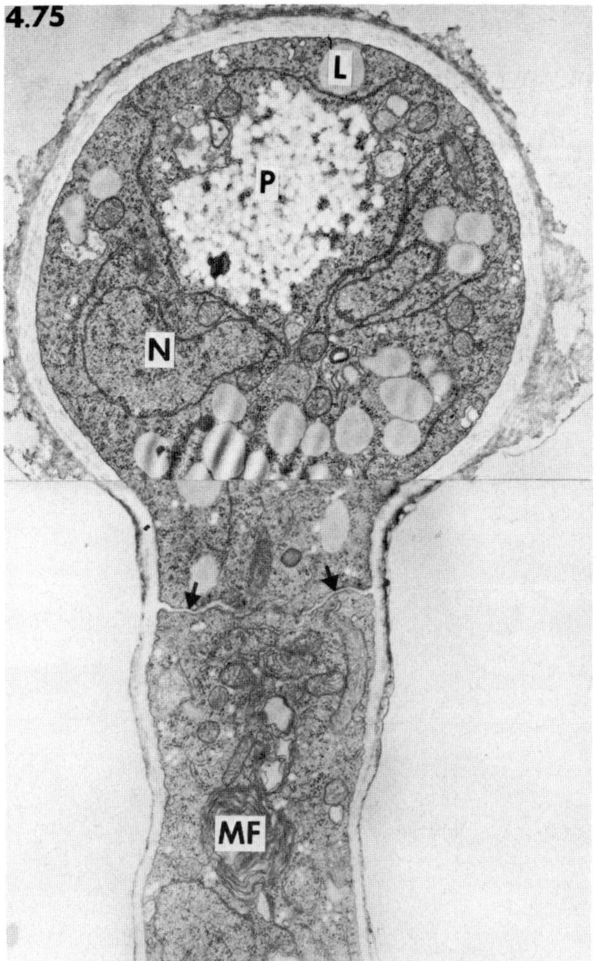

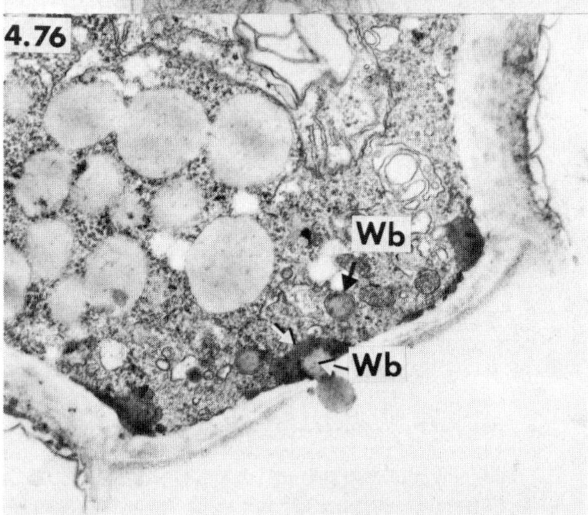

Figs. 4.75, 4.76 *Scopulariopsis brevicaulis.* Early stage of septal development at base of differentiated conidium (4.75). After conidial secession, septal pore is plugged by Woronin body (Wb) which is partially encompassed by electron-opaque deposit (arrow in 4.76). Arrows in 4.75 Indicate wall of young septum. L, lipid droplets; MF, myelin figure; N, nucleus; P, insoluble polysaccharide; Wb, Woronin body. ×19,800 and 28,750, respectively.

4.77

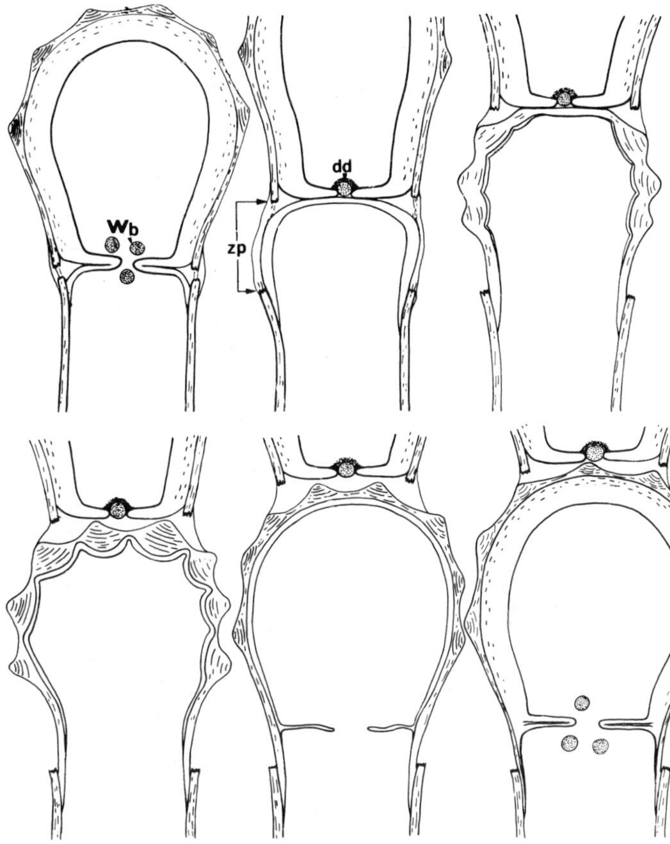

Fig. 4.77 Diagrammatic interpretation of wall differentiation during annellidic development in *Scopulariopsis brevicaulis.* Note that the first-formed conidium is holoblastic, while all subsequently formed conidia are enteroblastic. dd, electron-opaque deposit surrounding septal plug; Wb, Woronin body; zp, zone of proliferation.

vesicles which fuse with the plasmalemma and release their contents (wall precursors, mural enzymes, etc.) to the growing wall. During conidial secession (Fig. 4.75), the septal pore is usually plugged by a Woronin body which becomes partially surrounded by an electron-opaque deposit (Fig. 4.76). Similar observations have been reported for *Trichurus spiralis* Hasselbring (Hammill, 1977b).

Revised concept of annellidic development and comparison with phialidic development

On the basis of these data, a diagrammatic interpretation of annellidic development in *Scopulariopsis brevicaulis* is presented in Fig. 4.77A–F. Developmental similarities between annellides and phialides (Fig. 4.38A–I) are recognized. The primary conidium in each case is holoblastic. The septum at the base of the conidium is double layered, and each layer performs a separate function during conidial maturation and proliferation of the annellide. Similar differentiation of septal wall layers was demonstrated during

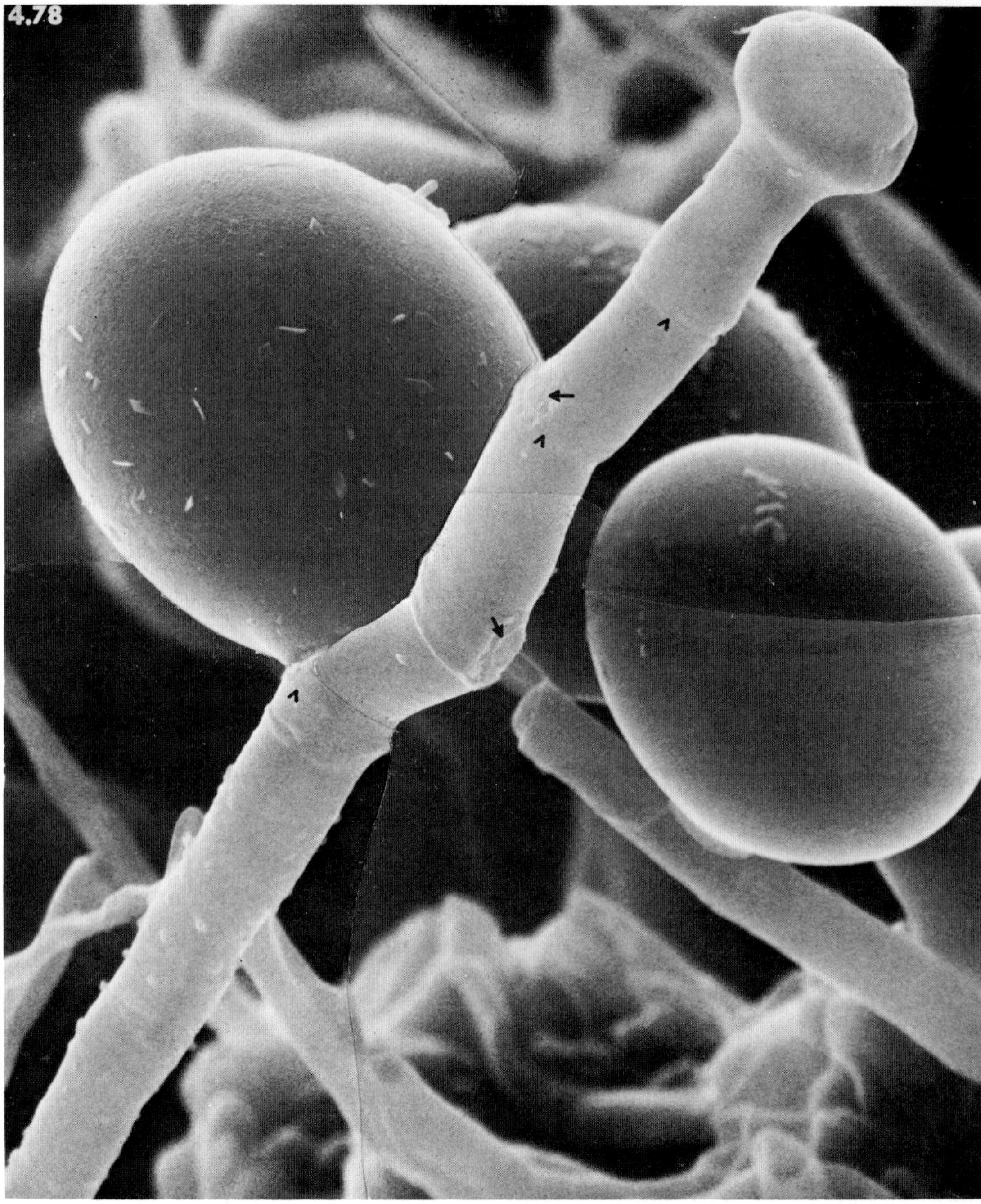

Fig. 4.78 *Acrogenospora sphaerocephala.* Proliferated conidiogenous cell showing scars of seceded conidia (arrows), laterally attached mature conidium, and conidial initial at apex. Several annellations (arrowheads) are visible along the conidiogenous cell. ×2,900.

4.79

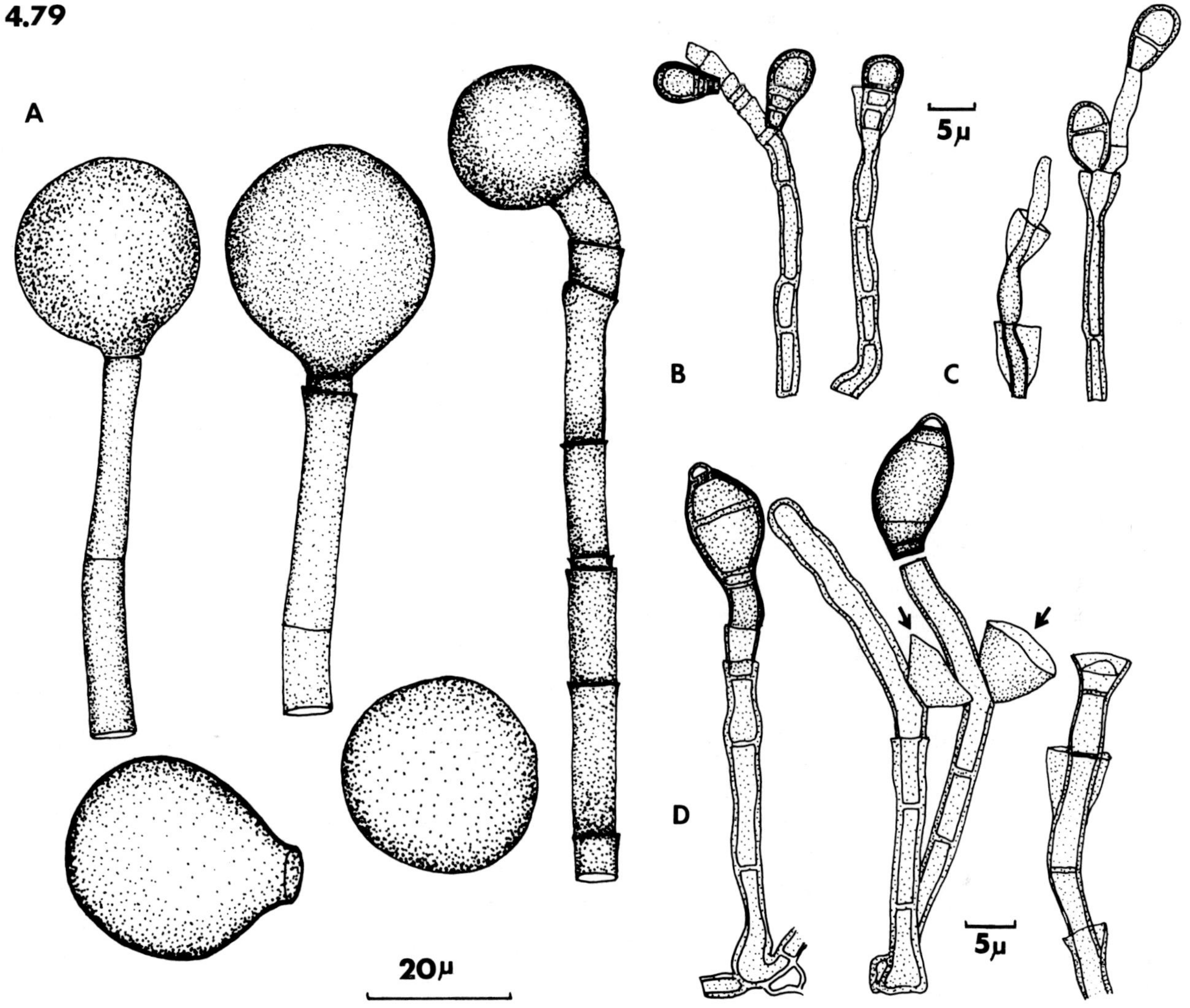

Fig. 4.79A *Acrogenospora sphaerocephala.* Conidia and annellated conidiogenous cells (after M. B. Ellis, 1971b).

Fig. 4.79B, C *Endophragmia* spp. Conidia, conidiogenous cells and prominent cup-shaped annellations, or abstriction scars of *E. bisepta* (B) and *E. unisepta* (C) (after M. B. Ellis, 1971b).

Fig. 4.79D *Endophragmiopsis pirozynskii.* Also demonstrates cup-shaped abstriction scars.

phialidic development in *Trichoderma saturnisporum* (Hammill, 1974a) and *Termitaria snyderi* (Khan and Aldrich, 1973). Formation of secondary conidia from both annellides and phialides is enteroblastic. This is, in fact, a reinterpretation of earlier known facts (Cole and Kendrick 1969a, b; Cole and Aldrich, 1971a; Kendrick, 1971a) but is a significant point in the consideration of developmental categories of Deuteromycetes (Chapter 8). The major differences between phialidic and annellidic development are as follows:

1. (a) The outer *annellide* wall layer usually ruptures at the base of the first-formed conidium, adjacent to the double layered septum leaving an annellation (Fig. 4.77A), while (b) the outer *phialide* wall layer may rupture at any point above the basal septum, leaving a collarette.
2. (a) The *annellide* undergoes percurrent proliferation after maturation of each conidium, leaving a series of annellations at its apex (Fig. 4.77B–F), while (b) no detectable elongation of the *phialide* occurs during formation of successive conidia, but circumscissile abstriction scars are sometimes left within the collarette (Fig. 4.38D, E).
3. (a) The septal pore at the base of each conidium which secedes from the *annellide* is sealed with a Woronin body (Fig. 4.77B, C), while (b) the septum at the base of each *phialoconidium* is complete; Woronin bodies are absent (Fig. 4.38D–H).

However, variations in annellidic development are known to be associated with ontogenetic features (1) and (2), and

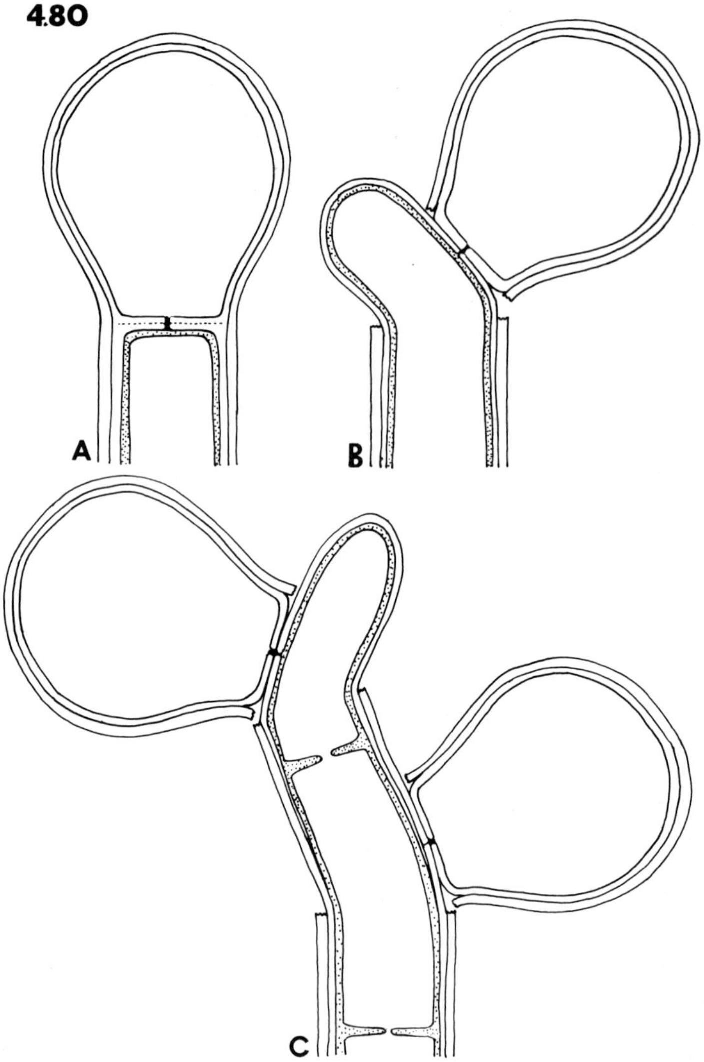

Fig. 4.80 Diagrammatic interpretation of wall differentiation during conidiogenesis and conidiogenous cell proliferation in *Acrogenospora sphaerocephala.*

additional thin-section examinations may also show that (3) is not consistent in distinguishing these two reproductive mechanisms in conidial fungi (R. Marchant, 1975; Hawes and Beckett, 1977b).

Variations in annellidic development

Hammill (1971) described variations in wall relations during annellidic development in *Scopulariopsis brevicaulis* and *S. koningii* (Oudem.) Vuill. For example, rupture of the outer conidiogenous cell wall may not always occur at the same level as the basal septum of the conidium but instead may occur above or below this point. In either case, the normal sequential formation of annellations at progressively higher levels on the conidiogenous cell neck as interpreted in Fig. 4.77 would not occur. A similar phenomenon has been reported from thin-section examinations of *Spilocaea pomi* Fr. ex Fr. (Reisinger, 1972; Hammill, 1973a; Corlett *et al.*, 1976).

Conidiogenesis in Acrogenospora

M. B. Ellis (1971b) illustrated distinct annellations on the conidiogenous cells of *Acrogenospora sphaerocephala* (Berk. & Br.) M. B. Ellis (Fig. 4.79A). However, Hammill (1972e) determined that percurrent proliferation of the conidiogenous cell did not involve extension growth of the lower layer of the double septum at the fertile apex, as illustrated for *Scopulariopsis* Bain. (Figs. 4.70, 4.71), but rather 'growth of intraconidiophore hyphae'. S. J. Hughes (1971b) considered this to be regeneration of a damaged fertile cell and not a process associated with normal conidiogenesis. Hammill (1972e) disagreed and provided evidence that endohyphae in *A. sphaerocephala* are 'concerned with normal production of successive conidia by a conidiophore'. On the basis of Hammill's (1972e) thin-section study and our SEM analysis (Fig. 4.78), we give a diagrammatic interpretation of wall differentiation in Fig. 4.80A–C. The proliferating endohypha causes the outer wall layer of the conidiogenous cell to rupture, leaving a distinct *annellation* (Fig. 4.80A, B) which may simultaneously, but not necessarily, cause the terminal conidium to secede. More often, the entire conidium is pushed off to one side and the endohypha continues to elongate laterally from the base of the conidium (Fig. 4.80B, C). In this case, the basal septum fails to undergo schizolysis. When one of these conidia finally secedes, a scar is left on the lateral surface of the conidiogenous cell just above the annellation (arrows in Fig. 4.78). The conidiogenous cell in Fig. 4.78 has produced at least four conidia and a conidial initial, which is still attached to the apex. Because of successive lateral proliferations of endohyphae, the conidiogenous cell assumes a geniculate shape. This should not be confused with sympodial proliferation as described in Chapter 3. Just below the conidial initial an annellation is present (arrowhead), but no lateral scar of a detached conidium. At this point, the conidiogenous cell proliferated through the scar of the previously formed terminal conidium, which apparently seceded with ease. Both percurrent and sympodial proliferation may occur during development of the same conidiogenous cell. Careful examination of Ellis's (1971b) illustrations of species of *Endophragmia* Duvernoy & Maire (Fig. 4.79B, C) indicates a similar developmental process except that the annellations are often prominent cup-shaped scars. The explanation of this phenomenon reminds one of Hammill's (1971) observation of *Scopulariopsis* in which the annellation may form above the level of the basal septum of the conidium. In *Endophragmia bisepta* M. B. Ellis (Fig. 4.79B) and *E. unisepta* M. B. Ellis (Fig. 4.79C), a circumscissile rupture of the outer conidial wall at the approximate mid-region of the cell occurs during secession. In *Endophragmiopsis pirozynskii* M. B. Ellis (Fig. 4.79D) a similar secession mechanism occurs and conidia are displaced laterally by a sequence of events comparable to those in *Acrogenospora sphaerocephala*. Formation of intrahyphal hyphae is common in the fungi (Miller and Anderson, 1961; Chan and Stephen, 1967; Calonge, 1968) and it is probable that several genera of annellated Deuteromycetes develop in this manner.

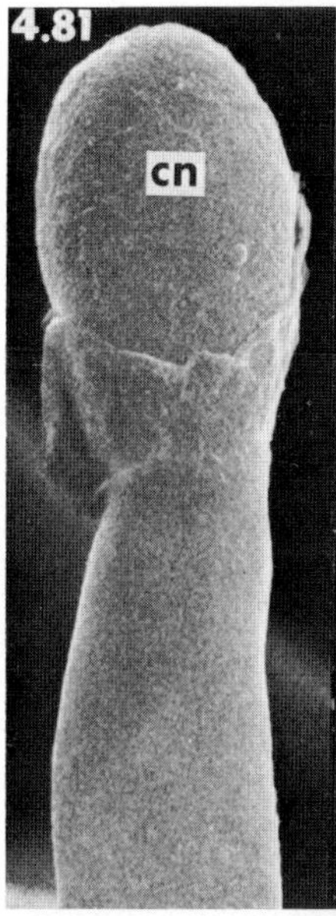

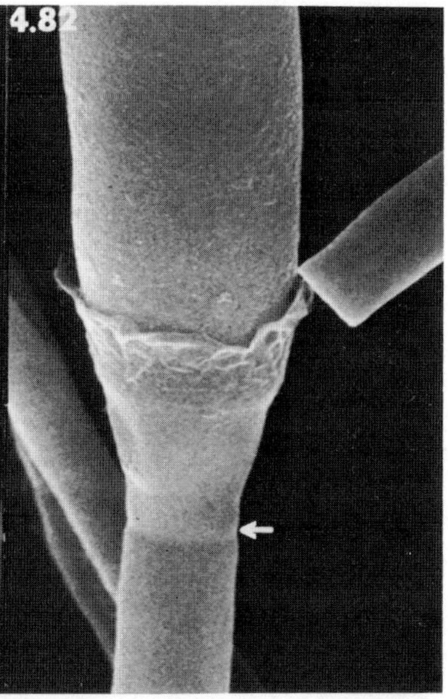

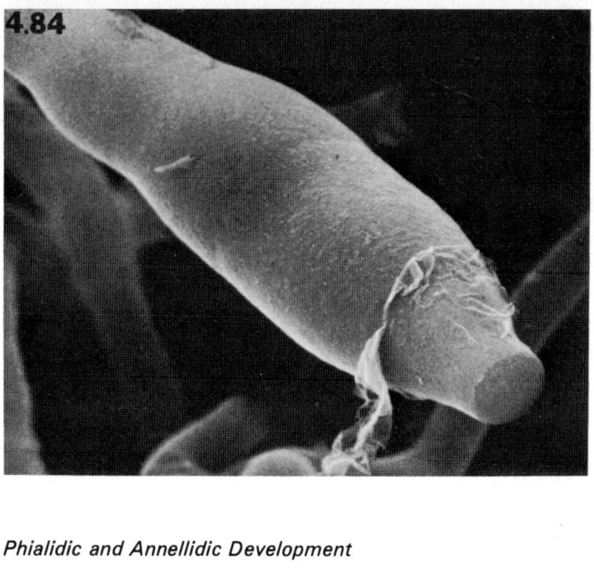

4.85

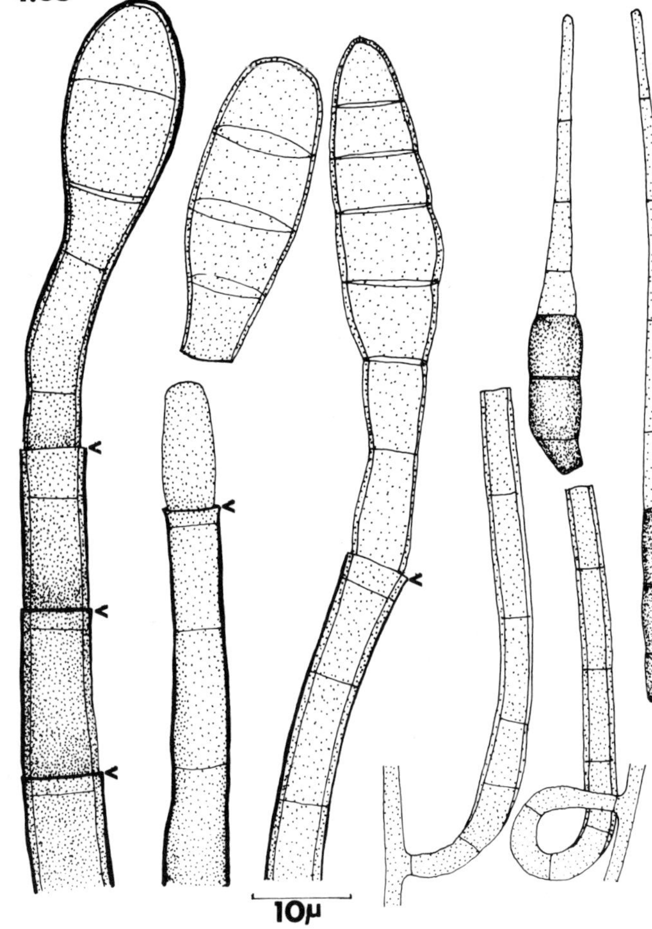

Figs. 4.81–4.84 *Sporidesmium ehrenbergii.* Conidial initial (4.81) and mature conidia attached to conidiogenous cell apex (4.82, 4.83), showing ring of ruptured outer wall encompassing base of conidium. Arrow in 4.82 locates transverse septum at conidial base. Schizolytic separation of conidium and fertile cell occurs below collar formed by ruptured, outer wall (4.84). cn, conidial initial. ×4,290, 2,780, 927 and 2,200, respectively.

Fig. 4.85 *Sporidesmium ehrenbergii.* Stages of conidial development and percurrent proliferation of conidiogenous cell which leaves annellations (arrowheads) on the surface of the latter.

Conidiogenesis in Sporidesmium

Each terminal conidium of *Sporidesmium ehrenbergii* M. B. Ellis has a broad basal attachment to the conidiogenous cell apex, and early in development, the outer cell wall ruptures, leaving a *collarette-like* ring of wall material at the base of the initial (Fig. 4.81) which persists (Figs. 4.82, 4.83). The basal septum of the conidium forms below the circumscissile rupture (arrow in Fig. 4.82) and when secession finally occurs, the *collarette* remains as part of the base of the detached cell (Fig. 4.84). Subsequently, the conidiogenous cell undergoes percurrent proliferation leaving an annellation on its outer surface (Fig. 4.85).

Convergence of phialidic and annellidic concepts

These subtle variations in wall differentiation during conidial and conidiogenous cell formation can easily lead to misinterpretations of development, particularly when fungi are examined only under the light microscope. This has been well demonstrated by study of a member of *Cryptosporiopsis* Bubak & Kabat (Sutton and Sandhu, 1969) and subsequent investigations of conidiogenesis in species of *Conioscypha* Höhnel (Shearer and Motta, 1973; Shearer, 1973; van Emden and Veenbaas-Rijks, 1973) and *Ascochyta* Lib. (Boerema and Bollen, 1975). A diagrammatic interpretation of conidiogenesis in *Cryptosporiopsis* spp. is presented in Fig. 4.86A–E. The primary conidium is holoblastic and later ruptures the outer conidiogenous cell wall and secedes schizolytically (Fig. 4.86A, B). The subsequent production of a basipetal series of conidia is associated with successive, short, percurrent proliferations of the conidiogenous cell. The annellations which remain after secession of conidia are hidden within the collarette (Fig. 4.86C–E). Although these *endogenous annellations* are separated and more distinct than the circumscissile scars found within the collarette of certain phialides (Figs. 4.36, 4.38E), it is questionable whether this developmental difference is sufficient justification to consider these conidiogenous cells annellides, as proposed by Sutton and Sandhu (1969). Before commenting further on this point, we will examine another example of essentially the same mode of development demonstrated by *Conioscypha varia* Shearer. On the basis of the time-lapse sequence (Fig. 4.87A–C) presented by van Emden and Veenbaas-Rijks (1973), one would logically interpret this mechanism of conidiogenesis as phialidic. The first-formed conidium is holoblastic (Fig. 4.88) and subsequent conidia are enteroblastic. But thin-sections of *Conioscypha lignicola* Höhnel, presented by Shearer and Motta (1973), demonstrate that the conidiogenous cell undergoes a slight percurrent proliferation before giving rise to the next conidium. However, rupture of the outer wall layer encompassing the first and subsequent conidia occurs at the apex of the newly formed cell rather than adjacent to the basal septum. Woronin bodies present in the cytoplasm at the fertile apex may function in plugging the septal pore.

In summary, ultrastructural studies have demonstrated that a developmental continuum exists between groups of phialide- and annellide-forming species, as previously suggested by Hammill (1973b) and Harvey (1974b). On the basis of present data, it is difficult to formulate distinct concepts of phialidic and annellidic development. Several earlier characters suggested for distinguishing between these two modes of conidiogenesis have proved to be inconsistent. For example, Morgan-Jones *et al.* (1972) proposed that the processes can be distinguished by differences in septal wall formation at the base of their seceding conidia. However, ultrastructural examinations have revealed similarities in this aspect of development. It has been suggested that conidia produced by annellides are *holoblastic* while those arising from phialides are *enteroblastic*. However, wall differentiation during these two modes of conidiogenesis are instead very similar.

4.86

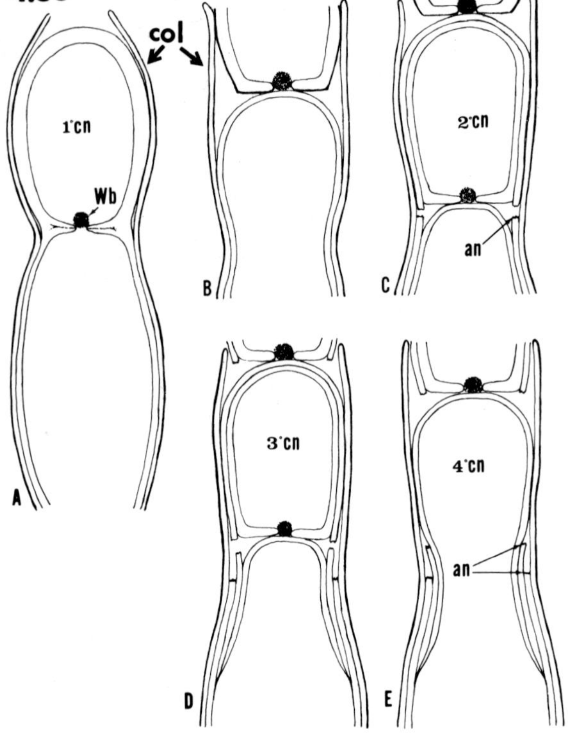

Fig. 4.86A–E Diagrammatic interpretation of wall differentiation during conidial and conidiogenous cell development in *Cryptosporiopsis.* an, annellation; col, collarette; 1°cn, primary conidium; 2°cn, secondary conidium; 3°cn, tertiary conidium; 4°cn, fourth conidium.

On the other hand, time-lapse light-microscopic studies have demonstrated that in phialides the conidiogenous locus is apparently fixed at the apex of the fertile cell, while in annellides it is located at successively higher levels due to the percurrent proliferation of the fertile cell. This raises a rather significant point concerning the functional aspects of a classification of the Deuteromycetes based on developmental characters. If such a classification is going to be useful to a large number of mycologists, surely the characters delimiting major groups of fungi must be detectable by light-microscopic investigations and not totally dependent on ultrastructural examinations. This does not suggest that ultrastructural studies have failed to provide useful information in designing such a classification. On the contrary, examinations of the fine structure of conidia and conidiogenous cells have made us more aware of the plasticity of morphogenesis in the conidial fungi and thereby cautious of establishing restrictive developmental concepts. We have come full circle in our consideration of distinguishing characters of annellidic and phialidic development: the best differentiating criterion for most species is surface morphology; annellides have annellations, phialides have collarettes. Both kinds of conidiogenous cells give rise to a basipetal succession of conidia. Camera lucida drawings of

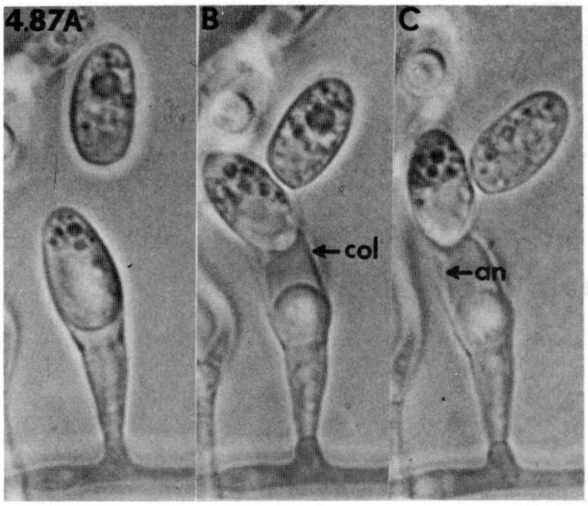

Fig. 4.87A–C *Conioscypha varia.* Time-lapse sequence of basipetal succession of enteroblastic conidium ontogeny. An endogenous annelation (an) is visible within the collarette (col). ×1,780 (after van Emden and Veenbaas-Rijks, 1973).

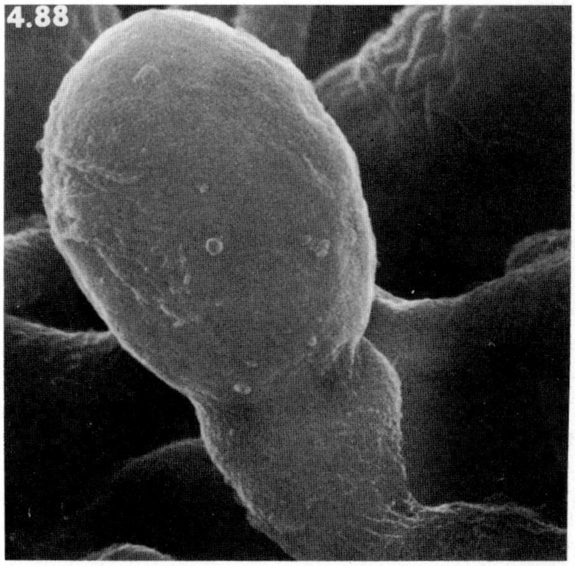

Fig. 4.88. *Conioscypha varia.* Primary, holoblastic conidium attached to apex of cup-shaped conidiogenous cell. ×9,700.

annellides of several species are presented in Fig. 4.89 demonstrating annellated, fertile apices. Annellations may sometimes be suspected to occur, because of the elongated nature of the conidiogenous cell apex, but are not visible with the light microscope, necessitating ultrastructural examinations. For example, the presence of annellations in *Exophiala jeanselmei* (Langeron) McGinnis & Padhye is only suggested by light microscopy (Wang, 1966), but confirmed by the SEM (Cole, 1978; Figs. 4.90–4.92) and thin-sections. Additional subcategories of these two major groups can be defined by characters derived from both

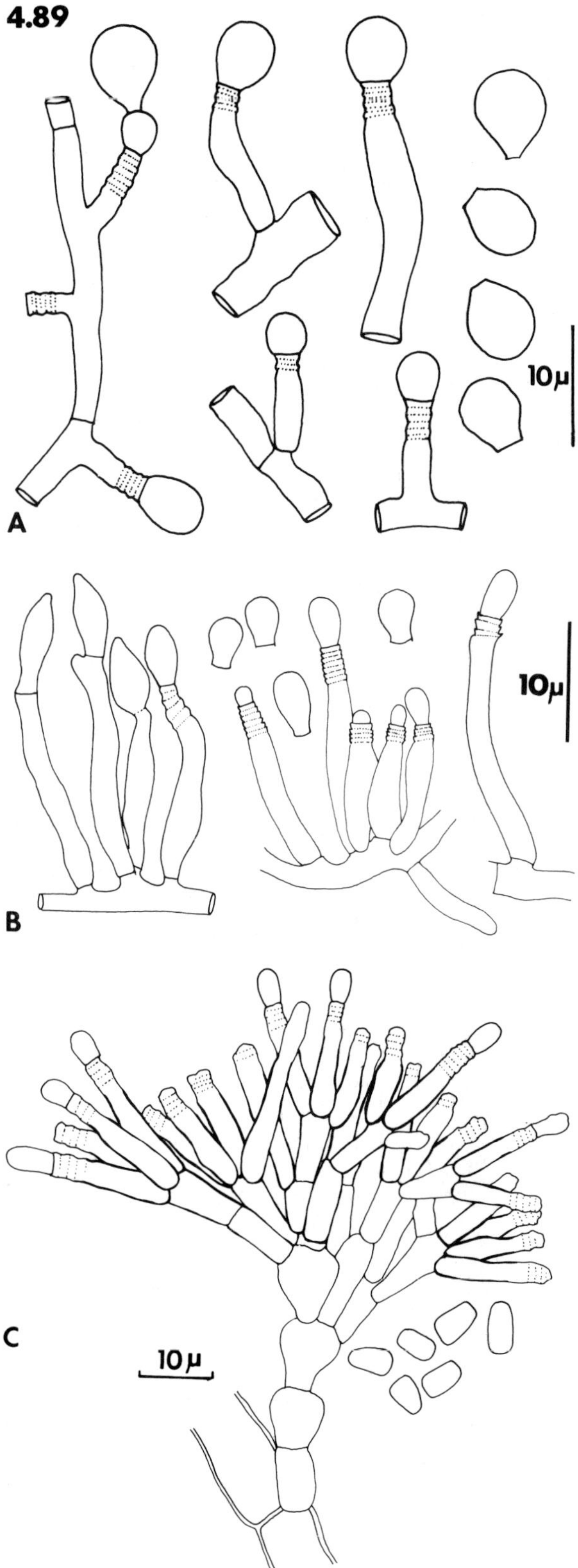

Fig. 4.89A–C Annellated conidiogenous cells. (A) *Oospora halophila* v. Beyma, (B) *Ambrosiozyma cicatricosa* (Scott & van der Walt) van der Walt, (C) *Leptographium lundbergii* Lagerberg & Melin.

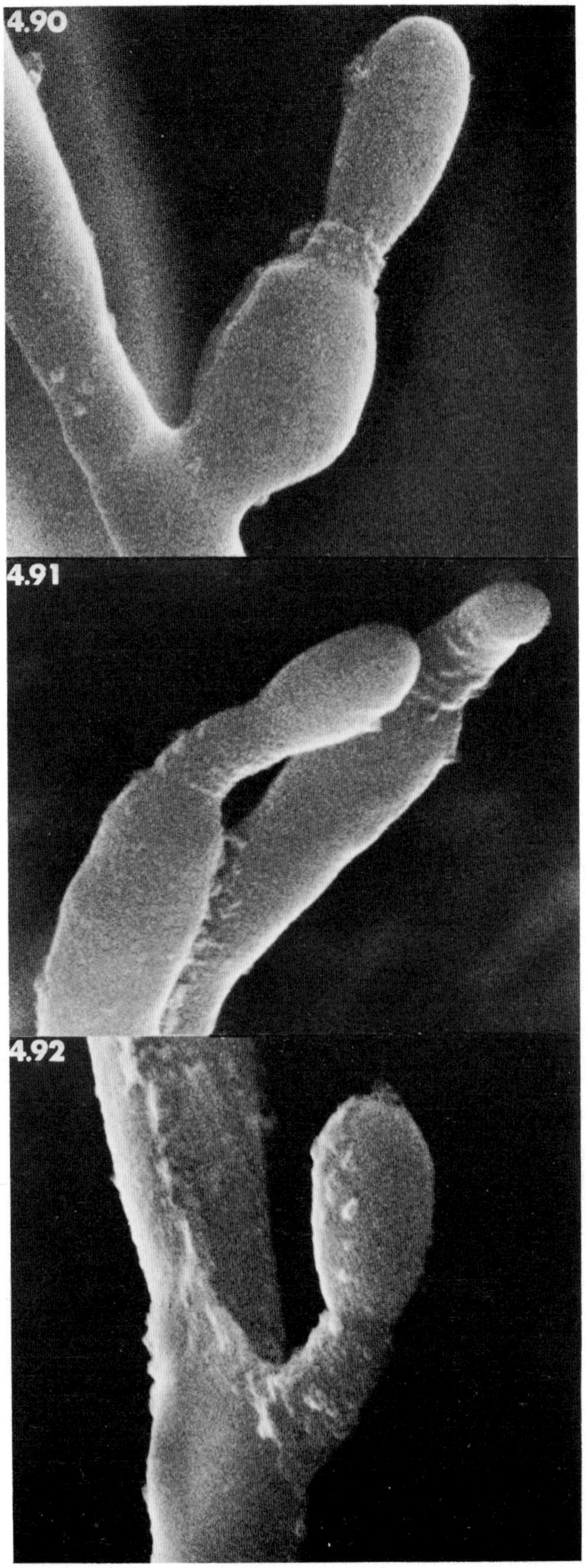

Figs. 4.90–4.92 *Exophiala jeanselmei.* Annellated conidiogenous cell apices and attached conidia. Note reduced annellide in 4.92. ×11,550, 13,200 and 13,900 respectively.

light-microscopic and thin-section investigations. For example, we would elect to place *Conioscypha varia, C. lignicola* and *Cryptosporiopsis* in a separate subdivision of phialidic development based on the absence of exogenous annellations, but the occurrence of distinct, endogenous proliferations of the conidiogenous cells (Figs. 4.86–4.88). Similarly, a subcategory of the annellidic group of conidial fungi can be designed to accommodate *Acrogenospora* (Figs. 4.79A, 4.80), *Endophragmia* (Fig. 4.79B) and *Endophragmiopsis* (Fig. 4.79C). This discussion, however, is incomplete without consideration of the third kind of conidiogenous cell which produces a basipetal succession of conidia. An analysis of retrogressive conidiogenous cell development forms the basis for discussion in Chapter 5.

Annellated yeasts

The imperfect yeasts are generally separated from other members of the Deuteromycetes because of their distinctive budding mechanisms of reproduction. However, several species have annellations. For example, von Arx (1967) reported that annellated cells were produced by *Eremascus fertilis* Stoppel. Streiblová (1971) demonstrated annellations in *Endomyces magnusii* Ludwig, *Schizosaccharomyces octosporus* Beijerinck, and *Saccharomycodes ludwigii* Hansen, using fluorescence microscopy. She described the wall architecture and differentiation in these same fungi, and it is evident from these studies that developmental similarities exist between this group of yeasts and those members of the conidial fungi whose fertile cells undergo percurrent proliferation. S. J. Hughes (1971b) concluded that the multiple scars of apiculate yeasts (Streiblová, 1971) are equivalent to annellations in the hyphomycetous fungi. Kreger-van Rij and Veenhuis (1971b) examined thin-sections of budding in bipolar yeasts, including *S. ludwigii, Hanseniaspora uvarum* (Neihaus) Shehata, *Wickerhamia fluorescens* Soneda, *Nadsonia elongata* Konokotina, and *Schizoblastosporion starkeyi-henricii* Ciferii, and concluded that, 'In all species a cross wall was formed between mother cell and bud which consisted of an electron-light layer between two layers of more electron-dense material. The cells separated along the light layer.' The complex process of cell wall growth in *Schizosaccharomyces pombe* Lindner (Fig. 4.93) has been extensively investigated by Streiblová and Wolf (1972) using techniques of thin-sectioning, fluorescence microscopy, and time-lapse cinemicrography. A summary of these different budding processes which involve percurrent proliferation is presented in Fig. 4.95. Budding from the yeastlike cells of *Sarcinomyces crustaceus* Linder (= *Cladosporium werneckii* Horta) (Fig. 4.94) also involves successive proliferations of the fertile apices and results in the formation of circumscissile scars, or annellations (Cole, 1978). The ultrastructural similarities between this mode of budding in *Sarcinomyces crustaceus* and annellidic development in other hyphomycetous fungi have been demonstrated by Gustafson *et al.* (1975).

Fig. 4.93 *Schizosaccharomyces pombe.* Replicas of annellated yeast cells. ×7,840.

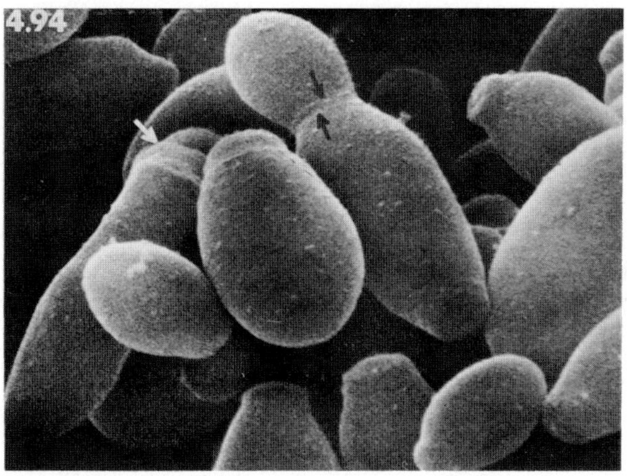

Fig. 4.94 *Sarcinomyces crustaceus* (= *Cladosporium werneckii*). Annellated cells in process of budding. Arrows locate annellations. ×7,000.

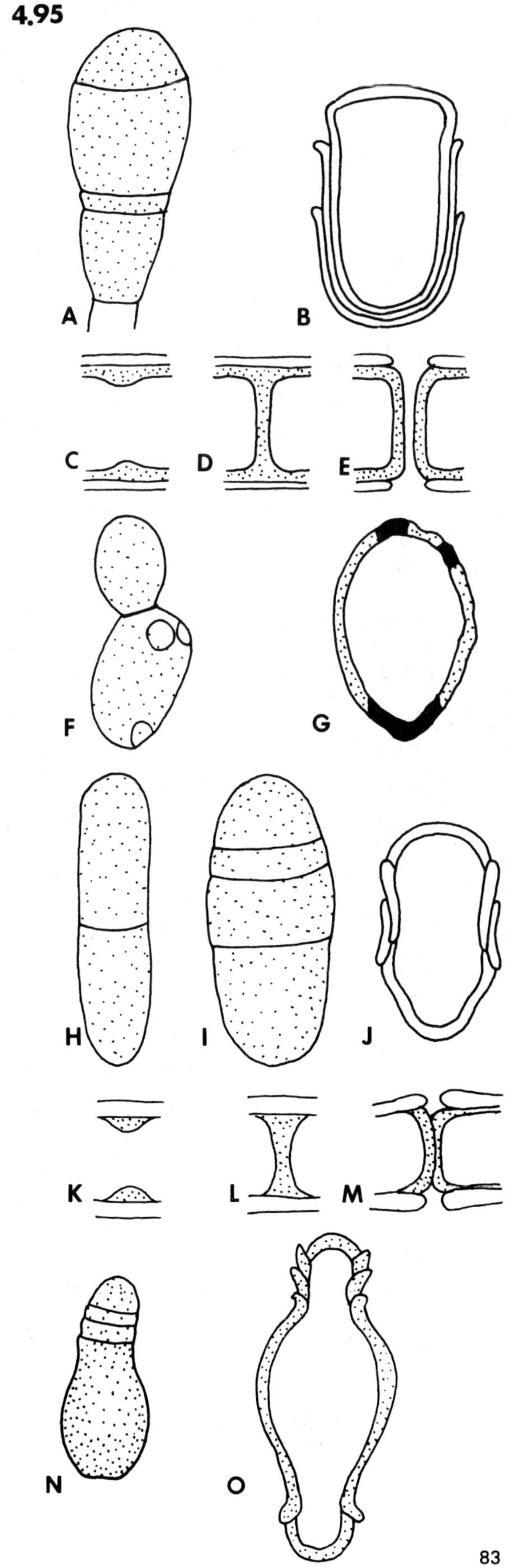

Fig. 4.95 Diagrammatic interpretation of wall relations in several species of annellated yeasts. (A–E) *Endomyces magnusii,* (F, G) *Saccharomyces cerevisiae* Hansen, (H–M) *Schizosaccharomyces octosporus,* (N, O) *Saccharomycodes ludwigii* (after Streiblová, *1971).*

5 Development of Retrogressive Conidiogenous Cells

The common developmental features of this group of fungi are the basipetal formation of blastic conidia at the apex of a conidiogenous cell, which becomes progressively shorter as conidiogenesis proceeds. The conidiogenous cells may be difficult to distinguish from vegetative hyphae, as in *Basipetospora rubra* Cole & Kendrick (Cole and Kendrick 1968a) and *Trichothecium roseum* (Pers.) Link ex S. F. Gray (Kendrick and Cole, 1969), or well differentiated, as in *Cladobotryum varium* Nees (Cole and Kendrick, 1971).

Development in *Cladobotryum varium*

A time-lapse sequence of conidium ontogeny in *C. varium* is presented in Fig. 5.1A–F. The young conidiogenous cells are morphologically similar to phialides—subulate (Fig. 5.1A), becoming swollen apically as the first conidium is differentiated (Fig. 5.1B). The second conidial initial (arrowhead in Fig. 5.1C) appears as a swelling below the primary conidium. Adjacent conidia typically adhere in loose chains, but the growth conditions of our culture chamber (Cole and Kendrick, 1968b) apparently prevented chains from forming. As the basipetal succession of conidia continues (Fig. 5.1D–F), a marked shortening, or retrogression of the conidiogenous cell is evident, witnessed by comparing distances between the reference point (arrow) and the apex of the fertile cell in Fig. 5.1A and F.

This shortening process can be suspected when conidiogenous cells from cultures of increasing age are prepared for the scanning electron microscope (SEM) (e.g., 5 days, Fig. 5.2; 15 days, Fig. 5.3; and 30 days, Fig. 5.4). As expected, increases in the width of both the fertile apex and bases of the conidia occur as retrogression proceeds. These features are also visible with the light microscope and help to identify this kind of development (Fig. 5.5A, B).

Two adjacent conidia of *C. varium* at different developmental stages in Fig. 5.6 show circular abstriction scars (ab) at their apices which represent sites of attachment of previously formed conidia. The rings of wall material at the base of the more mature conidium (r^1, r^2, Fig. 5.6) are important clues for defining the process of retrogression. A slight separation (arrowhead) between the conidium and conidiogenous cell indicates that secession has begun. The thickened rings of wall material at the base of the conidial initials in Fig. 5.7 (arrows) are similar to collarettes of phialides and correspond to r^1 in Fig. 5.6. The lower ring (r^2) in Fig. 5.6 is the new fertile cell apex. When a conidium secedes, a

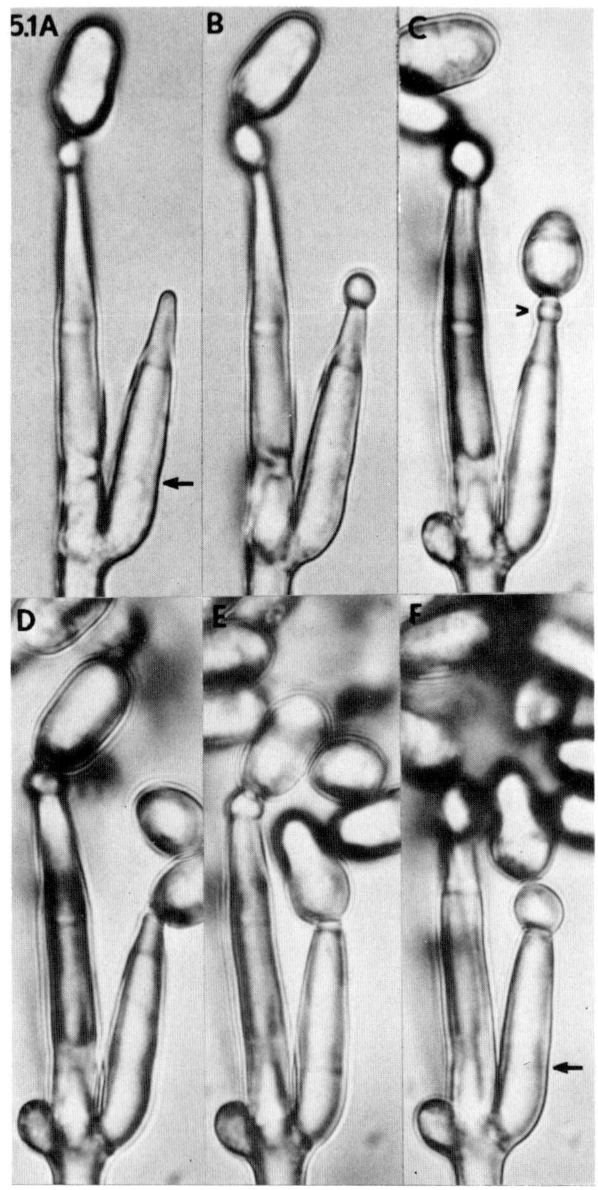

Fig. 5.1A–F *Cladobotryum varium.* Time-lapse sequence of retrogressive conidiogenous cell development. Arrows in A and F indicate common reference points at the base of the fertile cell. Arrowhead in C locates conidial initial. T = 0, $\frac{1}{2}$, 4, 8$\frac{3}{4}$, 15, 23$\frac{1}{4}$ hours, respectively. ×1,380.

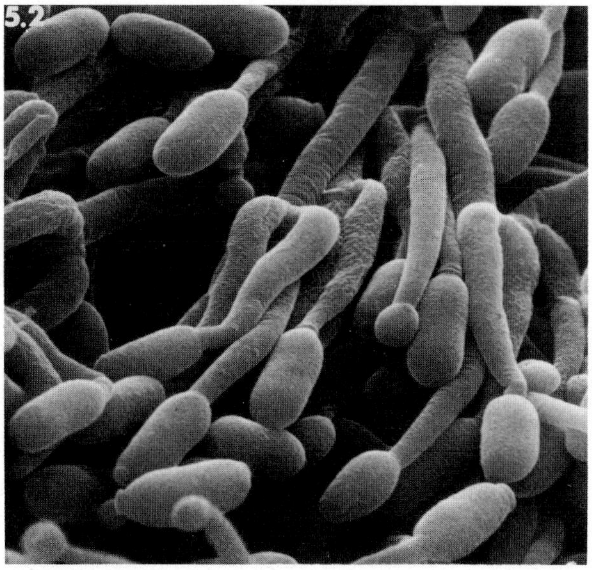

Fig. 5.2 *Cladobotryum varium.* Conidiogenous cells and terminal conidia from 5 day old culture. ×1,800.

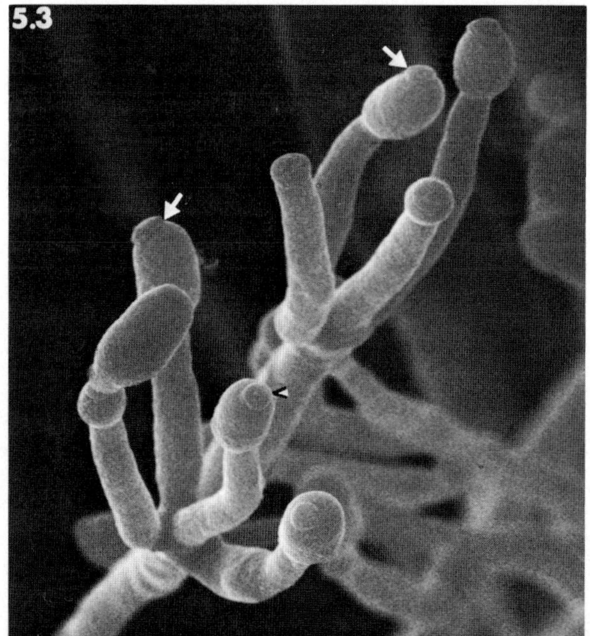

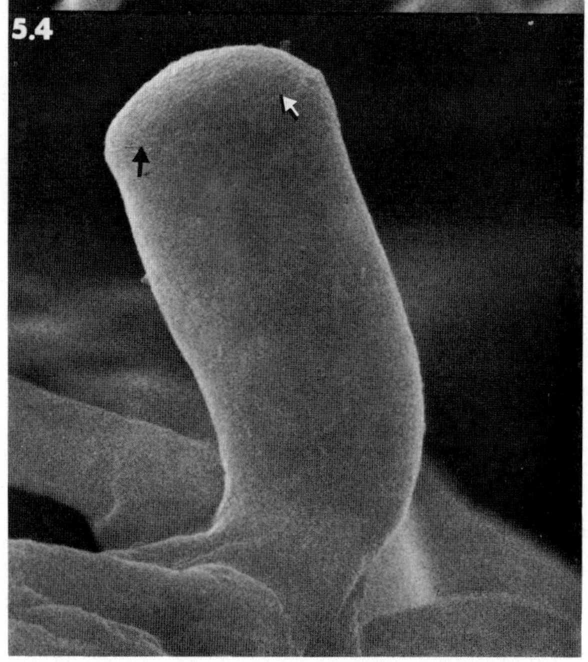

Figs. 5.3, 5.4 *Cladobotryum varium.* Conidiogenous cells from 15 day (5.3) and 30 day (5.4) cultures. Arrows in 5.3 indicate abstriction scars at apex of attached conidia. Arrows in 5.4 locate circumscissile rupture of outer conidiogenous cell wall. ×1,980 and 6,820, respectively.

5.5

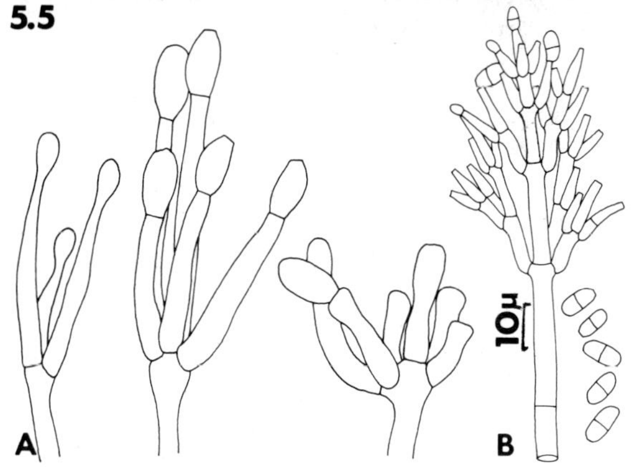

Fig. 5.5A *Cladobotryum varium.* Conidia and conidiogenous cells from cultures of increasing age (5, 15 and 30 days, left to right).

Fig. 5.5B *Cladobotryum verticillatum* (Link ex S. F. Gray) Hughes.

thickened band is present at its base (arrows in Figs. 5.6, 5.8). Freeze-fractures of seceded conidia (Fig. 5.9) reveal this as a band of rodlet fascicles which are generally oriented circumferentially. Another layer of rodlet fascicles, arranged more or less parallel to the long axis, is shown above and below the raised, circumferential band suggesting that the latter is not contiguous with the wall of the new conidium but is part of the conidiogenous cell wall remaining attached to the conidium when it secedes. Thin-sections have confirmed this interpretation (Figs. 5.10–5.13). A *collarette* (arrows in Fig. 5.10) is present at the base of the conidial initial. The outer wall of the young conidium is continuous with the inner wall layer of the conidiogenous cell, confirming that conidiogenesis in *C. varium* is enteroblastic. The slightly tangential section in Fig. 5.10 has passed through part of the apical abstriction scar. A collarette is also visible in Fig. 5.11 (arrows) which illustrates a very early stage of conidium ontogeny. A similar stage was demonstrated by the SEM in Fig. 5.4. The slightly flattened appearance of the fertile apex

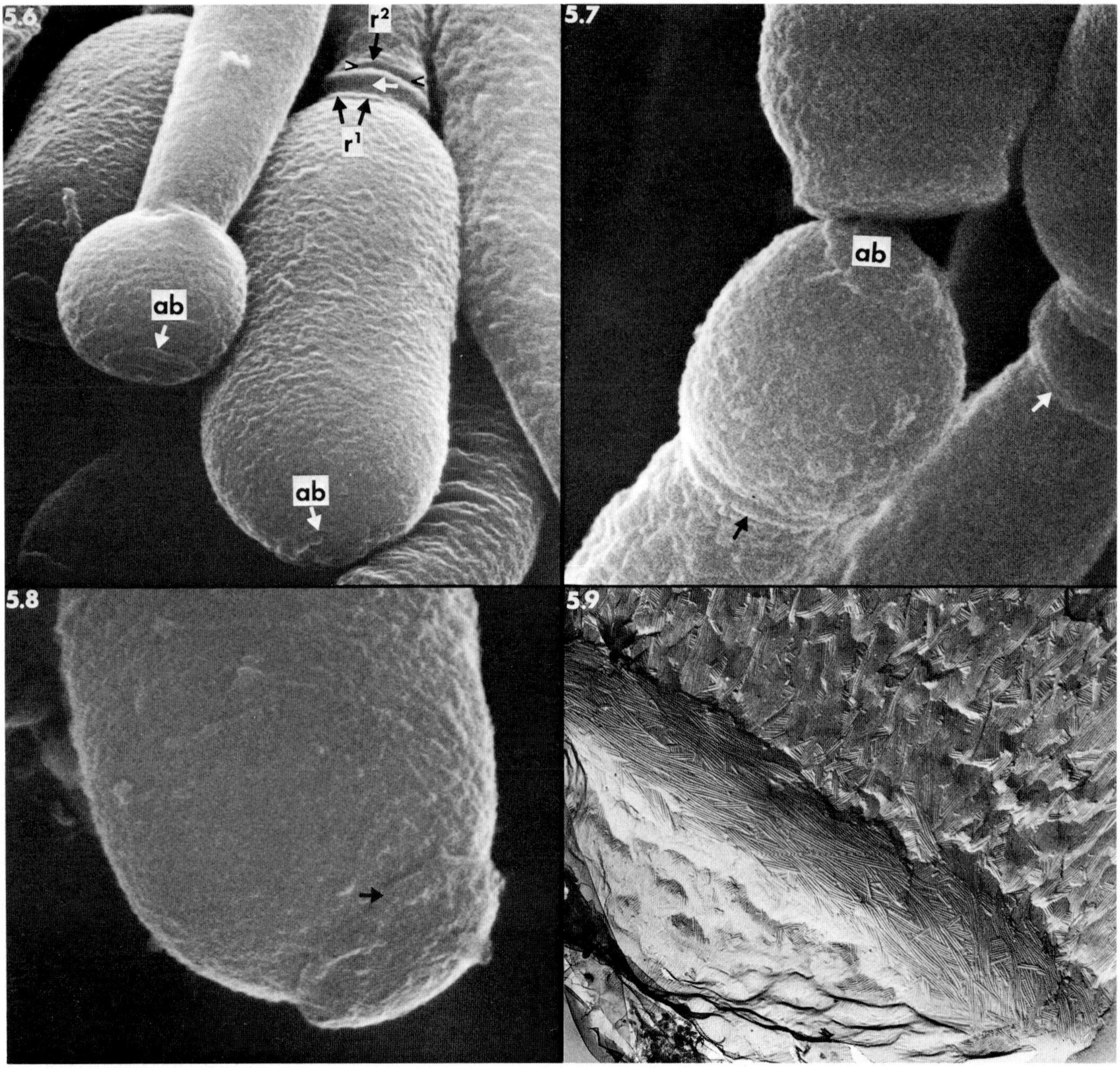

Figs. 5.6–5.9 *Cladobotryum varium.* Stages of conidial development (5.6, 5.7) and ultrastructure of conidial base (5.8, 5.9). Arrow in 5.6 represents same band of wall material at base of conidium as shown in 5.8 (arrow) and 5.9 (circumferential band of rodlets). Arrowheads in 5.6 locate zone of separation between conidial base and conidiogenous cell apex. Arrows in 5.7 indicate circumscissile rupture (*collarette*) of outer conidiogenous cell wall. ab, abstriction scar; r^1, ring of wall material corresponding to *collarette* shown in 5.7; r^2, conidiogenous cell apex. ×9,555, 11,550, 15,400 and 35,000, respectively.

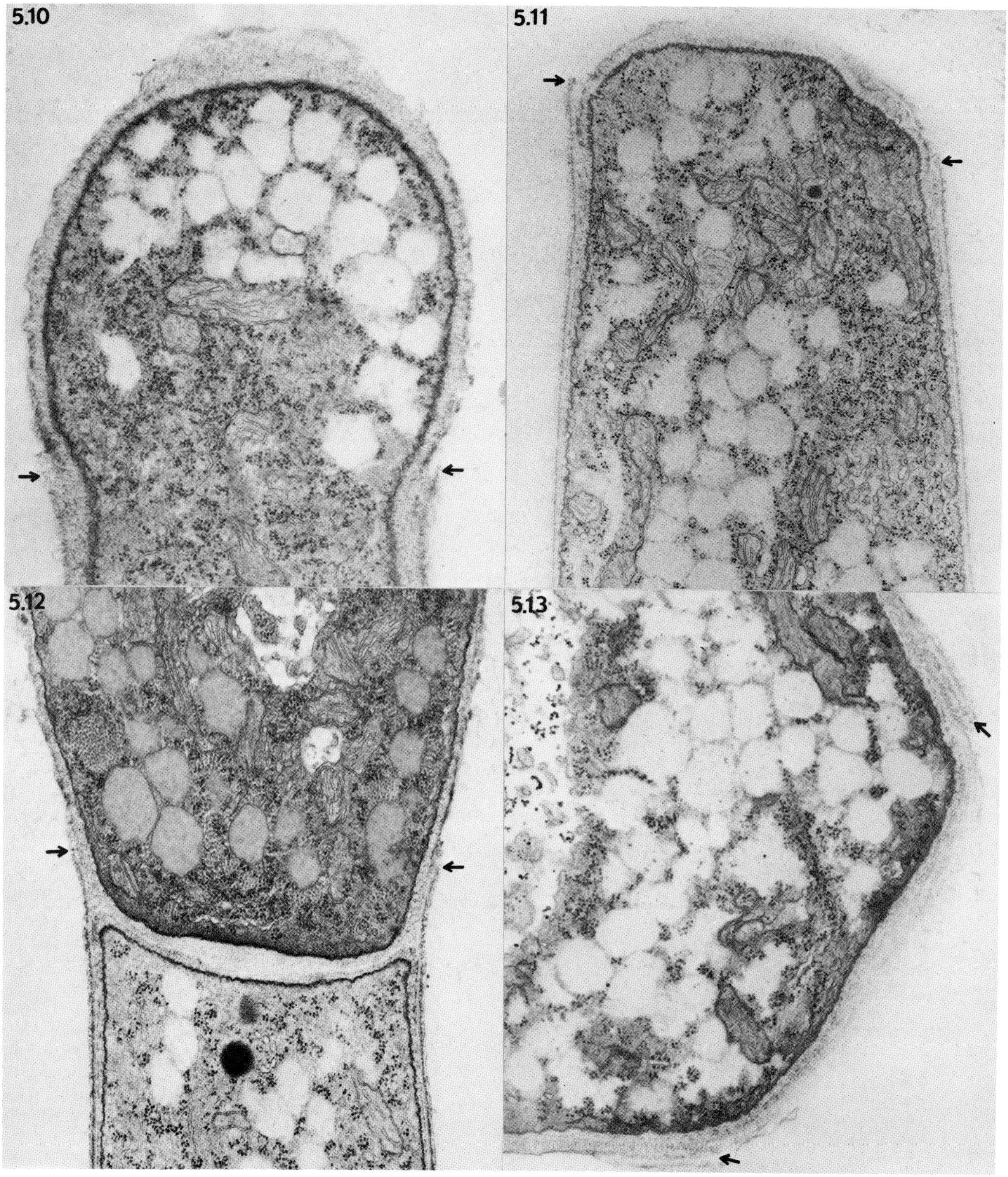

Figs. 5.10–5.13 *Cladobotryum varium.* Thin-sections of stages of conidiogenesis. Arrows in 5.10, 5.11 and 5.12 locate tip of *collarette* at conidiogenous cell apex. Note that basal septum of conidium lies below tip of *collarette* in 5.12 and the latter is fused to inner conidial wall. Arrows in 5.13 indicate circumferential band of wall at conidial base. ×32,000, 24,000, 32,000 and 32,000, respectively.

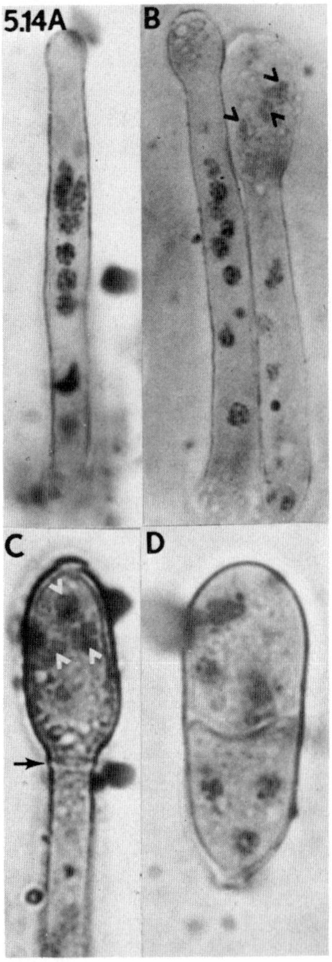

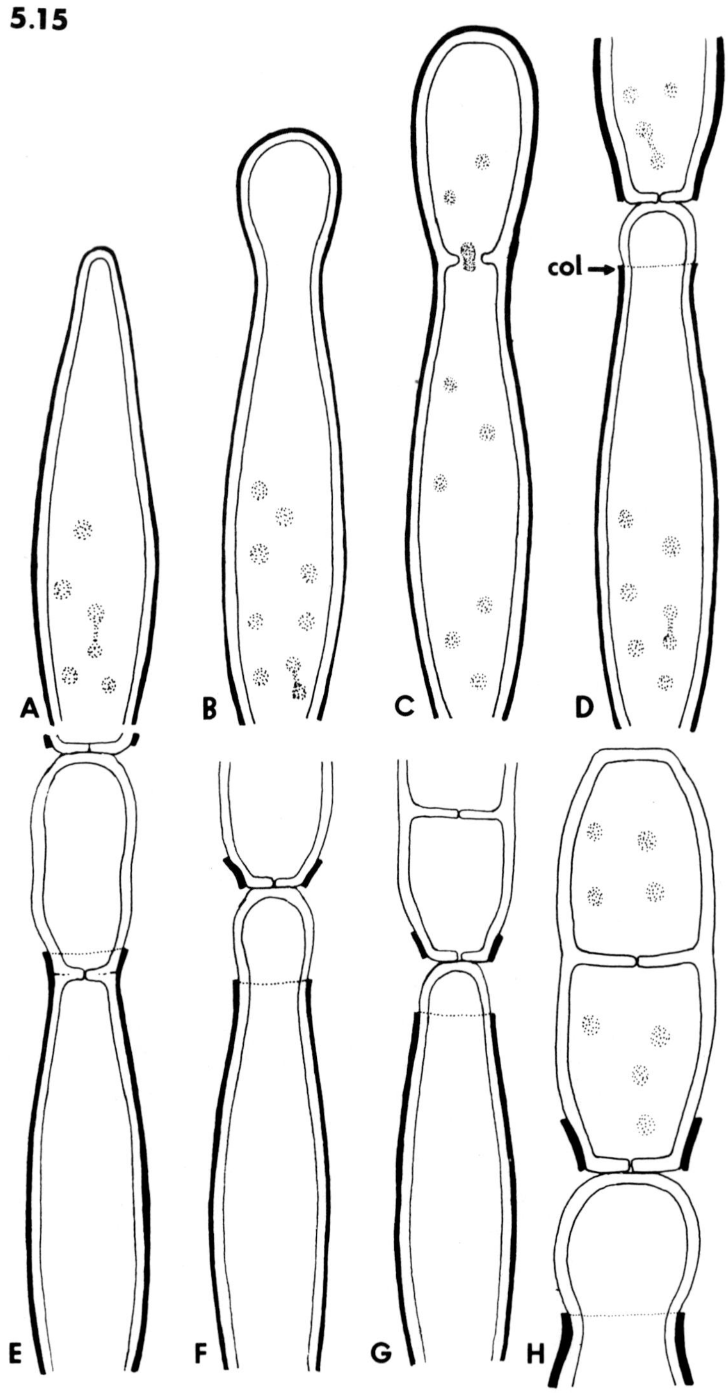

Fig. 5.14A–D *Cladobotryum varium.* Giemsa-aceto-orcein stained nuclei of conidiogenous cells and conidia. Nuclei located in subapical region below conidial initial (A). Arrowheads in (B) and (C) locate nuclei within differentiated conidia. Arrow in (C) indicates basal septum of conidium. A, B, ×2,400 and C, D, ×3,200.

Fig. 5.15A–H Diagrammatic interpretations of retrogressive conidiogenous cell development in *Cladobotryum varium.* The first-formed conidium is holoblastic (A–C). The conidiogenous cell shortens due to loss of segments of its outer wall attached to the base of seceding conidia (E–H). col, collarette.

is probably due to its attachment to the previously formed conidium. The feature responsible for retrogression of the conidiogenous cell is revealed in Fig. 5.12. In this section, a septum delimits the base of a conidium from the apex of the fertile cell. The tip of the collarette (arrows) is located above the septum. However, the conidiogenous cell wall comprising the collarette is fused to the inner conidial wall. When the septum undergoes schizolysis the ruptured outer wall of the fertile cell above the septum remains attached to the base of the conidium (Figs. 5.8, 5.9, 5.13). The retrogressive nature of the conidiogenous cell, therefore, is simply a function of the level of septum formation during secession. We have already discussed the flexibility that exists in the position of the basal septum during development of other kinds of blastic conidia (e.g., *Scopulariopsis brevicaulis* (Sacc.) Bain., Hammill, 1971).

Karyological studies of *C. varium* have demonstrated additional features of conidiogenesis which appear to be typical of blastic development (Fig. 5.14A–D). During initiation of conidia, nuclei are found in the subapical region of the conidiogenous cell, but not within the initial (Fig. 5.14A; left cell in Fig. 5.14B). Later accumulation of nuclei in the young conidium may be the result of nuclear division (Fig. 5.14A) and migration of nuclei from adjacent supporting cells. At about the time the septum is initiated at the base of the conidium, nuclei migrate towards the apex of the fertile cell and enter the still immature conidium (arrowheads, right cell in Fig. 5.14B). After completion of septal development but before the conidium has become two celled, its cytoplasm contains 3–5 nuclei (Fig. 5.14C). Most mature conidia, on the other hand, contain 8 nuclei, with an average of 4 in each cell (Fig. 5.14D). This is probably the result of 1 or 2 mitotic events occurring within the delimited conidium. The septal pore narrows quickly during formation of the basal septum, which soon restricts movement of cytoplasm between the two cells.

A diagrammatic summation of the features of retrogressive conidiogenous cell development is presented in Fig. 4.15A–H.

Development in *Trichothecium roseum*

During retrogressive conidiogenous cell development in *Trichothecium roseum,* successive conidia are formed from opposite sides of the fertile apex and remain together in characteristic chains (Figs. 5.16–5.24). Earlier developmental studies of this species, which produced several interpretations of conidiogenesis, have been presented by Ingold (1956), Nicot and Leduc (1957) and Kendrick and Cole (1969). The time-lapse sequence in Fig. 5.16A–D shows the formation of a basipetal succession of conidia as the conidiogenous cell shortens.

The conidiogenous cell is indistinguishable from the vegetative hypha until the first conidium is initiated. This primary conidium develops holoblastically as a terminal swelling of the fertile hypha (Fig. 5.17), in contrast to all subsequent, laterally proliferated conidia. An early stage of secondary conidium development is shown in Fig. 5.18. The outer conidiogenous cell wall has apparently ruptured, leaving a *collarette* (arrow) through which the initial has emerged. The primary conidium has remained firmly attached to the fertile apex due to continuity of wall material (Fig. 5.18) rather than a mucilaginous connection as previously suggested (Kendrick and Cole, 1969). The secondary conidium in Fig. 5.19 is also still attached to the conidiogenous cell apex. A deep groove near its base (arrow) represents the abstriction scar of the previously formed conidium. This scar is more clearly shown on the surface of the seceded conidium in Fig. 5.20. Shadowed replicas of conidia have revealed that a septal pore is present at the centre of the abstriction scar, which is suggested in Fig. 5.20 (arrow). Cytoplasmic continuity may be maintained between the conidiogenous cell and developing conidium, even after the next propagule is initiated (e.g., Fig. 5.18). The constriction at the base of the conidium in Fig. 5.19 (arrowhead) represents the region where conidial secession will later occur. Therefore, each secondary conidium which secedes from the chain has both a lateral abstriction scar, representing the point of detachment of the previously formed conidium, and a terminal scar exposed after separation from the subsequently formed conidium, or conidiogenous cell apex. Camera lucida drawings in Fig. 5.21 illustrate the arrangement of the conidial chain in *T. roseum.*

Developmental similarities are evident between *Cladobotryum varium* and *T. roseum.* For example, thin-sections have confirmed the presence of a ruptured outer wall layer (*collarette*) at the base of secondary conidial initials in *T. roseum* (arrows in Fig. 5.22) and, as in *C. varium,* therefore, conidiogenesis is enteroblastic. The ruptured outer wall remains firmly attached to the inner wall layer and the basal septum of the developing conidial initial forms below the tip of the collarette (arrowhead in Fig. 5.22). In Fig. 5.23, a collarette is shown (arrows) at the base of a laterally oriented conidial initial. This section also passed tangentially through the base of the previously formed conidium. A comparable stage of conidiogenesis was illustrated by the SEM in Fig. 5.18.

A diagrammatic interpretation of conidiogenesis in *T. roseum*, illustrating aspects of wall differentiation, is presented in Fig. 5.24A–D. The wall relations established during delimination of the second conidium (Fig. 5.24C) and initiation of the third (Fig. 5.24D) illustrate the retrogressive mechanism of conidiogenesis. The septum at the base of the second conidium forms below both the ruptured conidiogenous cell wall and the region of lateral attachment to the primary conidium. Secession of these two conidia results in loss of segments of the fertile hypha.

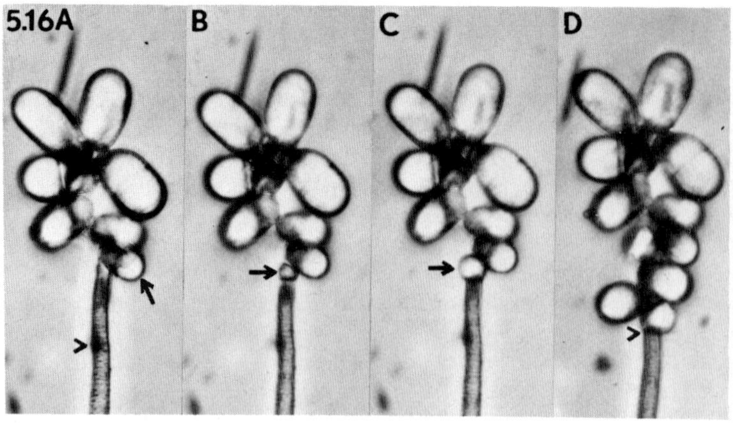

Fig. 5.16A–D *Trichothecium roseum.* Time-lapse sequence of conidial and conidiogenous cell development. Arrow in (A) indicates initial at apex of fertile cell and base of conidial chain. Arrows in (B) and (C) locate next conidial initial which grows asymmetrically in opposite direction. Three new conidia have formed in (D). Arrowheads in (A) and (D) represent common reference points. T = $8\frac{1}{4}$, $8\frac{3}{4}$, $10\frac{1}{4}$, $16\frac{1}{4}$ hours, respectively. ×1,480.

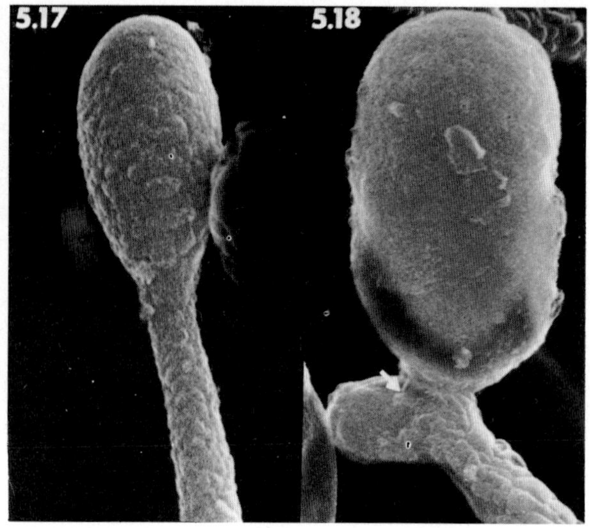

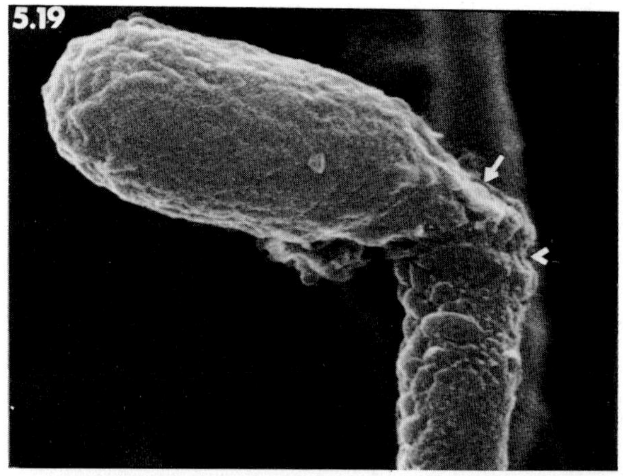

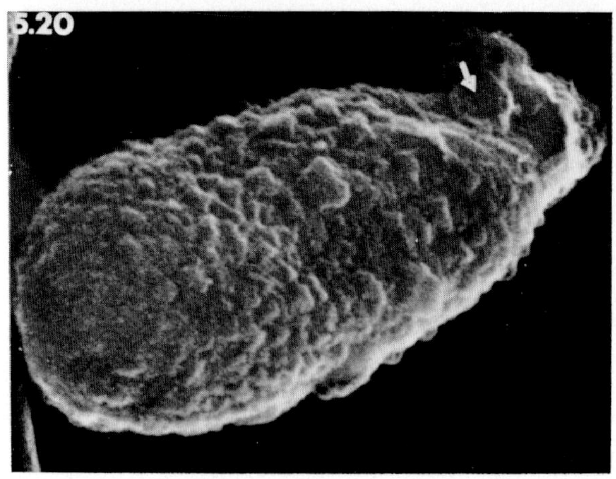

Figs. 5.17–5.20 *Trichothecium roseum.* Stages of conidial and conidiogenous cell development. Primary, holoblastic conidium (5.17) and second conidial initial which has emerged through ruptured outer wall (*collarette*) of fertile cell (arrow in 5.18) below primary conidium. Secondary conidium attached to fertile cell apex (5.19) showing abstriction scar (arrow) of previously formed conidium and region where later secession from conidiogenous cell apex will occur (arrowhead). Seceded conidium (5.20) showing scar of previously attached conidium. Arrow in 5.20 locates septal pore. ×6,035, 5,270, 6,864 and 9,000, respectively.

5.21

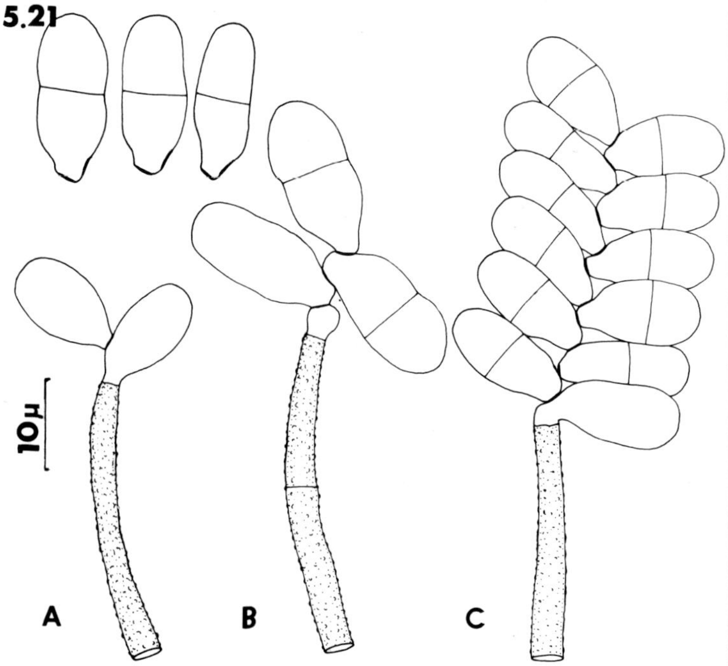

Fig. 5.21A–C *Trichothecium roseum.* Arrangement of conidia in chains. Seceded conidia showing abstriction scars.

5.22

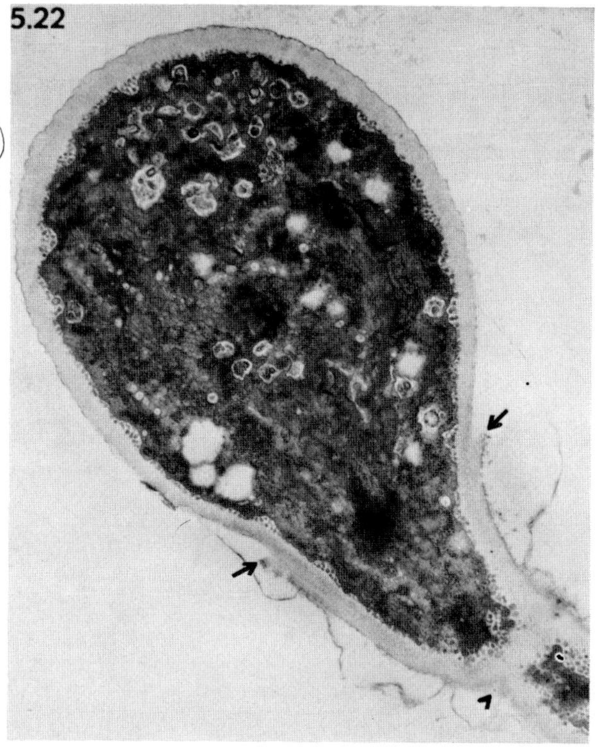

5.23

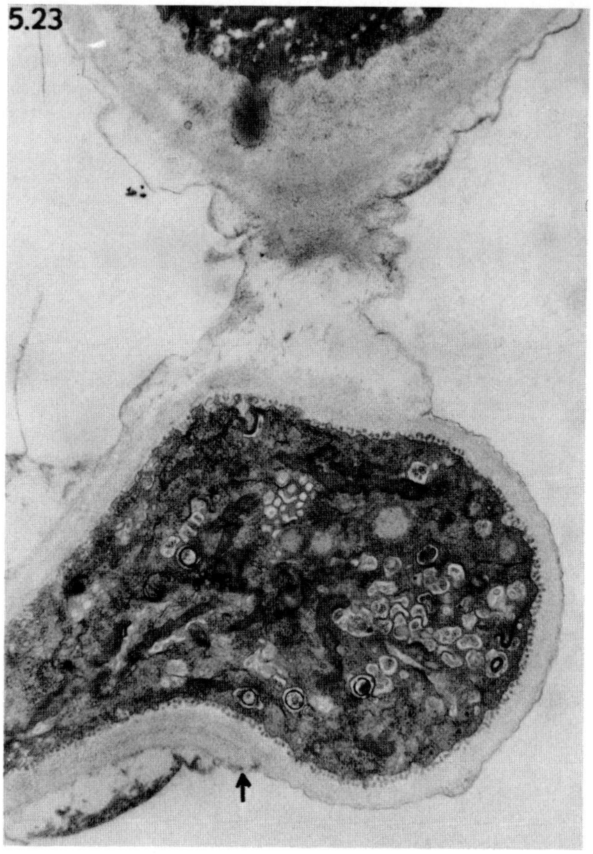

Figs. 5.22, 5.23 *Trichothecium roseum.* Thin-sections of conidial initials showing ruptured, outer conidiogenous cell wall (arrows in 5.22 and 5.23). Basal septum (arrowhead in 5.22) is located below *collarette.* ×21,600 and 22,200 respectively.

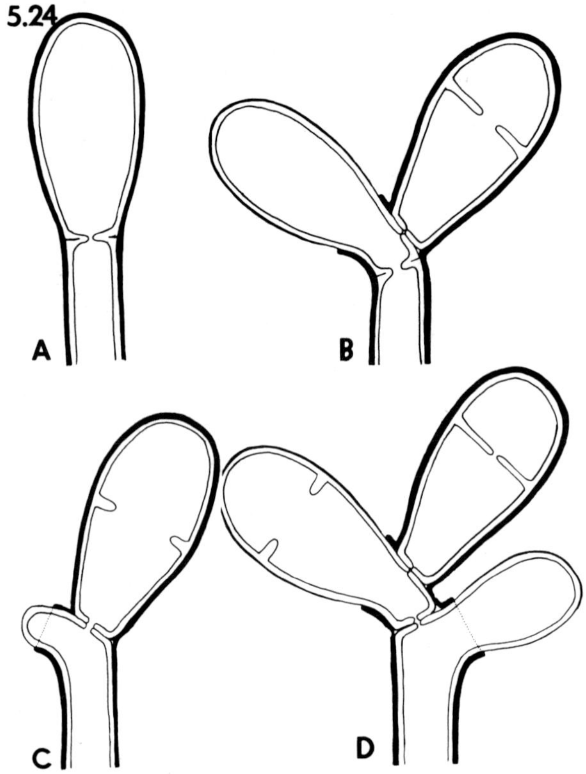

Fig. 5.24A–D Diagrammatic interpretation of conidiogenesis in *Trichothecium roseum.* Primary conidium is holoblastic (A). Secondary conidia develop laterally below base of attached conidium (C, D). Conidiogenous cell shortens due to loss of segments of outer wall as conidia secede (D).

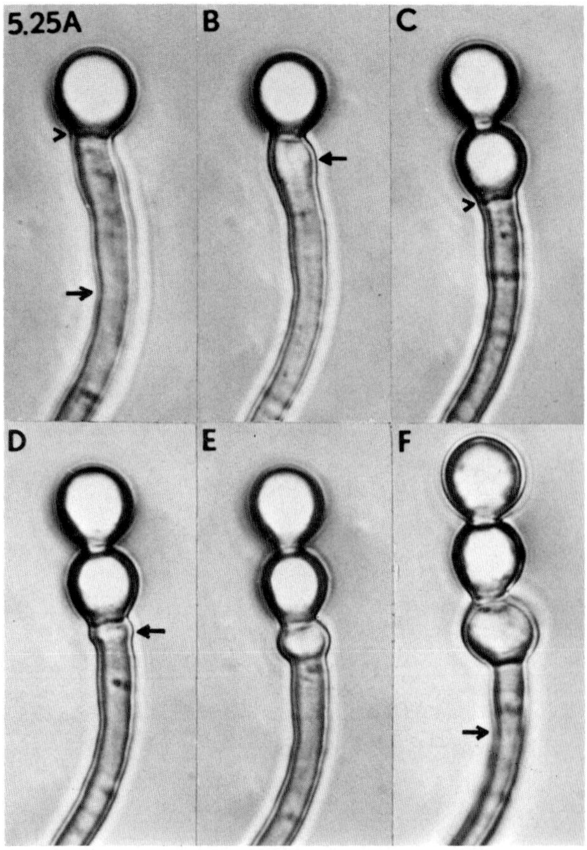

Fig. 5.25A–F *Basipetospora rubra.* Time-lapse sequence of conidial and conidiogenous cell development. Arrowheads in (A) and (C) locate basal septa. Arrows in (B) and (D) indicate conidial initials. Arrows in (A) and (F) represent common reference points. T = $\frac{1}{2}$, $6\frac{1}{2}$, $10\frac{1}{4}$, $12\frac{1}{2}$, $13\frac{3}{4}$, $15\frac{1}{2}$ hours, respectively.

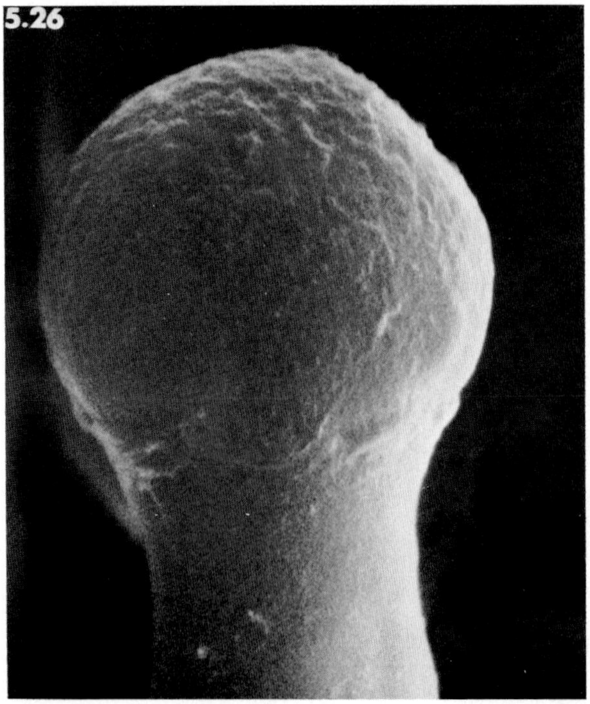

Fig. 5.26 *Basipetospora rubra.* Holoblastic, primary conidium attached to conidiogenous cell apex. ×12,650.

Development in *Basipetospora rubra*

The time-lapse sequence in Fig. 5.25A–F demonstrates the mechanism of retrogressive conidiogenous cell development in *Basipetospora rubra*, the anamorph of *Monascus ruber* van Tieghem. During formation of three conidia there has been a substantial shortening of the fertile hypha. The details of wall differentiation during this developmental process, however, differ from *C. varium* and *T. roseum.*

The conidiogenous cell is first differentiated from a vegetative hypha with the swelling of its apex and development of the primary, holoblastic conidium (Figs. 5.25A, 5.26). In contrast to retrogressive development in the two previous examples, the basipetal succession of secondary conidia in *B. rubra* is holoblastic (Fig. 5.27). The thin, outer wall layer of the fertile hypha is ruptured only at the time of conidial secession (arrow in Fig. 5.27), revealing the thick, inner wall layer below. An interrupted secession in Fig. 5.28 shows thin strands of outer wall material still uniting the two adjacent cells.

A diagrammatic interpretation of wall differentiation in this species is presented in Fig. 5.29A–C. The outer conidiogenous cell apparently remains extensible, encompassing the chain of cells. After differentiation of the primary conidium (Fig. 5.29A), a segment of the fertile hypha below its basal septum swells and differentiates into the next propagule (Fig. 5.29B). Each conidium, therefore, incorporates a portion of the pre-existing conidiogenous cell wall into its own wall as well as undergoing considerable growth by secondary intussusception of new wall material. The conidial chain thereby elongates simultaneously with retrogression of the fertile hypha (cf. Fig. 5.29A, C).

Other examples of retrogressive development

Several anamorphs of the Hypocreales also demonstrate retrogressive conidiogenous cell development. Tubaki (1975) presented a time-lapse sequence of conidiogenesis in *Hypomyces trichothecoides* Tubaki which is essentially identical to our illustrations of *Trichothecium roseum.* Although the author stated that the fertile hypha does not shorten during formation of successive conidia, his photomicrographs indicate the opposite. It is also suggested that the anamorphs of *Heleococcum japonense* Tubaki (Tubaki, 1967) and *Hypomyces polyporinus* Peck demonstrate conidiogenesis similar to that of *T. roseum.*

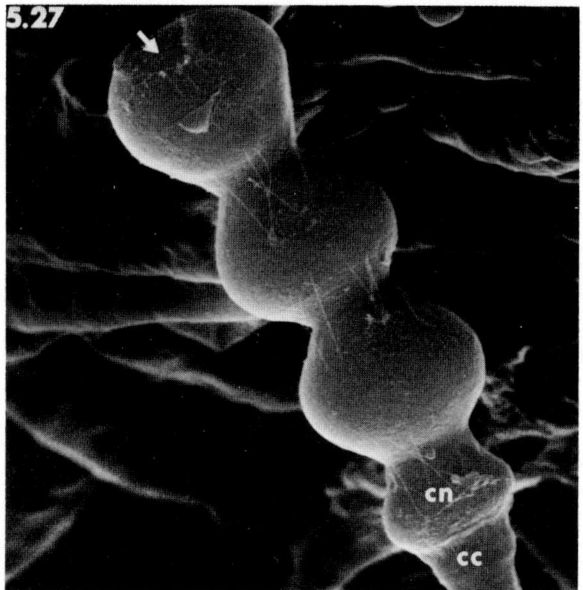

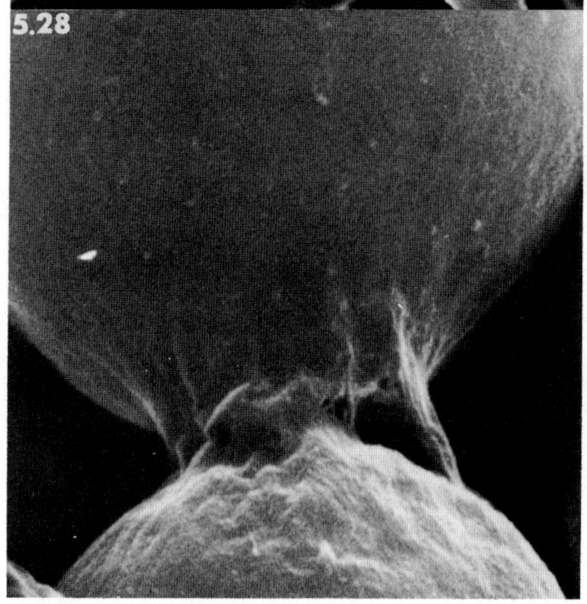

Figs. 5.27, 5.28 *Basipetospora rubra.* Conidial chain (5.27) showing continuous outer wall layer of fertile hypha between adjacent cells. During conidial secession, this outer wall ruptures (5.28), revealing the inner conidial wall (arrow in 5.27). cn, conidial initial; cc, conidiogenous cell. ×3,080 and 12,600, respectively.

5.29

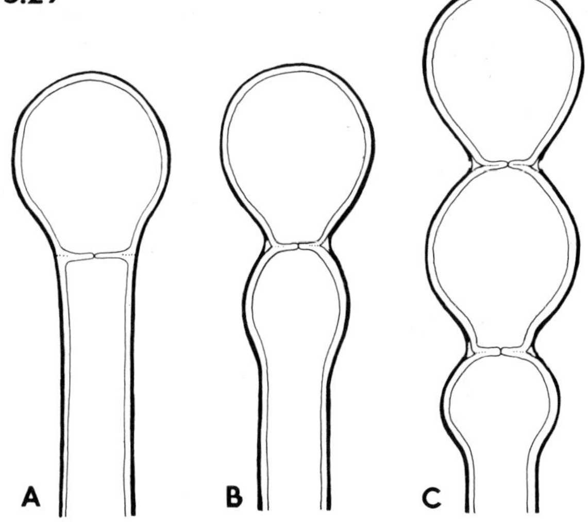

Fig. 5.29 Diagrammatic interpretation of retrogressive development in *Basipetospora rubra.*

However, among the Hypocreales, which accommodates approximately 100 genera (Rogerson, 1970), the phialide is the most common and almost universal kind of conidiogenous cell. Of interest in this respect is that similarities clearly exist between phialidic and retrogressive conidiogenous cell development which have already been alluded to in this chapter. The following discussion considers these overlapping, developmental concepts.

Comparison of retrogressive, phialidic and annellidic ontogeny

One unifying developmental feature of these three kinds of conidiogenous cells is that each produces a basipetal succession of conidia. Similarities in wall differentiation also exist—the primary conidium is holoblastic. The first-formed conidium at the apex of these conidiogenous cells is represented diagrammatically by Fig. 5.30A. The outer wall which encompasses the developing conidium is apparently capable of stretching, or is *unset* (Da Riva Ricci and Kendrick, 1972). On the other hand, the region of this same wall layer below the fertile apex is nonextensible, or *set*. Except in *Basipetospora rubra*, this relationship of set and unset regions of the fertile cell wall persists during formation of successive conidia. The inner conidial wall layer continues to grow and thicken as the conidium develops. Differentiation of the two-layered basal septum and its subsequent schizolytic separation initiates the sequence of events leading to formation of the second conidium.

The developmental characteristics of secondary conidium formation from phialides (Fig. 5.30B–D), annellides (Fig. 5.30E–G) and retrogressive conidiogenous cells (Fig. 5.30H–K), produce morphological differences which have been used to distinguish between these three modes of ontogeny. In phialidic development, the outer conidiogenous cell wall usually reaches its limit of extensibility at some stage during differentiation of the first conidium, and growth of the lower layer of the double septum (secondary conidium initiation) causes the outer wall to rupture (Fig. 5.30B), forming a collarette at the phialide apex (e.g., *Phialophora lagerbergii* (Melin and Nannf.), Chapter 4). The collarette no longer participates directly in the formation of secondary conidia and is represented in Fig. 5.30B–D as part of the *set*, outer wall layer. Figure 5.30C, D illustrates the wall differentiation associated with formation of the third (Fig. 5.30C) and fourth conidia (Fig. 5.30D). The presence of endogenous wall layers at the fertile apex of the phialide, which accumulate as successive conidia develop and secede, is characteristic of one kind of phialidic development (Chapter 4, Fig. 4.38).

The annellide has exogenous rings of wall material at its fertile apex which represent scars left as successive conidia secede (Fig. 5.30E–G). As in phialidic development, the outer wall layer encompassing the first-formed conidium reaches its limit of extensibility during differentiation of the basal septum. Rupture of the outer wall in this case, however, usually occurs at the same level as the double septum (Fig. 5.30E).

This difference in the location of the rupture of the outer wall during phialidic and annellidic development appears to depend mainly on the degree of structural association between the inner conidial and outer conidiogenous cell wall layers during conidiogenesis. In the case of phialides with well-differentiated collarettes, these two layers are *not* bound firmly together above the conidiogenous locus, permitting secondary conidia to slip easily out of the collarette (Fig. 5.30B–D). On the other hand, these same layers at the annellide apex are usually bound firmly together and a circumscissile tear of the outer wall occurs at the zone of greatest weakness, which is usually in the region of the double layered septum adjacent to the zone of schizolysis (arrowheads in Fig. 5.30E). The annellide subsequently undergoes percurrent proliferation before its apex swells and differentiates into the next conidium. The same process of wall differentiation occurs during formation of secondary conidia. Annellations (Fig. 5.30F, G) result from sequential, percurrent proliferation of the conidiogenous cell and basipetal formation of successive terminal conidia. As previously discussed in Chapter 4, intermediary developmental patterns between these typical phialidic and annellidic modes are recognized (e.g., *Conioscypha* Höhnel and *Cryptosporiopsis* Bubák & Krieger). Clearly, a merging of the two developmental concepts is apparent, even though the phialide and annellide can usually be morphologically distinguished.

The retrogressive conidiogenous cell shows additional variations in wall differentiation during secondary conidium formation. In one case (Fig. 5.30H, I), exemplified by *Basipetospora rubra*, a restricted region of the outer wall layer directly below the septum of the first-formed conidium becomes *unset* and swells as growth of the corresponding region of the inner wall occurs (Fig. 5.30H). The outer wall above and below this meristematic zone remains set. The demarcation of these set and unset regions is probably rather gradual. As a result of this mode of wall differentiation, a basipetal chain of conidia is formed bound together by the continuous outer wall and involving the progressive shortening of the fertile hypha (Fig. 5.30I). In the second example of retrogressive development (Fig. 5.30J, K), represented by *Cladobotryum varium*, the *set* outer wall is ruptured at the base of the first-formed conidium opposite the double-layered septum, leaving a collarette similar in appearance to that of a phialide (Fig. 5.30J). However, unlike phialidic development, the outer and inner walls remain fused in the region of the ruptured wall, which is more like annellidic development. The second conidium is initiated by growth of the lower layer of the basal septum of the primary conidium (Fig. 5.30J, E). However, this developmental process is distinguished from annellidic ontogeny in that the septum of each secondary conidium forms *below* the collarette and secession results in loss of a segment of the conidiogenous cell (Fig. 5.30K).

5.30

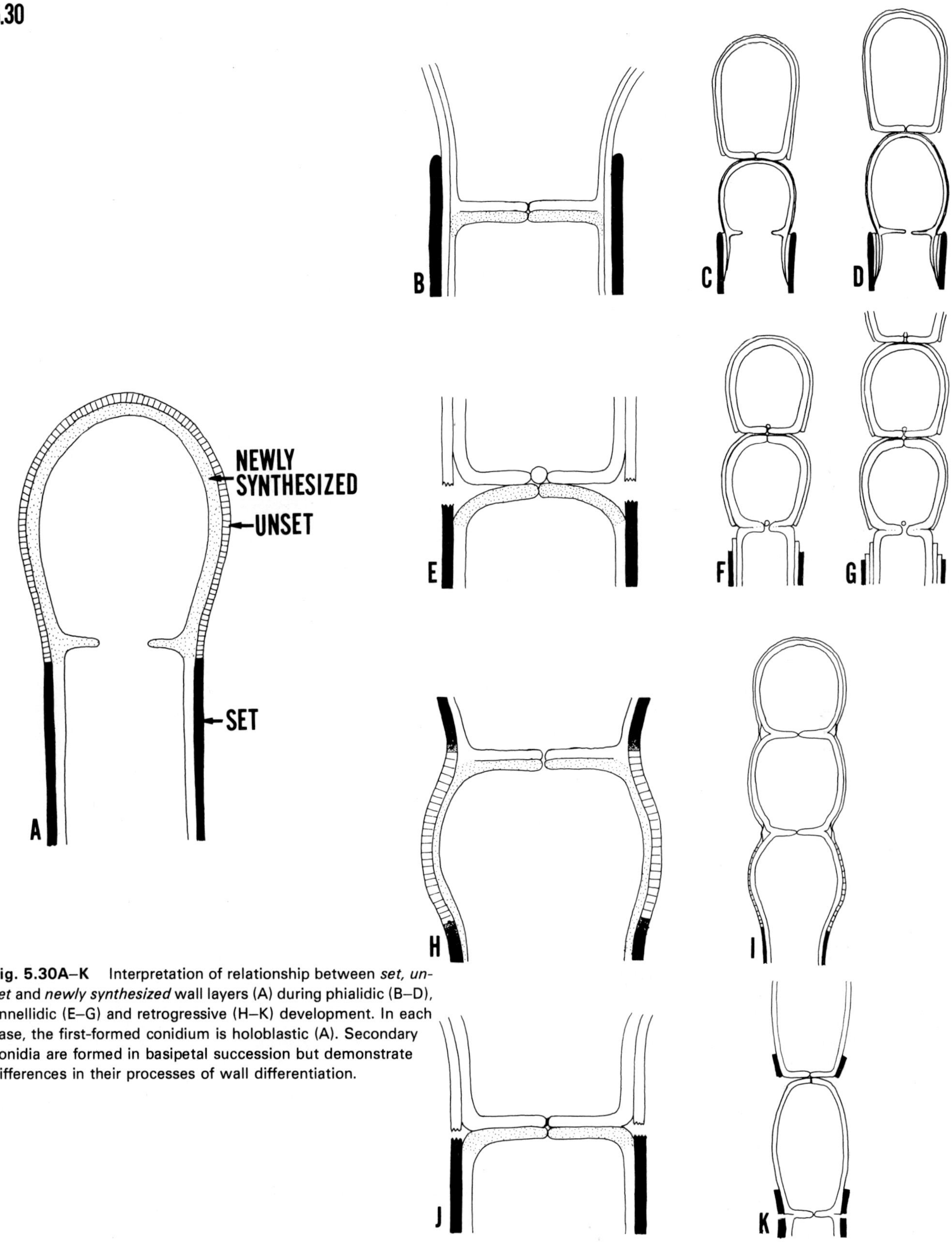

Fig. 5.30A–K Interpretation of relationship between *set, unset* and *newly synthesized* wall layers (A) during phialidic (B–D), annellidic (E–G) and retrogressive (H–K) development. In each case, the first-formed conidium is holoblastic (A). Secondary conidia are formed in basipetal succession but demonstrate differences in their processes of wall differentiation.

6 Development of Basauxic Conidiogenous Cells

S. J. Hughes (1953) formed a separate section in his experimental classification of conidial fungi to accommodate the unique mode of development which he observed in species of *Spegazzinia* Sacc., *Arthrinium* Kunze ex Fr. and *Dictyoarthrinium* Hughes. He coined the term 'basauxic' to describe their fertile cells which 'elongate by a basal growing point'.

Development in *Spegazzinia tessarthra*

The time-lapse sequence of photomicrographs of *Spegazzinia tessarthra* (Berk. & Curt.) Sacc. in Fig. 6.1A–J reveals that at least two distinct kinds of conidia and conidiogenous cells are produced by the same mycelium (Cole, 1974). The spherical initial in Fig. 6.1A, which has arisen blastically from a short lateral branch of the vegetative hypha, becomes echinulate and darkly pigmented (Fig. 6.1B–D), and is later pushed upward by growth of a basal filament (Fig. 6.1E). The elongated filament appears to have ruptured the cup-shaped cell at its base (arrowhead in Fig. 6.1E). The erect filament which supports the terminal conidium is, by definition, a conidiogenous cell. The fertile branch which initially gave rise to the conidium (arrowhead in Fig. 6.1B), in retrospect, should be referred to as a conidiophore. The residual cup which encloses the base of the elongated conidiogenous cell has been called a *conidiophore mother-cell* (S. J. Hughes, 1953). The three adjacent mother-cells in Fig. 6.1E have lost their conidiogenous cells.

Figure 6.1F–J demonstrates the ontogeny of another kind of conidium produced by the same mycelium. In Fig. 6.1F, two young conidiophore mother-cells (arrows) have arisen laterally from a vegetative hypha. Each fertile cell gives rise to a smooth-walled, cruciately septate conidium and in Fig. 6.1J, the structural relationships between the cup-shaped conidiophore mother-cells (arrowheads), erect conidiogenous cells (arrow) which have undergone extension growth, and apical, smooth-walled conidia are demonstrated. The conidiogenous cells supporting the smooth conidia are considerably shorter than the erect, fertile hypha in Fig. 6.1E. Conidial secession commonly occurs by schizolysis at the base of the conidium, but occasionally the propagule and conidiogenous cell adhere while the latter separates at its base from the conidiophore mother-cell.

The scanning electron microscope (SEM) has clarified some aspects of basauxic development in *S. tessarthra* (Figs. 6.2, 6.3). Most mother-cells in Fig. 6.2 have lost their

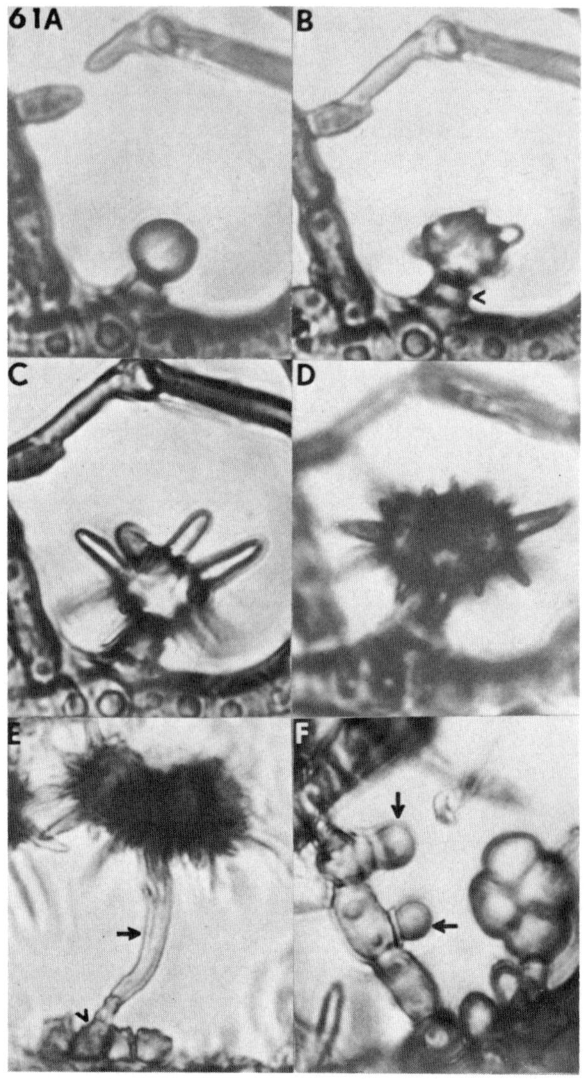

Fig. 6.1A–J *Spegazzinia tessarthra.* Time-lapse sequences of holoblastic ontogeny of echinulate (A–D) and smooth-walled (G–I) conidia and basauxic development of respective conidiogenous cells (arrows in E and J). The cup-shaped structures at the base of the conidiogenous cells in (E) and (J) (arrowheads) are the conidiophore mother-cells. Young mother-cells are indicated by arrows in (F). Arrowhead in (B) represents unruptured mother-cell. A–D, ×2,000; T = 0, $2\frac{1}{2}$, $5\frac{1}{2}$, $16\frac{1}{2}$ hours, respectively. E, ×1,800; (F–I), ×2,000; G–I, T = 0, $\frac{1}{4}$, $2\frac{1}{2}$ hours, respectively. J, ×1,800.

endogenous fertile hypha and remain as empty cups. The intact, rough-walled conidiogenous cell broadens apically and is attached to one of the four cells of an echinulate condium. The conidial spines have branched during their development and show considerable variation in length. Seceded, smooth-walled conidia are visible beneath the conidiogenous cell. In Fig. 6.3, one of these same smooth-walled conidia is supported by a short fertile cell (cc) covered with raised excrescences. The base of the fertile filament is also enclosed by a cup-shaped mother-cell, and its apex is attached to one of the four conidial cells. This interrelationship between conidial shape and length of the conidiogenous cell is maintained, although some developmental variation has been reported (Cole, 1974). Thin-section examinations are required to define the process of wall differentiation during conidium ontogeny. Basauxically developing conidiogenous cells may be ideal for examining the cytological events associated with intercalary growth and secondary intussusception in deuteromycetous fungi.

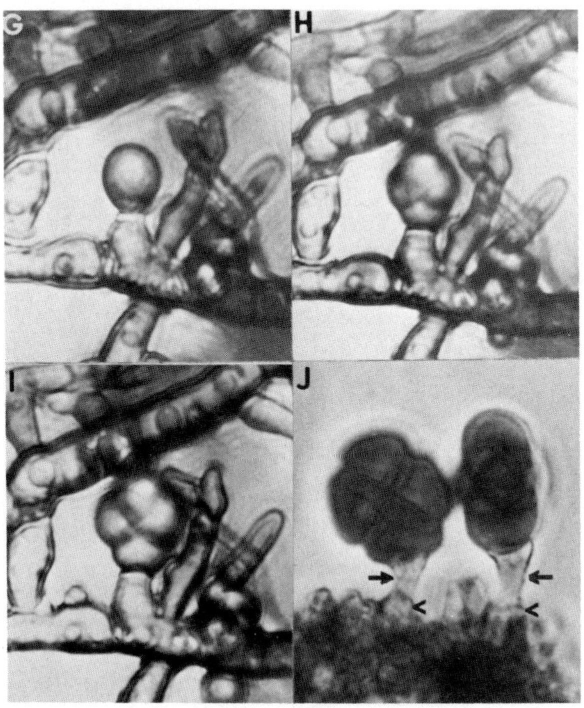

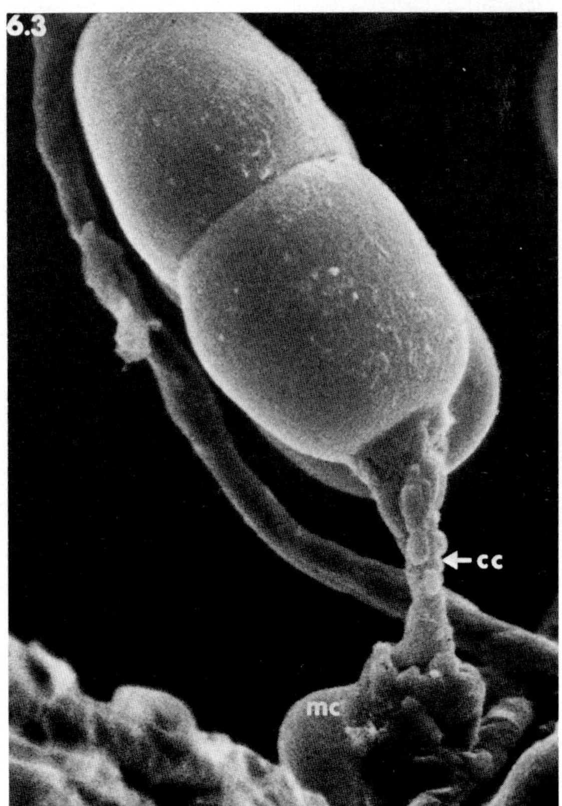

Figs. 6.2, 6.3 *Spegazzinia tessarthra.* Echinulate and smooth-walled conidia attached to apices of basauxically developed conidiogenous cells. In each case the elongated fertile hyphae have emerged from a cup-shaped conidiophore mother-cell at their base. Several conidiophores from which the conidiogenous cells have seceded are shown in 6.2 (arrows). The apically-tapered conidiogenous cells are attached to only one of the four cells of the conidia (6.2). Note difference in orientation of two kinds of conidia. cc, conidiogenous cell; mc, conidiophore mother-cell. ×2,400 and 5,700, repectively.

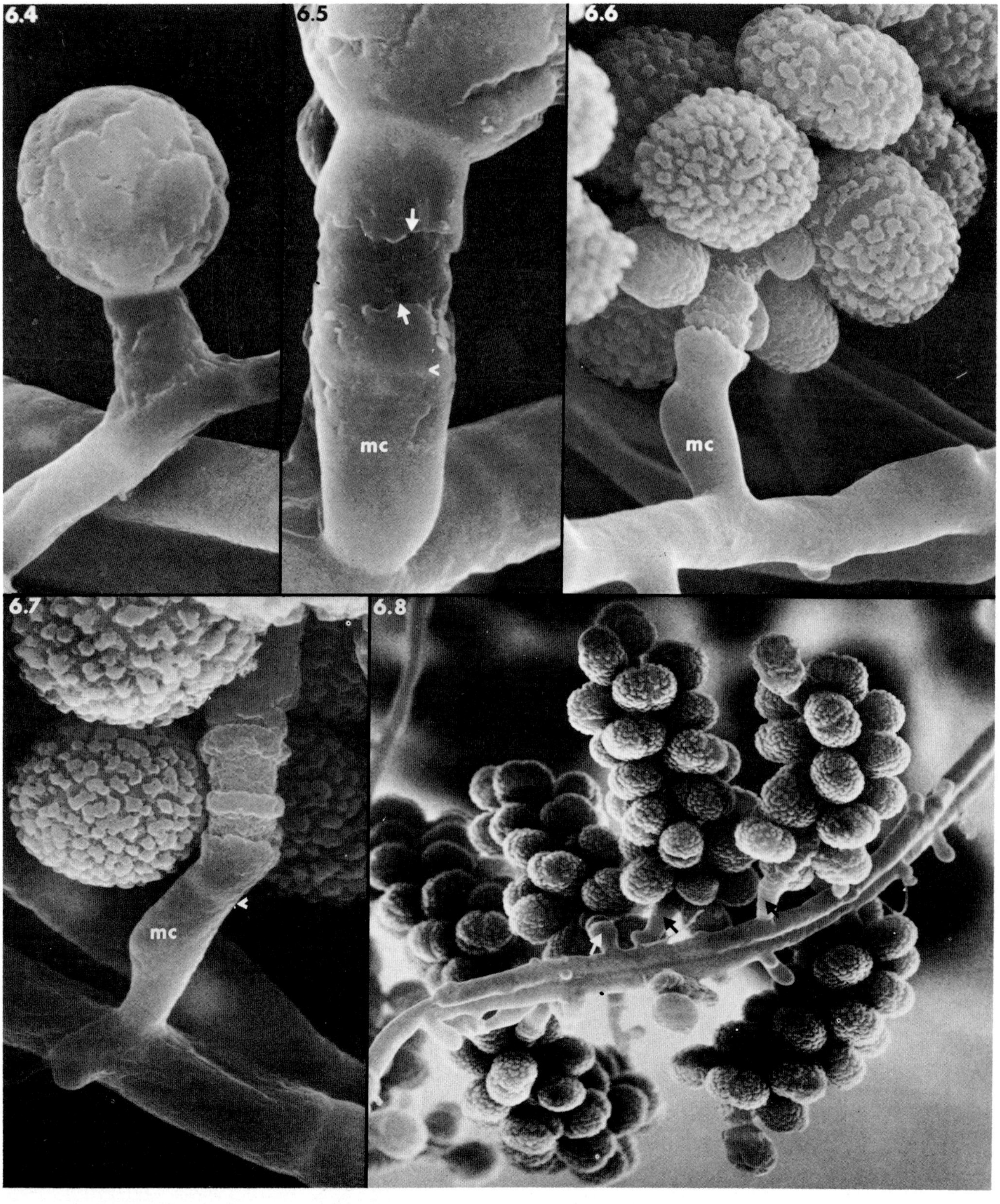

Figs. 6.4–6.8 *Dictyoarthrinium sacchari.* Stages of basauxic conidiogenous cell development and basipetal succession of holoblastic conidia. Formation of primary holoblastic conidium at conidiophore apex (6.4) is followed by rupture of outer wall layer (arrows in 6.5) which is associated with intercalary growth of inner wall of conidiogenous cell. A vase-shaped mother-cell (mc) remains. Arrowheads in 6.5 and 6.7 locate thickened septa inside mother-cells. Arrows in 6.8 indicate conidiophore mother-cells. ×4,300, 7,300, 3,100, 4,000 and 1,100, respectively.

Development in *Dictyoarthrinium sacchari*

Our second example of basauxic development is the dematiaceous Hyphomycete, *Dictyoarthrinium sacchari* (Stevenson) Damon (Figs. 6.4–6.9). As above, the primary conidium forms holoblastically at the apex of a short lateral branch of a vegetative hypha (Fig. 6.4). Basauxic growth of the inner wall layer occurs a short distance below the base of the terminal conidium and causes a circumferential rupture of the apparently nonextensible outer wall layer of the lateral branch (arrows in Fig. 6.5), leaving a vase-shaped conidiophore mother-cell. The exposed endogenous wall continues to elongate and the intercalary meristem becomes differentiated as the conidiogenous cell, giving rise to a basipetal succession of conidia. Each new conidium forms just above the ruptured apex of the mother-cell (Fig. 6.6). Thick-walled, darkly pigmented transverse septa are formed at regular intervals along the length of the conidiogenous cell (Fig. 6.7). Young septa are visible within the ruptured wall of the conidiophore mother-cell in Figs. 6.5 and 6.7 (arrowheads). It is suggested that these thickened crosswalls later emerge from the mother-cells as a result of elongation of the fertile hypha. Secondary intussusception associated with basauxic conidiogenous cell development, therefore, occurs within the conidiophore mother-cell. Lateral conidia arise as holoblastic outgrowths of the fertile hyphal wall between the thickened septa (Figs. 6.7, 6.9) forming dense clusters of propagules along the length of the erect conidiogenous cells (Figs. 6.8, 6.9). In both *D. sacchari* (Fig. 6.8) and *Spegazzinia tessarthra* (Fig. 6.1E), conidiophore mother-cells are closely juxtaposed along the surface of vegetative hyphae. The conidia of *D. sacchari* secede from the fertile cells leaving behind short, inconspicuous, truncate denticles. Occasionally the entire conidiogenous cell may separate from the conidiophore mother-cell. Figure 6.9 is a series of camera lucida drawings illustrating similar stages of basauxic, conidiogenous cell development in *D. sacchari.*

Development in *Arthrinium* spp.

The doliiform conidiophore of the *Arthrinium* state of *Apiospora montagnei* Sacc. produces a holoblastic, apical conidium which is at first spherical (Fig. 6.10), but later becomes lenticular (Fig. 6.11). As the conidium matures, the neck of the conidiophore narrows slightly (Fig. 6.11). A circumferential rupture of the outer wall is visible in Fig. 6.12 (arrow) which represents an early stage in basauxic growth of the inner conidiogenous cell wall (cf. *Dictyoarthrinium sacchari*, Fig. 6.5). Campbell (1975) has presented thin-sections of this species illustrating a conidiophore mother-cell through which the conidiogenous cell emerges (Fig. 6.13). The latter gives rise to lateral, holoblastic conidia as the erect, fertile filament continues to elongate (Fig. 6.14). The mature conidiogenous cell is up to 50 μm long, flexuous and threadlike and bears a large number of conidia, the youngest usually found just above the collar of the conidiophore mother-cell in the region of elongation. Seceded conidia leave short, truncate denticles on the surface of the fertile hypha. Campbell (1975) also demonstrated that the circumferential ridge of the mature, lenticular conidium (Fig. 6.12,

6.9

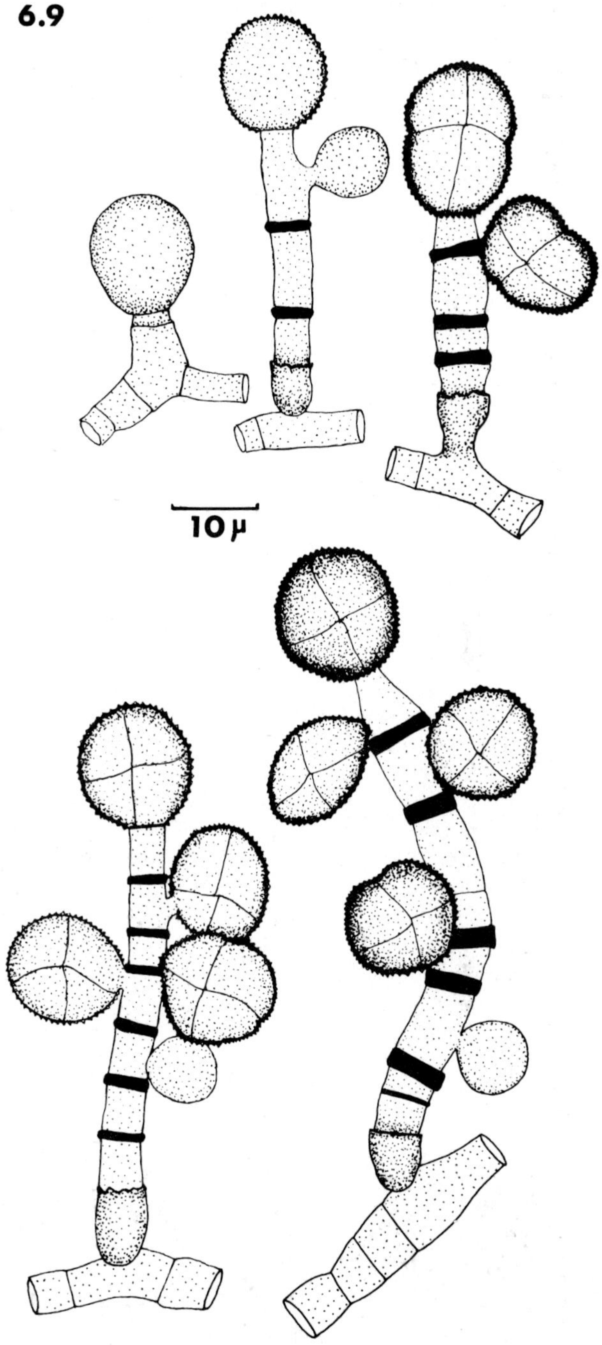

Fig. 6.9 *Dictyoarthrinium sacchari.* Stages of basauxic conidiogenous cell development and basipetal succession of holoblastic conidia.

arrowhead) is composed of electron-transparent wall material which is continuous with the inner conidial wall layer. This thin, outer wall of the conidium is ruptured during conidial germination resulting in separation of the two halves of the bivalvate conidium (Figs. 6.15A, 6.16E).

Drawings of different stages of conidiogenesis in this species are illustrated in Fig. 6.15A. Other species of *Arthrinium* (Fig. 6.15B–D) produce prominent, transverse septa at regular intervals along basauxically developed conidiogenous cells (e.g., *A. sporophleum* Kunze, Fig. 6.15B; *A. spegazzinii* Subram., Fig. 6.15C; *A. kamtschaticum* Tanz. & Woronich., Fig. 6.15D).

A diagrammatic interpretation of wall differentiation in *Arthrinium,* based on ultrastructural aspects of the *Arthrinium* state of *Apiospora montagnei*, is presented in Fig. 6.16A–E. The developmental features of this species are characteristic of the basauxic mode of ontogeny and we suggest that the process of wall differentiation is representative of other species examined in this chapter. The primary conidium develops holoblastically at the apex of the conidiophore (Fig. 6.16A–C). Intercalary growth of the inner wall layer, below the basal septum of the terminal conidium, initiates basauxic development of the conidiogenous cell and causes the outer conidiophore wall to rupture (Fig. 6.16C). Elongation of the conidiogenous cell continues, with new wall intussusception presumably occurring within the conidiophore mother-cell just below its ruptured apex (Fig. 6.16D). Above the ruptured apex, concomitant with extension of the fertile hypha, holoblastic development of successive conidia occurs from alternate sides of the fertile hypha.

M. B. Ellis (1971b) illustrated several genera, in addition to those described above, which undergo basauxic development during conidiogenesis in his account of dematiaceous Hyphomycetes (Fig. 6.17). *Cordella johnstonii* M. B. Ellis produces lenticular conidia and prominent setae, 200–300 μm long (Fig. 6.17A). The conidiophores and elongated conidiogenous cells of *Pteroconium pterospermum* (Cooke & Massee) Grove (Fig. 6.17B) arise from sporodochia and produce lobed or dentate conidia. In species of *Endocalyx* Berk. & Br., not illustrated here, prominent, stilboid or cupulate mother-cells are produced which contain monoblastic or polyblastic conidiogenous cells. Ellis (1971b) also illustrated fertile cells of *E. thwaitesii* Berk. & Br. which are capable of undergoing sympodial proliferation in addition to basauxic development. We observed this same phenomenon in *Dictyoarthrinium sacchari.*

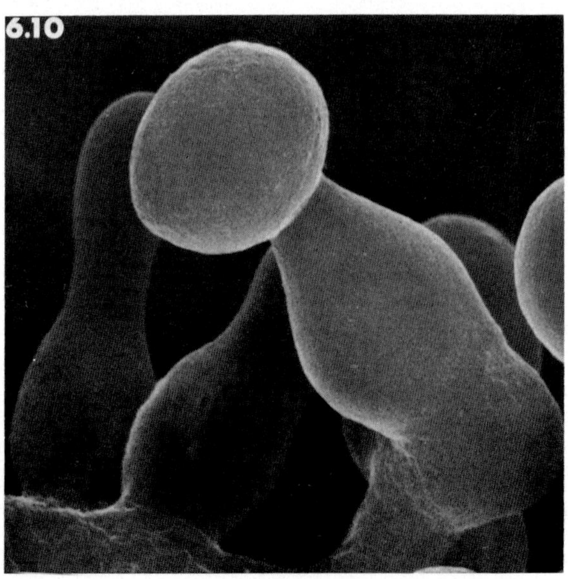

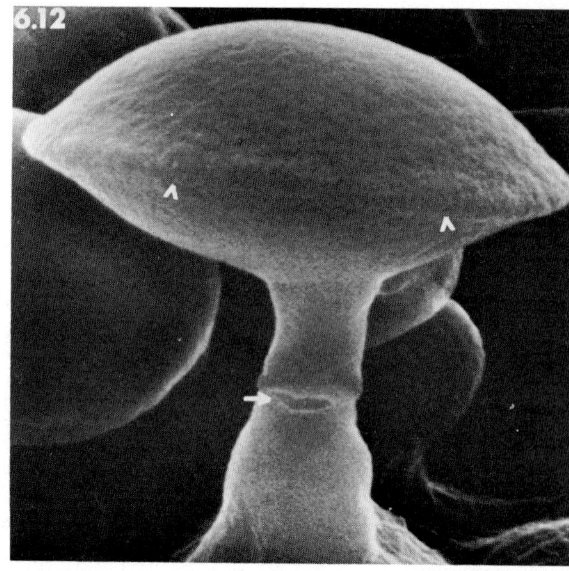

Figs. 6.10–6.12 *Arthrinium* state of *Apiospora montagnei.* Globose conidial initial (6.10) and mature, lenticular, primary conidium at apex of doliiform conidiophore (6.11). Arrow in 6.12 indicates rupture of outer conidiophore wall representing early stage of basauxic development. Arrowheads in 6.12 represent thickened ridge of mature conidium. ×6,300, 11,700 and 9,700 respectively.

Figs. 6.13, 6.14 *Arthrinium* state of *Apiospora montagnei.* Thin-sections through ruptured apex of conidiophore mother-cell (mc) and filamentous conidiogenous cell (cc). Holoblastic development is revealed in 6.14. Arrows in 6.13 indicate ruptured apex of mother-cell. ×36,200 and 24,000, respectively (from Campbell, 1975).

6.13

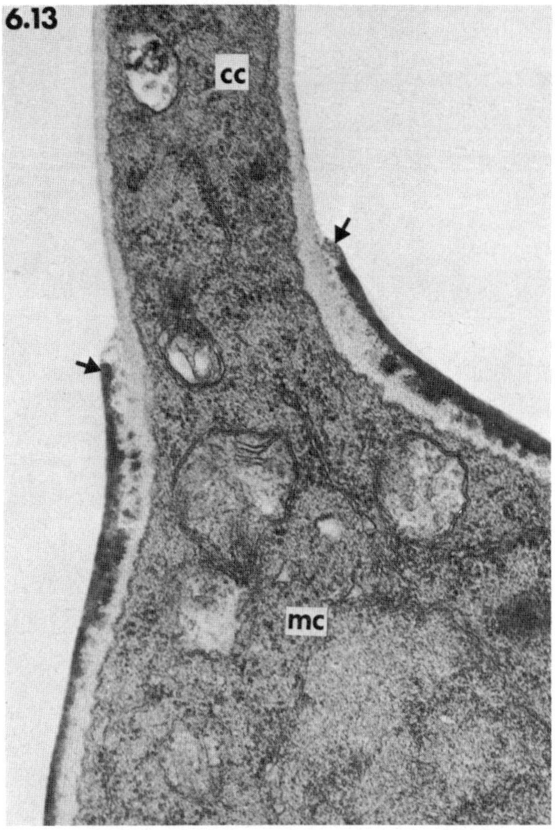

6.14

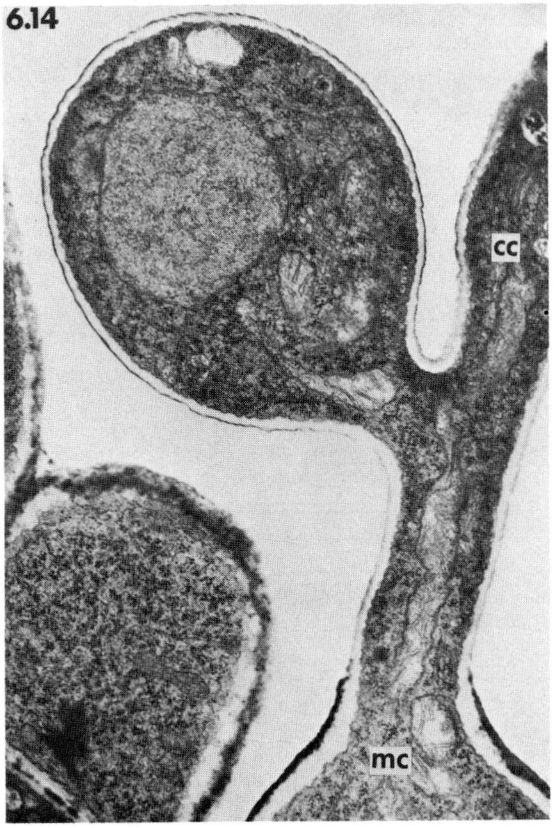

6.15

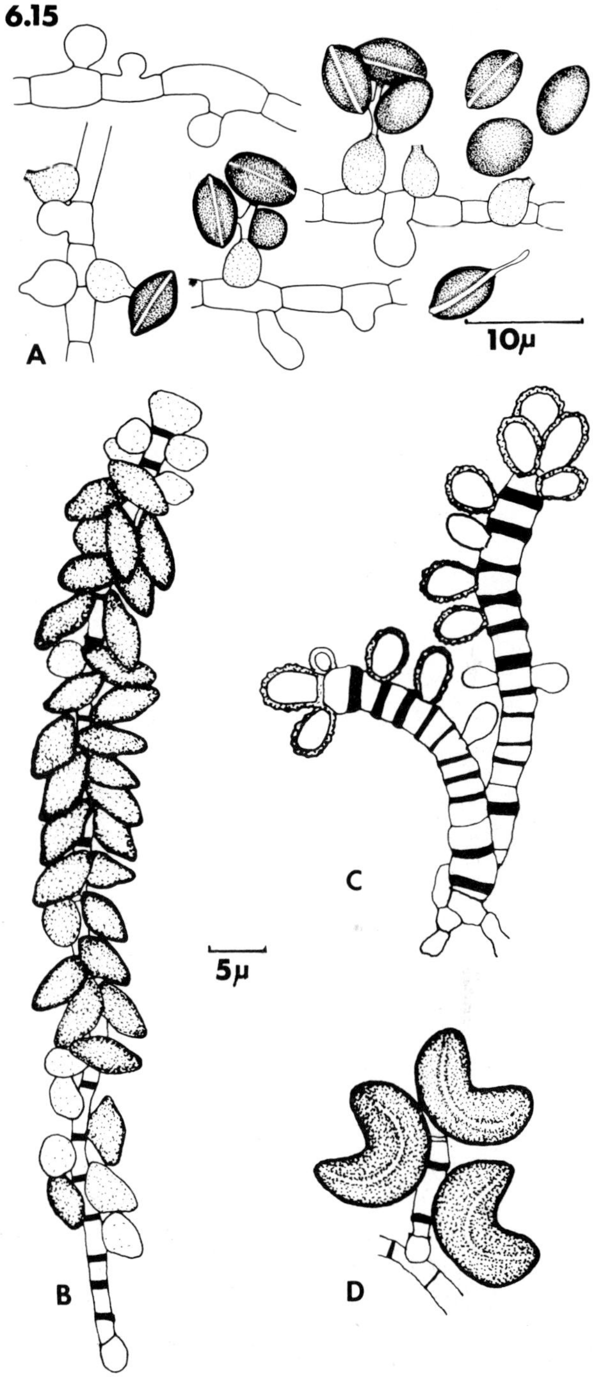

Fig. 6.15A *Arthrinium* state of *Apiospora montagnei*. Stages of basauxic conidiogenous cell development and basipetal succession of holoblastic conidia.

Fig. 6.15B *Arthrinium sporophleum* (after M. B. Ellis, 1971b).

Fig. 6.15C *Arthrinium spegazzinii* (after M. B. Ellis, 1971b).

Fig. 6.15D *Arthrinium kamtschaticum* (after M. B. Ellis, 1971b).

6.16

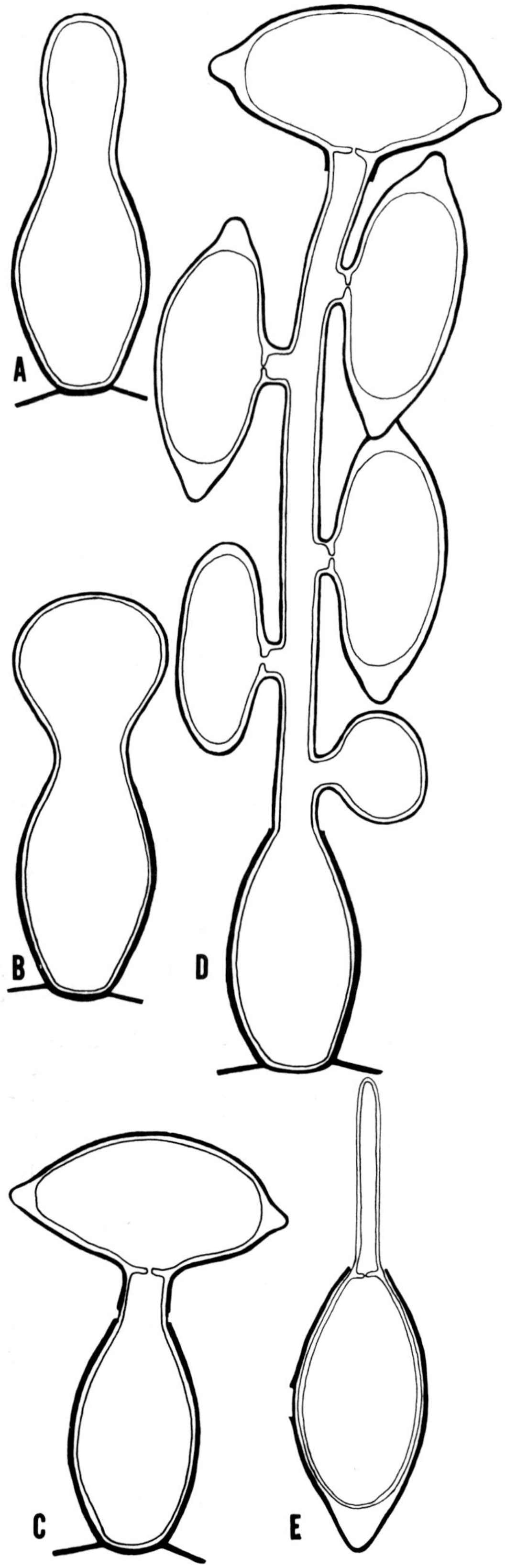

Fig. 6.16A–D Diagrammatic interpretation of basauxic conidiogenous cell development, basipetal succession of holoblastic conidia, and conidial germination (E) in *Arthrinium* state of *Apiospora montagnei.*

6.17

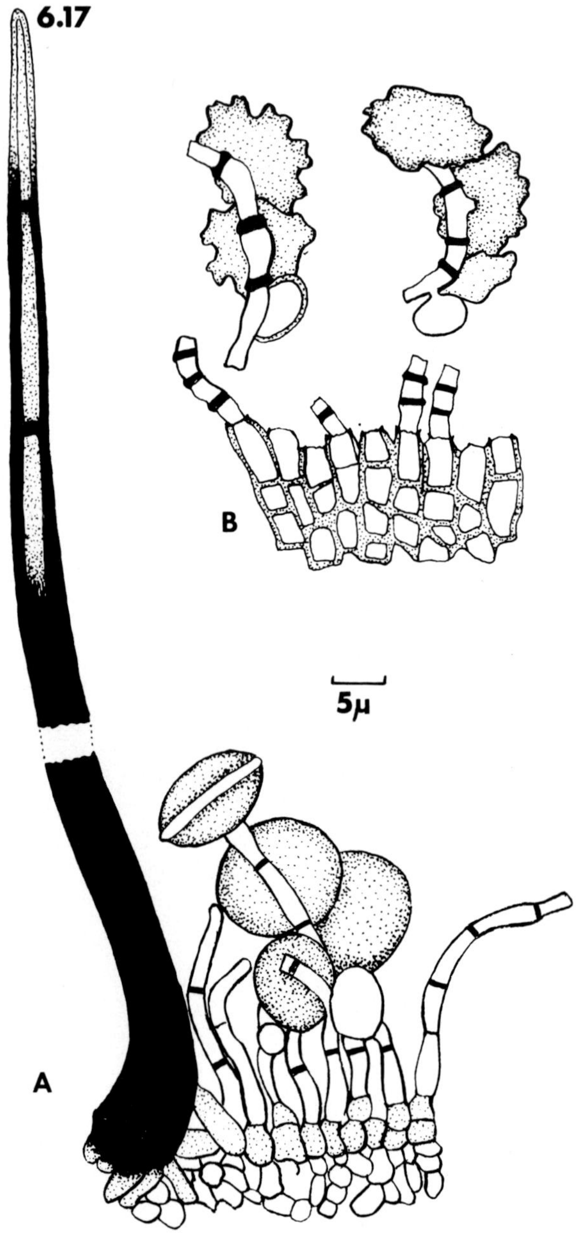

Fig. 6.17A *Cordella johnstonii* (after M. B. Ellis, 1971b).

Fig. 6.17B *Pteroconium pterospermum* (after M. B. Ellis, 1971b).

Variations in basauxic development

Conidiogenesis in Wallemia sebi

Controversy exists over the interpretation of ontogeny in this species. Madelin and Dorabjee (1974) described it as a special mode of phialidic development while several others have interpreted the process of conidiogenesis as arthric development (Barron, 1968; von Arx, 1970; M. B. Ellis, 1971b; Hashmi and Morgan-Jones, 1973). In some ways, the mode of conidium ontogeny in *Wallemia sebi* (Fr.) v. Arx is reminiscent of S. J. Hughes's (1953) concept of Section V in his developmental classification of Deuteromycetes which he defined as follows: 'Conidia developing in a gradually maturing basipetal series and originating by the meristematic growth of the apical region of the conidiophore in such a way that the chain of conidia merges imperceptibly with the conidiophore which gives rise to the chain'. Hughes referred to these conidia as 'meristem arthrospores' and cited several examples of fungi which demonstrate this developmental process, including species of *Coniosporium* Link ex Fr., *Trimmatostroma* Corda, *Phragmotrichum* Kunze apud Kunze & Schmidt, and the anamorphs of *Hysterium insidens* Schw. and *Erysiphe polygoni* DC. However, the concept of thallic-arthric development is based on disarticulation of a fertile hypha which has terminated extension growth (Chapters 1 and 7). In *W. sebi* the fertile hypha continues to elongate during conidium formation. The term 'meristem arthrospore' is, therefore, ambiguous because of the contradictory concepts of thallic-arthric development and meristematic activity associated with blastic ontogeny.

It is our contention that *W. sebi* produces chains of blastic conidia which mature in basipetal succession, by differentiation of a basauxically developing fertile hypha. In a time-lapse photomicrographic analysis of conidium ontogeny in this species, Madelin and Dorabjee (1974) demonstrated that a tapered, smooth-walled conidiophore gives rise to the verrucose fertile hypha (Fig. 6.18) which elongates by intercalary growth at a zone just above the conidiophore apex (Fig. 6.22A–I). Unlike other examples of basauxic development, the outer wall layer remains continuous in the region of the basal meristem. However, the junction between the non-extensible conidiophore and elongating fertile hypha is visible as a circumscissile constriction of the outer wall (arrow in Fig. 6.19) and could easily be misinterpreted as the collarette of a phialide. As the fertile hypha elongates, it undergoes septation, first separating into long segments (Fig. 6.20) which later subdivide into four cells (Fig. 6.22A–E). The latter are originally cylindrical but later swell (Fig. 6.21) as the septa between adjacent cells separate schizolytically, resulting in disarticulation of the chain (Fig. 6.22F–I).

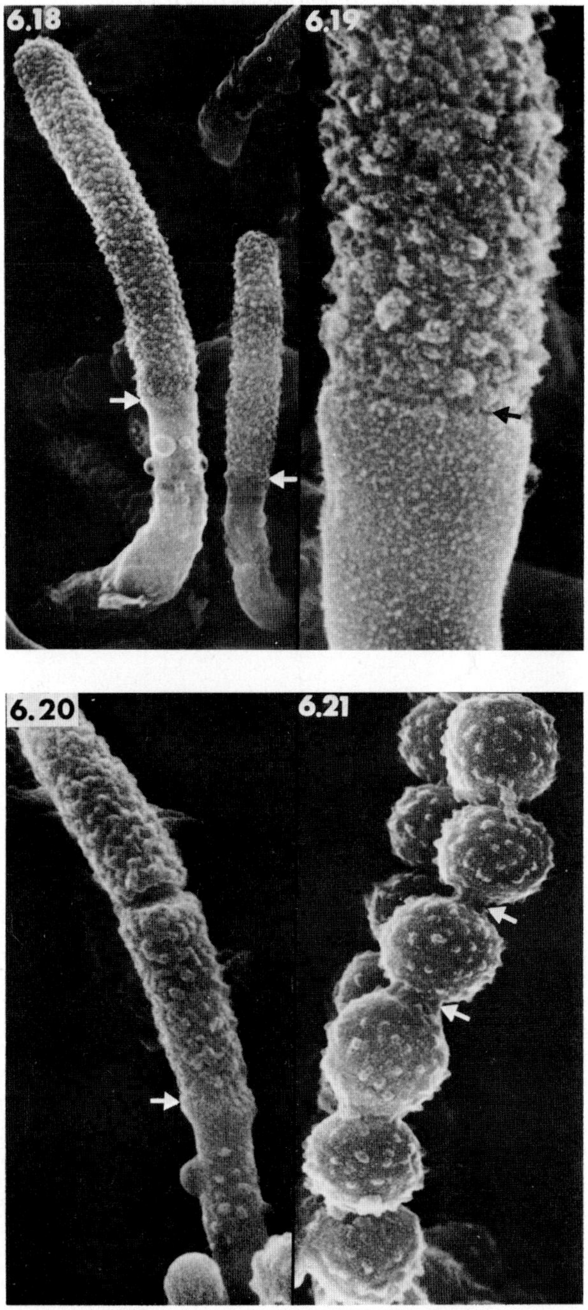

Figs. 6.18–6.21 *Wallemia sebi.* Stages of conidial formation and basauxic development of fertile hypha. Arrows in 6.18, 6.19 and 6.20 locate junction between smooth-walled conidiophore and verrucose, fertile hypha. Arrows in 6.21 indicate connectives between mature conidia. ×4,000, 14,000, 6,300 and 5,000, respectively.

Conidiogenesis in Erysiphe

Kendrick (1971c) presented a diagrammatic interpretation of development in the anamorph of *Erysiphe polygoni*, adopting S. J. Hughes's (1953) concept of a gradually maturing conidial chain and active-growing point at the base of the fertile hypha (Fig. 6.23A–D). The anamorph of *E. graminis*

6.22

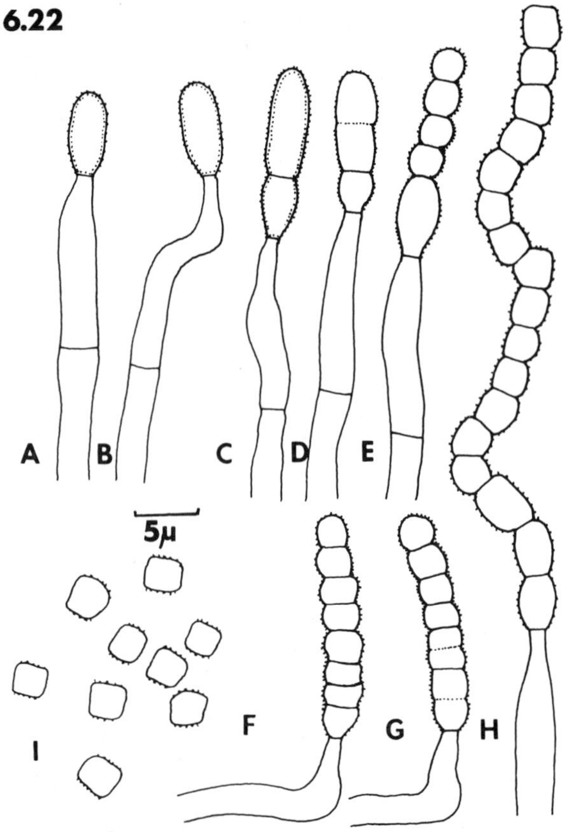

Fig. 6.22A–I *Wallemia sebi.* Camera lucida drawings of successive stages of basauxic development and conidiogenesis.

Fig. 6.23A–D Diagrammatic interpretation of basauxic development and conidial formation in the anamorph of *Erysiphe polygoni* (after Kendrick, 1971c).

6.23

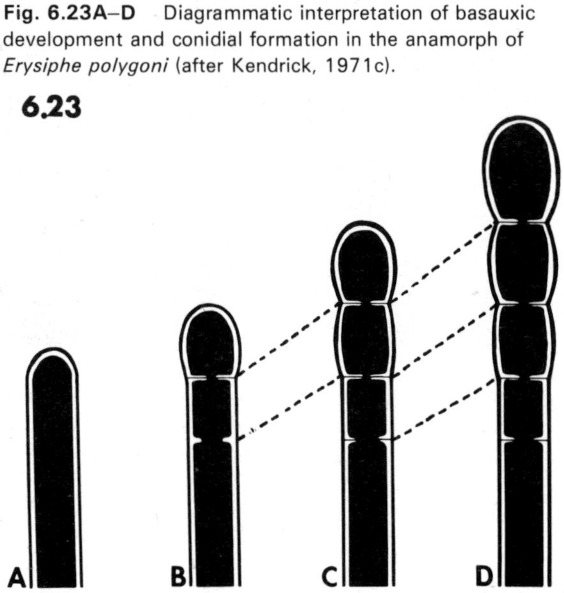

Fig. 6.24 Anamorph of *Erysiphe graminis.* Montage of scanning electron micrographs which shows globose conidiophore mother-cell (mc), basipetally maturing chain of holoblastic conidia, and basal meristematic zone of fertile hypha (MZ) which does not differentiate into conidia. Arrowheads locate septa. ×1,820.

DC. has a basipetally maturing chain of conidia and a short, globose conidiophore mother-cell at the base of the fertile hypha (Figs. 6.24, 6.27). The latter arises holoblastically from the conidiophore (Fig. 6.25) and during early development, probably elongates by polar-tip growth like a vegetative hypha. As the fertile hypha continues to elongate, a septum forms just above the conidiophore apex (arrowhead in Fig. 6.26) and then additional septa develop in acropetal succession along the fertile hypha (arrowheads in Fig. 6.24). At this stage, the apical cell begins to swell, signalling the differentiation of the primary conidium (Fig. 6.26). Some extension growth occurs as each conidium matures, in addition to the secondary intussusception associated with the basal region of the fertile hypha which has not differentiated into conidia (Fig. 6.24). This region can be identified even after a long conidial chain has developed (Fig. 6.27). The outer wall of the fertile hypha finally ruptures as the septa of adjacent conidia separate schizolytically (Fig. 6.28). Few developmental and ultrastructural studies have been conducted on this species and other members of Hughes's (1953) Section V (Plumb and Turner, 1972). These fungi may also prove to be well-suited for investigations of intercalary growth.

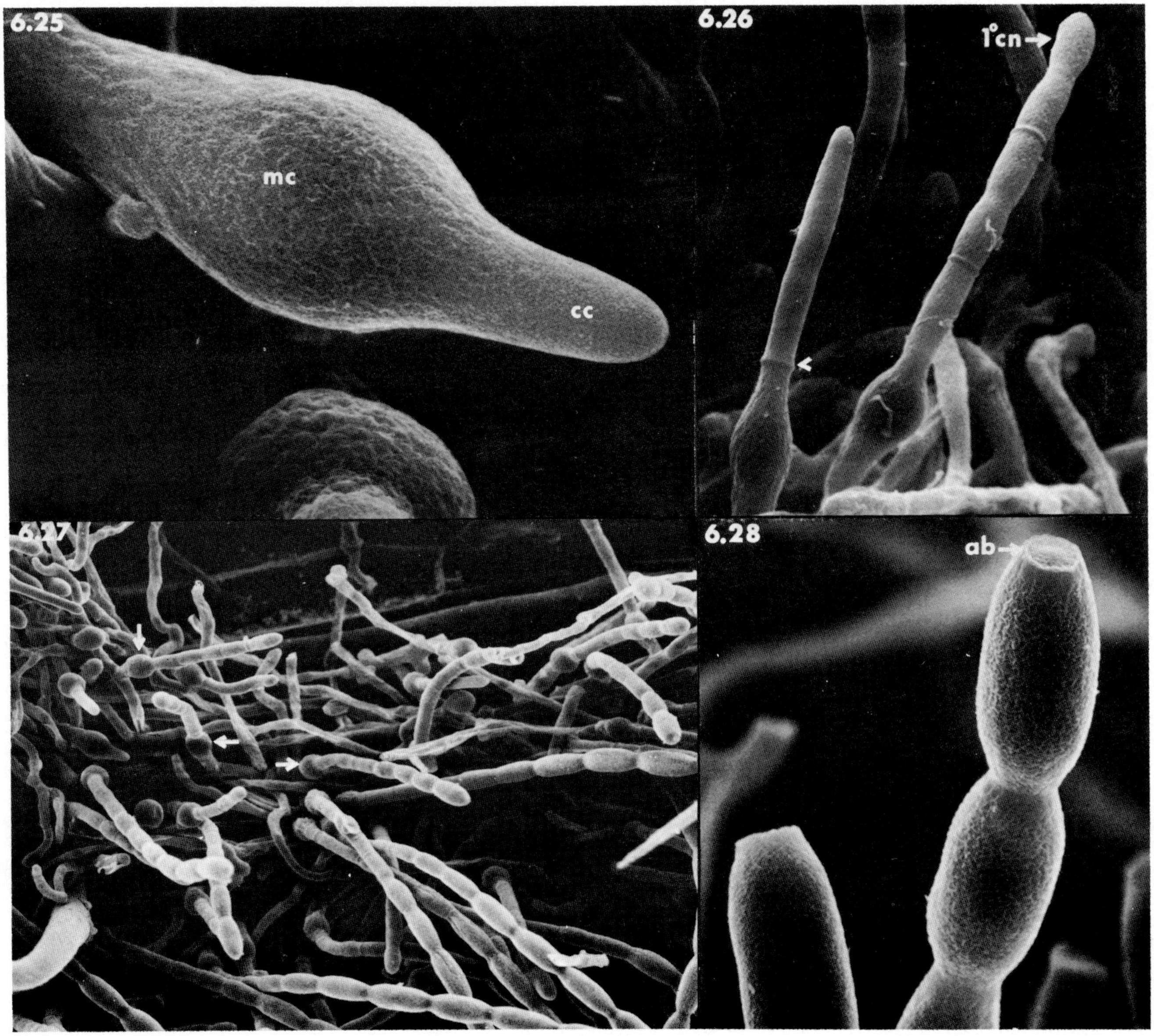

Figs. 6.25–6.28 Anamorph of *Erysiphe graminis.* Stages of basauxic development of fertile hypha from globose conidiophore mother-cell (6.25), acropetal succession of septation (6.26), basipetal succession of conidial differentiation (6.26, 6.27) and conidial secession (6.28). Note continuity of outer wall layer between conidiophore and fertile hypha, or conidiogenous cell (cc) in Fig. 6.25. Arrowhead in 6.26 indicates septum. Arrows in 6.27 locate conidiophore mother-cells at the base of each chain of conidia. 1°cn, primary conidium; ab, abstriction scar. ×1,820, 4,700, 13,795, 462 and 2,314, respectively.

7 Development of Thallic Conidia

Two basic kinds of thallic development were recognized in Chapter 1. Formation of holothallic conidia involves conversion of a terminal or intercalary segment of a fertile hypha into a thick-walled and usually pigmented propagule. On the other hand, thallic-arthric conidia form by conversion and fragmentation of a fertile hypha. In each case, the significant aspect of thallic development is the conversion of a pre-existing hyphal element into one or more propagules, in contrast to blastic conidium ontogeny in which the initial arises *de novo* as a blown-out part of the conidiogenous cell.

Holothallic conidia

In an earlier developmental study of *Microsporum canis* Bodin, one of us (Cole, 1976a) suggested that its conidia are holoblastic and conidiogenous cells are retrogressive. However, in a later cinephotomicrographic analysis of the same mechanism in *M. gypseum* (Bodin) Guiart & Grigorakis (Cole, 1977), it was apparent that conidiogenesis is holothallic. The time-lapse sequence in Fig. 7.1A–D follows early development of three conidia (arrows in Fig. 7.1A). The fertile branch in Fig. 7.1B (arrow) extends, then its apical segment swells (Fig. 7.1C) signalling initiation of conidial development. Some extension growth occurs as the swelling continues (Fig. 7.1D), presumably due to secondary intussusception associated with thickening of the conidial wall. The distal segment of the fertile hypha has thus been converted into a conidium and the outer hyphal wall layer has remained intact (i.e., holothallic development). The transverse septum which delimits the conidium forms after swelling of the hyphal apex (Fig. 7.2A, B), which contradicts one earlier concept of thallic development (Kendrick, 1971b). The timing of septation at the base of both blastic and thallic conidia is variable and cannot be used as a distinguishing developmental character for these two kinds of conidia. Just below the delimiting septum of the holothallic conidium in *M. gypseum* (arrow in Fig. 7.2B) another crosswall forms delimiting a short, cylindrical cell. Later, usually after the conidium has become septate, the protoplasm of this basal cell degenerates, and its thin, lateral wall collapses (arrow in Fig. 7.2C) releasing the macroconidium rhexolytically. Microconidia develop similarly.

Galgoczy (1975) also examined conidium ontogeny in *M. gypseum* by time-lapse photomicrography and his diagrammatic interpretation of conidiogenesis in this species is reproduced in Fig. 7.3A–H. We would slightly modify the diagram, on the basis of our observations, to indicate that some elongation of the developing holothallic initial occurs (Fig. 7.3B–D), and that the basal septum of the conidium (Fig. 7.3C) forms at a later stage. Galgoczy (1975) demonstrated similar conidiogenesis in other Dermatophytes, including *Microsporum fulvum* Uriburu, *M. audouinii* Gruby, *M. nanum* Fuentes, *Trichophyton rubrum* (Castell.) Sabour., *T. terrestre* Durie & Frey, *T. ajelloi* (Vanbreuseghem) Ajello, and *Epidermophyton floccosum* (Harz) Langer. & Milochevitch. The scanning electron microscope (SEM) micrographs of *Trichophyton ajelloi*, anamorph of *Arthroderma uncinatum* Dawson & Gentles, show clusters of macroconidia (Fig. 7.4) and adjacent microconidia (arrows in Fig. 7.4). The thin-walled basal cells (bc) are visible as darker regions of the micrographs where fewer 2° electrons are produced. A collapsed cell at the base of the mature macroconidium is shown in Fig. 7.5. The same process of rhexolytic secession occurs at the base of mature microconidia (Fig. 7.6).

The term *chlamydospore* was given a broad definition by the participants of the Kananaskis Conference (Kendrick, 1971b), because of its wide use in a variety of connotations. It was described as a 'thick-walled, thallic, terminal or intercalary spore'. Most chlamydospores of hyphomycetous fungi appear to be essentially holothallic. In an ultrastructural examination of chlamydospore wall differentiation in *Fusarium oxysporum* Schlecht. ex Fr., Griffiths (1973b) determined that 'the chlamydospore' (wall) 'is essentially two-layered, the outer layer consisting of the original hyphal wall and the inner representing secondary wall material incorporated' (secondary intussusception) 'between the hyphal wall and the plasmalemma'. Chlamydospores do not secede

Fig. 7.1A–D *Microsporum gypseum.* Time-lapse sequence of holothallic conidium ontogeny. Arrows in (A) represent three different developmental stages. Arrow in (B) indicates hyphal branch just prior to differentiation of holothallic conidium. T = 0, 30, 60, 85 minutes, respectively. ×550.

Fig. 7.2A–C *Microsporum gypseum.* Time-lapse sequence of septation and secession of holothallic conidium. Arrow in (B) locates septum at base of conidium and arrowhead indicates delimiting crosswall of basal cell. Arrow in (C) represents collapsed basal cell. T = 0, 15, 25 minutes, respectively. ×700.

Fig. 7.3A–H Diagrammatic interpretation of macroconidium development in *Microsporum gypseum* (after Galgoczy, 1975).

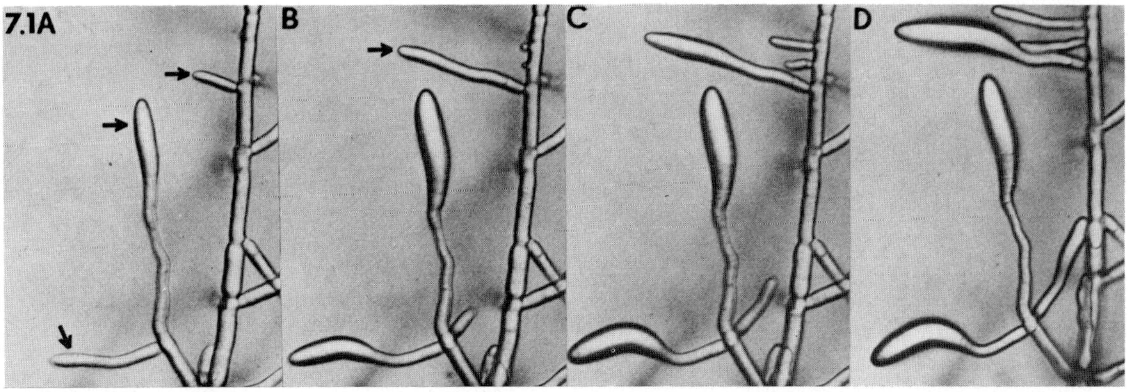
7.1A
B
C
D

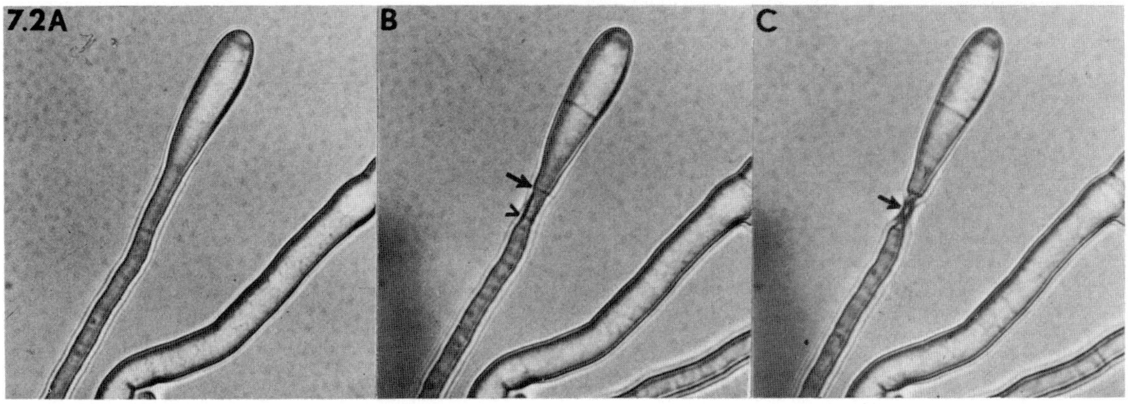
7.2A
B
C

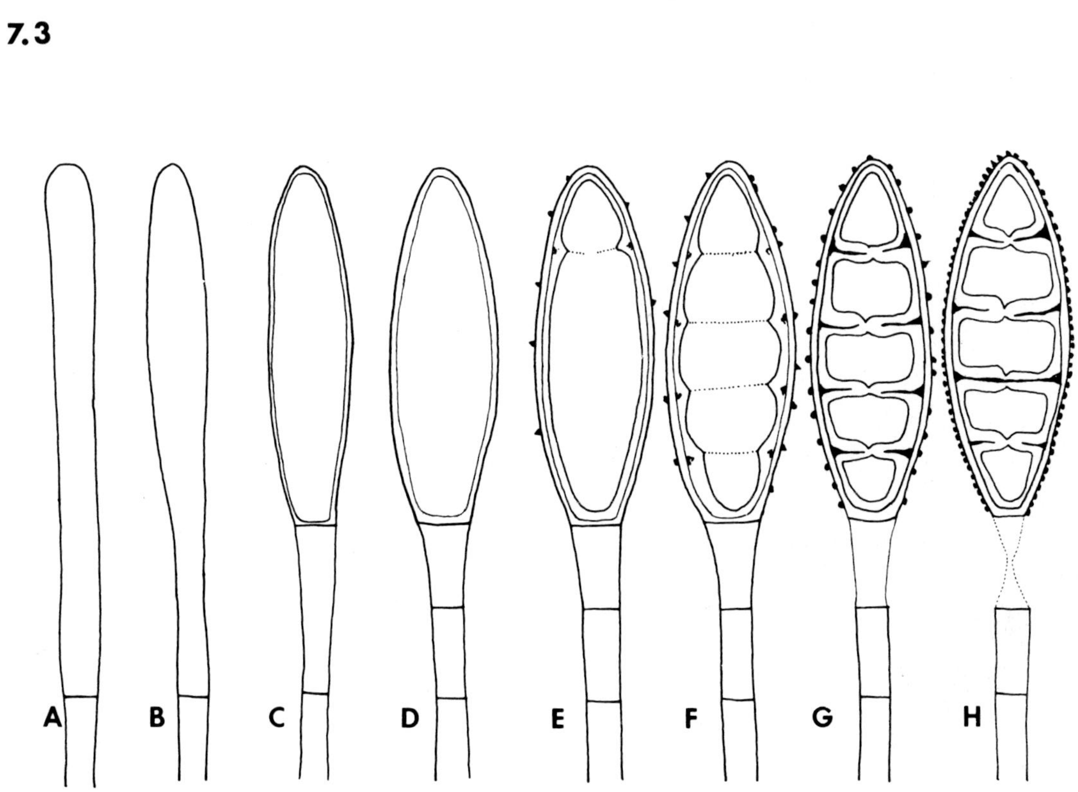
7.3
A
B
C
D
E
F
G
H

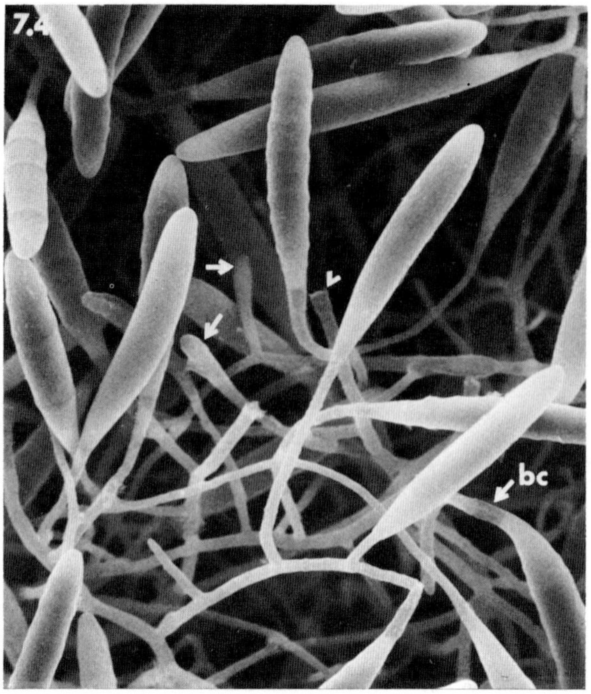

Figs. 7.4–7.6 *Trichophyton* state of *Arthroderma uncinatum.* Macroconidia in 7.4 show differentiated basal cell (bc) which later undergoes autolysis (7.5). Microconidia (arrows in 7.4) also show degenerated basal cells at time of secession (7.6). Arrowhead in 7.4 locates abstriction scar at apex of fertile hypha. Arrow in 7.5 indicates septum at fertile, hyphal apex. ×700, 5,700 and 6,800, respectively.

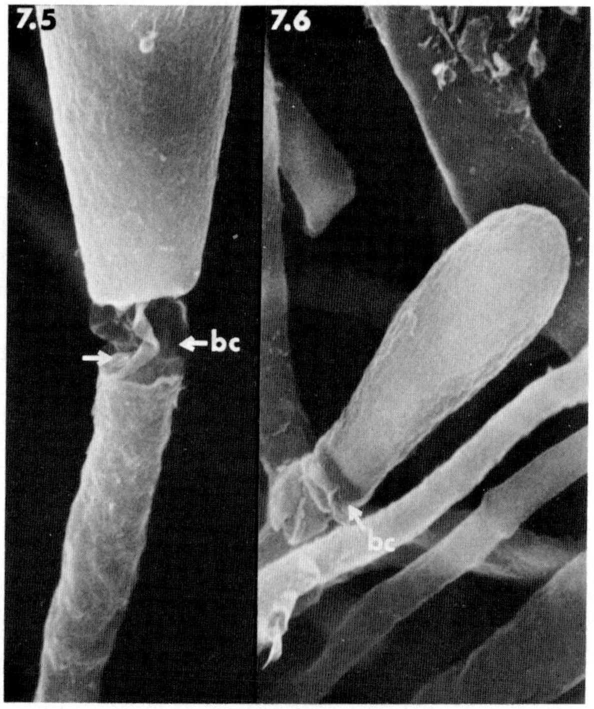

readily from their parent hyphae but are released only after lysis of the adjacent cell(s), as in rhexolytic secession in *M. gypseum* (Fig. 7.2A–C). Additional comparative studies of chlamydospores and holothallic conidia are necessary.

Thallic-arthric conidia

In the conversion of pre-existing hyphal elements into deciduous conidia, the fertile hyphae cease apical growth, become septate, and eventually disarticulate. As defined earlier (Chapter 1), we recognize two kinds of thallic-arthric ontogeny, holoarthric and enteroarthric, which are distinguished primarily on the basis of wall differentiation.

Holoarthric conidia

Geotrichum candidum Link ex Pers. epitomizes holoarthric development. Fertile and vegetative hyphae are difficult to distinguish in early stages of conidium ontogeny. However, when the fungus is grown in our thin culture chamber, the following developmental characteristics are recognized:

1. septation is at first acropetal (Fig. 7.7A–C), later random (Fig. 7.7D–F),
2. hyphal tip growth ceases before septa are formed in the apical region (Fig. 7.7A–C),
3. continued septation yields progressively shorter segments (Fig. 7.7C–F), and
4. septation ceases and the conidia enlarge slightly, their end walls visibly convex (Fig. 7.7E, F).

Conidial dimensions are variable in culture (Fig. 7.8). The width of conidia, at least in early development, is that of the hypha from which they are formed (Cole, 1975a). What determines conidial length is as yet unknown.

The acropetal and then random septation, rather than the originally postulated basipetal sequence (S. J. Hughes, 1953), must be associated with critical karyological events which assure that each disarticulating segment of the fertile hypha receives at least one nucleus. Further discussion of this aspect of arthric development is presented later in the chapter. The initial acropetal order of septation is logical when one considers that secondary intussusception associated with septal formation occurs concomitantly with hyphal tip growth and is more advanced progressively further back from the tip (Hunsley and Burnett, 1970; Bartnicki-Garcia, 1973). Cole and Kendrick (1969c) have demonstrated that a rapid succession of septation occurs in *G. candidum* shortly after growth ceases at the hyphal tip. Park and Robinson (1969) proposed that specific physiological factors are associated with conversion of hyphae from a vegetative to a sporulation phase and initiation of septal formation in this species. They suggested that conidiogenesis in *G. candidum* may be induced by a diffusible fungal metabolite from an older colony of the same species, or of another species. Conidium ontogeny is accompanied by metabolic change in the fertile hypha which leads to 'cessation of apical growth' and 'change in form and disposition of the mitochondria' contained within its cytoplasm.

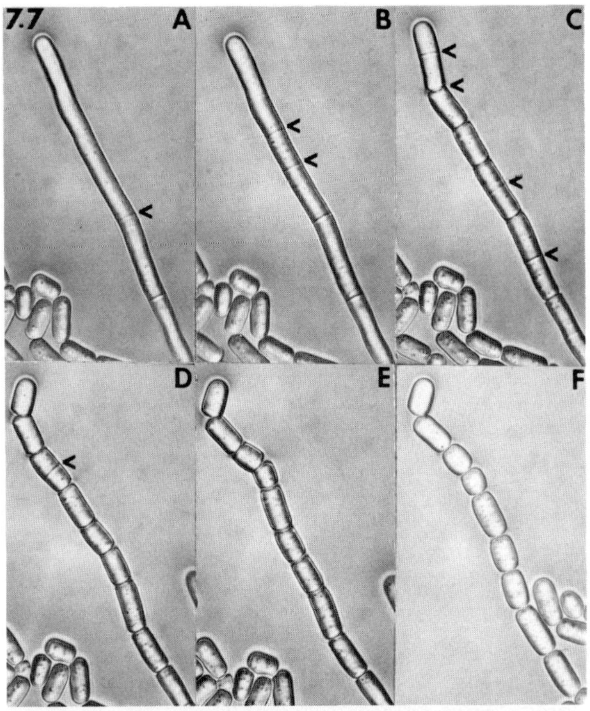

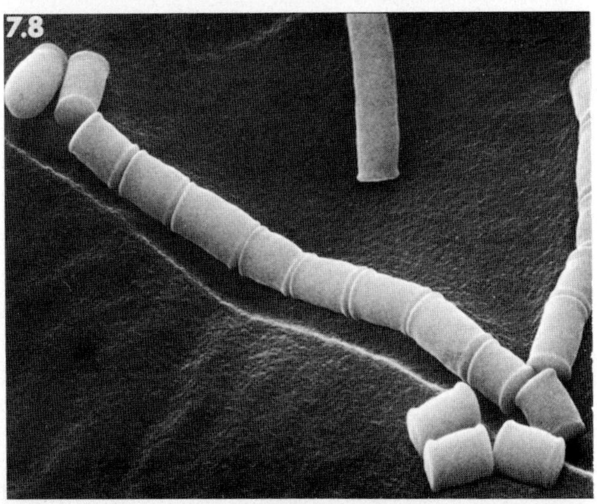

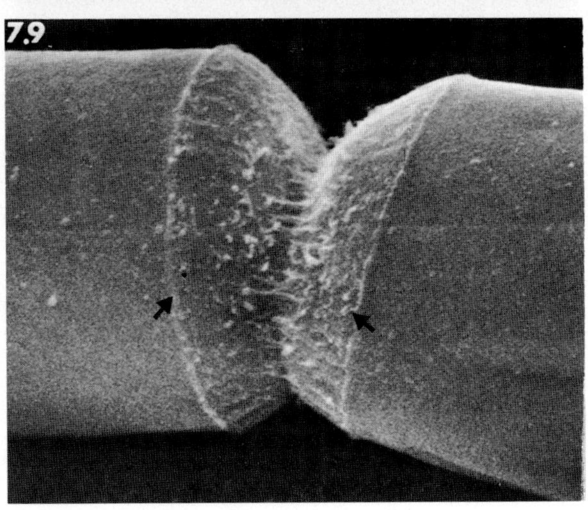

Fig. 7.7A–F *Geotrichum candidum.* Time-lapse sequence of holoarthric development. Arrowheads in (A–D) locate newly formed septa, demonstrating the initial acropetal and later random order of septation. T = 0, 15, 60, 105, 170, 1,140 minutes, respectively. ×500.

Figs. 7.8, 7.9 *Geotrichum candidum.* Fragmented fertile hypha (7.8) showing separation of septal wall layers of adjacent holoarthric conidia (7.9). Arrows in 7.9 indicate ruptured outer wall layer of adjacent conidia. ×2,100 and 10,800.

Renewal of vegetative hyphal growth may occur by providing the fungus with sufficient exogenous carbon and nitrogen sources (Trinci and Collinge, 1974b).

Associated with the swelling of differentiated conidia (Fig. 7.7E, F) is the centripetal separation of septa (schizolysis) and concomitant rupture of the outer hyphal wall layer, leaving a circumscissile scar at the ends of each conidium (arrows in Figs. 7.9, 7.10). Complete disarticulation is retarded by the presence of thin strands of wall material at the central convexity of septa (Figs. 7.9, 7.10) which maintains a weak union between adjacent end walls. Mature septa are perforated by numerous *micropores* (Wilsenach and Kessel, 1965c; Bracker, 1967), or *plasmadesmal canals* (Hashimoto *et al.*, 1973) which may permit translocation of ribosomes, ER and some cytoplasmic deposits (e.g., glycogen), but not nuclei, mitochondria, and other larger organelles between adjacent cells (Fig. 7.11). As conidia mature, the micropores are occluded with electron-opaque material (arrowhead in Fig. 7.10). Formation of multiperforate septa is not understood, though dissolution of pores or spoke-like growth of the hyphal crosswall have been suggested (Bracker, 1967; Hashimoto *et al.*, 1973).

The lipid droplets contained by the conidium in Fig. 7.10 probably serve as an energy reserve during cell differentiation. Conidia obtained from older cultures are more vacuolate. Vacuoles may be connected with cell dormancy since they are associated with reduced germination (Park and Robinson, 1970). Freeze-fractures of the cell membrane of mature conidia show many plasmalemma invaginations (Fig. 7.12). Cross fractures of the invaginations show that they are simply folds of the plasmalemma and are not connected with cytoplasmic organelles. These infoldings of the cell membrane are typical of conidia of many deuteromycetous fungi examined by the senior author (unpublished data), and yeasts reported by other investigators (Moor and Mühlethaler, 1963). Hess (1973) determined that plasmalemma invaginations are commonest in dormant conidia and showed that during germination, the cell membrane surrounding the germ tube is smooth except for faint scars representing previously invaginated regions. The invaginations may simply increase the surface area of the cell membrane in preparation for the initial, rapid membrane expansion during germ tube emergence before the synthesis of new cell membrane becomes synchronized with tube growth.

Trichophyton mentagrophytes (Robin) Blanchard, which typically forms holothallic macro- and microconidia (Chapter 1, Fig. 1.22), has been reported to produce chains

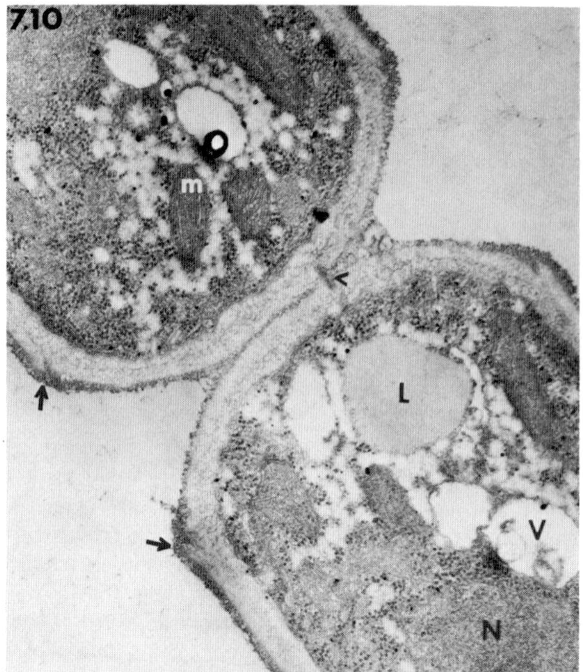

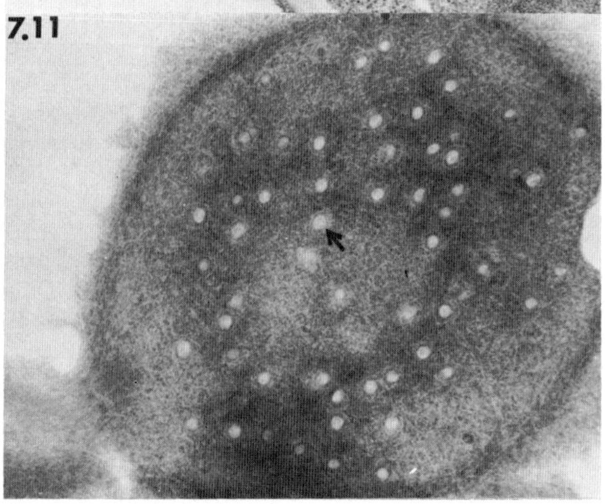

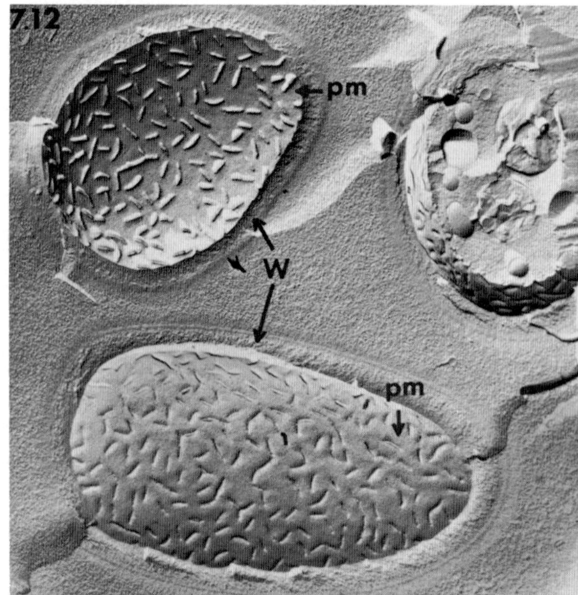

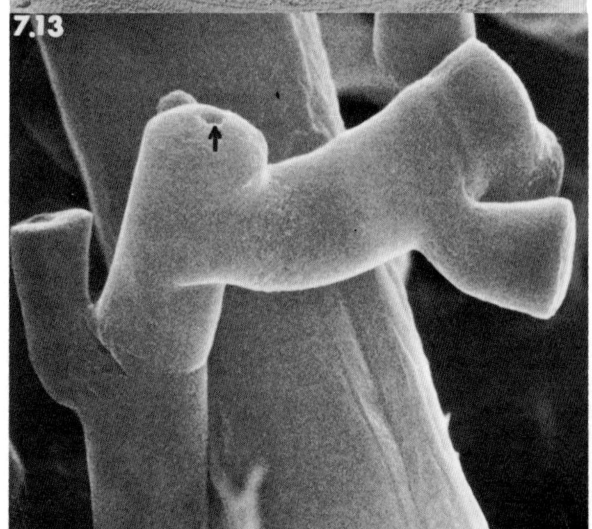

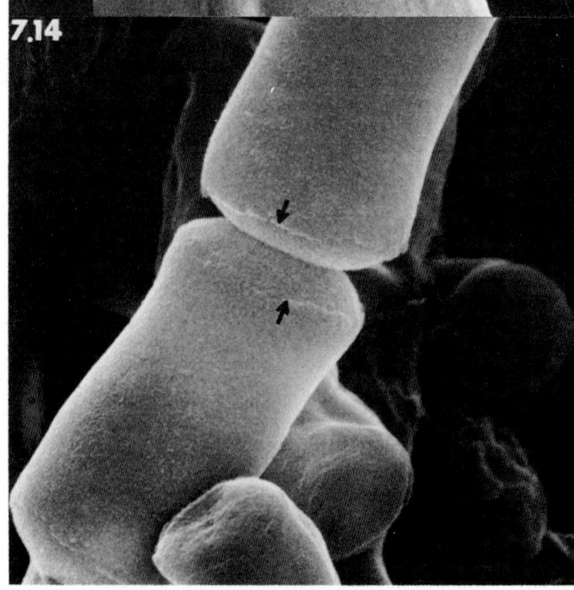

Figs. 7.10–7.12 *Geotrichum candidum.* Thin-sections reveal differentiation of wall layers of adjacent, seceding conidia (7.10) and multiperforate nature of conidial septum (7.11). Freeze-fractured conidia (7.12) show numerous plasmalemma invaginations. Upper cell shows concave fracture, lower is convex fracture. Arrows in 7.10 locate ruptured outer wall of fertile hypha. Arrowhead in 7.10 indicates occluded micropore. L, lipid droplet; m, mitochondrion; N, nucleus; pm, plasma membrane; W, conidial wall. ×28,600, 33,000, 10,800 (7.11, from Hashimoto *et al.*, 1973).

7.15

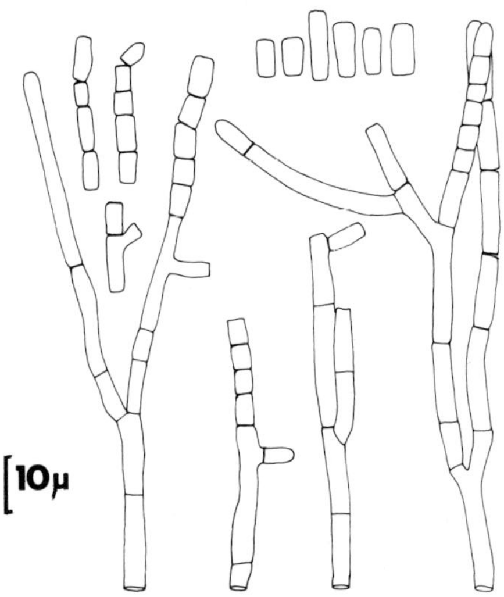

Figs. 7.13–7.15 Anamorph of *Peniophora gigantea.* Sympodially proliferated conidiophore (7.13, 7.15), the fertile hyphae of which have formed holoarthric conidia (7.14). Arrow in 7.13 indicates centrally located, septal pore. Arrows in 7.14 indicate ruptured outer wall of fertile hypha. 7.13, 7.14; ×6,700 and 8,600, respectively.

Fig. 7.16 *Sympodiella acicola.* Sympodially proliferated conidiophore and holoarthric conidia.

7.16

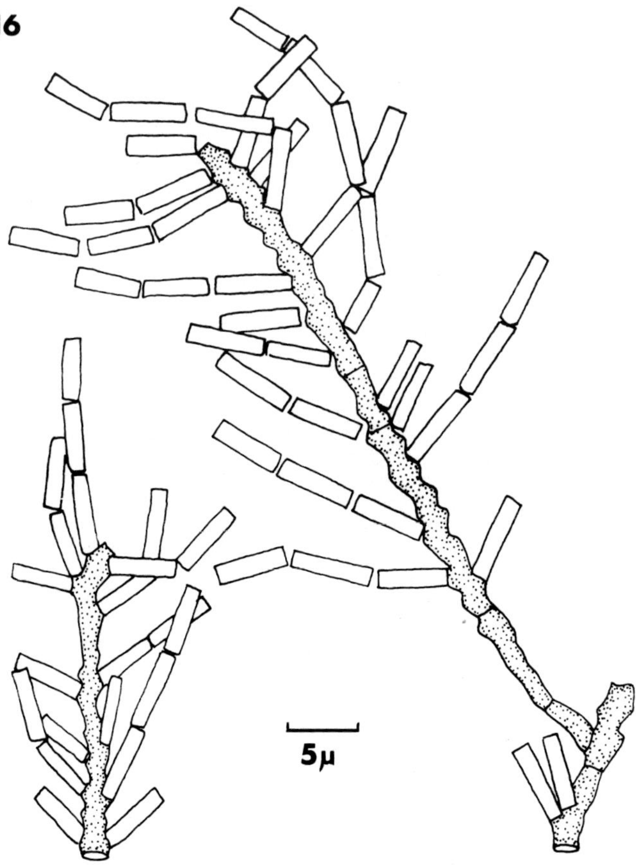

of holoarthric conidia when incubated in an atmosphere of 8% CO_2 (Bibel *et al.*, 1977). Wall differentiation is essentially the same as outlined above. It would be interesting to test the morphogenetic plasticity of other dermatophytic species under similar growth conditions.

Development of the anamorph of the Basidiomycete *Peniophora gigantea* (Fr.) Massee (Figs. 7.13–7.15) resembles that of *Geotrichum candidum* under the light microscope. Both species are holoarthric (Fig. 7.14), but the Basidiomycete has a centrally perforated septum (arrow in Fig. 7.13). The fertile hyphae demonstrate typical dolipore septa. The mechanism of schizolysis during holothallic conidium ontogeny in *Antromycopsis broussonetiae* Pat. & Trabut, anamorph of the Basidiomycete *Pleurotus cystidiosus* Miller (Pollack and Miller, 1976), has been examined by Moore (1977) who has presented an interpretation of ultrastructural changes associated with dolipore disjunction.

The fertile hyphae of *Peniophora gigantea* arise sympodially from the apex of the conidiophore (Figs. 7.13, 7.15). Similar development occurs in species of *Sympodiella* Kendrick (Fig. 7.16). Conidiogenesis in the type species, *S. acicola* is holoarthric rather than blastic in acropetal chains as originally suggested (Kendrick, 1958).

Staheliella nodosa van Emden demonstrates yet another variation of holoarthric development (Figs. 7.17–7.22). A cluster of short, branched hyphal initials appear at the apex of the conidiophore (Figs. 7.17–7.19). The initials continue to branch, producing a dense cluster of hyphae (Figs. 7.18, 7.19), which all elongate and cease growth synchronously. In Fig. 7.20, the fertile hyphae have become septate and the delimited cells differentiate into holoarthric conidia, which later secede by disarticulation of the chains (Figs. 7.21, 7.22). The crosswalls show no sign of a central pore (Fig. 7.21), suggesting multiperforate septa. Thin-sections are required to examine this possibility.

Conidium ontogeny in *Oidiodendron truncatum* (Robak) Barron involves a regular basipetal maturation of cells which form by septation and fragmentation of fertile hyphae. The time-lapse sequence in Fig. 7.23A–D illustrates that conidiogenesis involves the separation of adjacent cells while they remain linked in chains by narrow, cylindrical hyaline *connectives*. One of these connectives shows the beginning of circumscissile separation (arrow in Fig. 7.23D). The end walls of mature conidia are thickened and appear truncate. Species of *Oidiodendron* commonly form long, delicate chains of conidia with characteristic intercellular connectives (Fig. 7.24). The SEM has revealed an early stage of fragmentation of the fertile hypha of *O. truncatum* (Fig. 7.25), and a later stage (Fig. 7.26) in which the connective and zone of circumscissile separation (arrow) have differentiated between two adjacent conidia. Barron (1962) suggested that the connectives might be initially gelatinous deposits which later dry up. However, using the freeze-fracture technique, Cole (1973c) demonstrated that the connectives are actually part of the conidial wall (Fig. 7.27) and develop as a result of polar growth of the cells in the chain. A diagrammatic interpretation of wall differentiation during conidiogenesis in *O. truncatum* is presented in Fig. 7.28A, B. Elongation of the

7.17

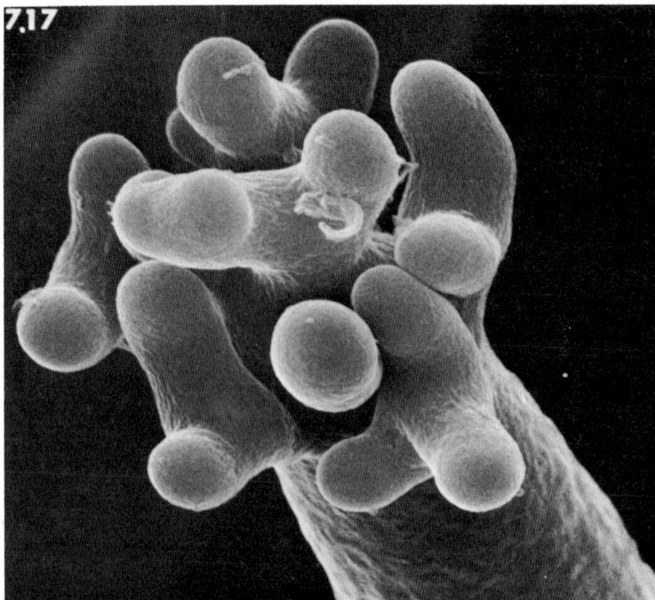

7.18

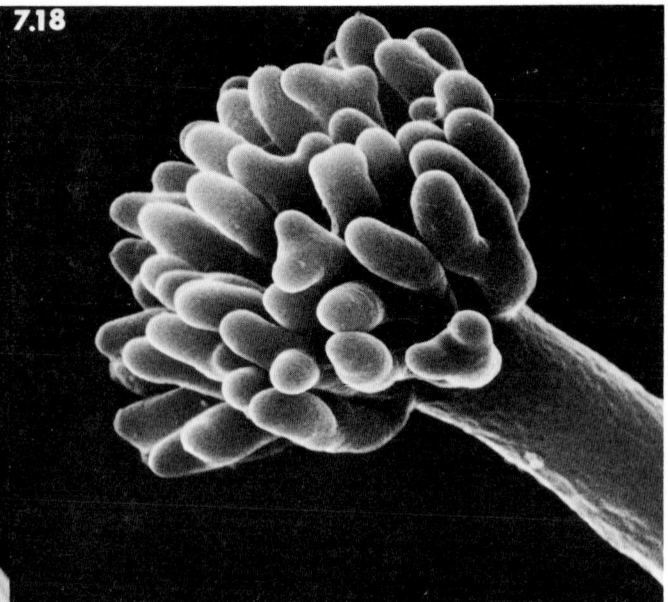

7.19

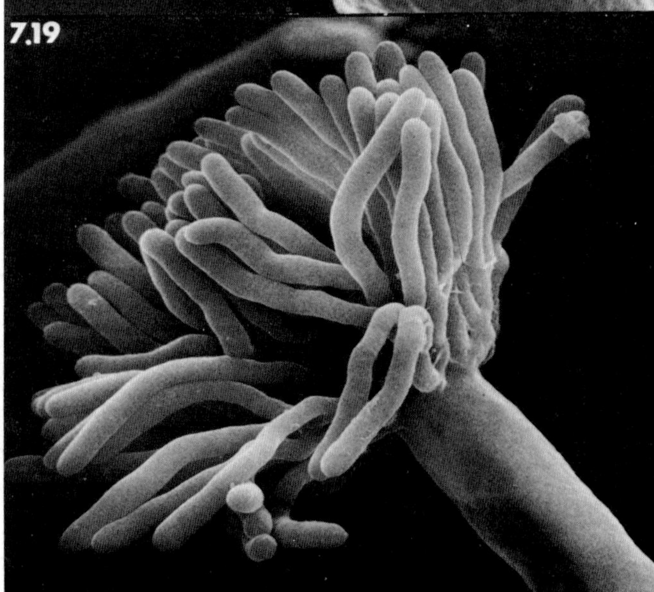

7.20

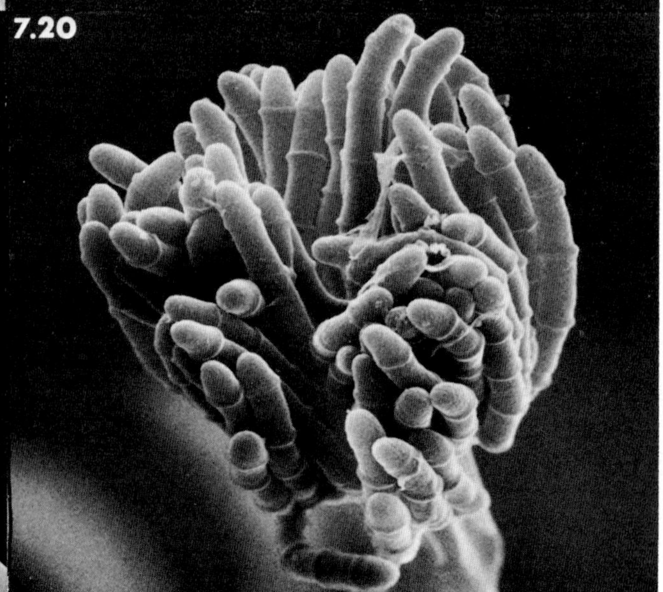

7.21

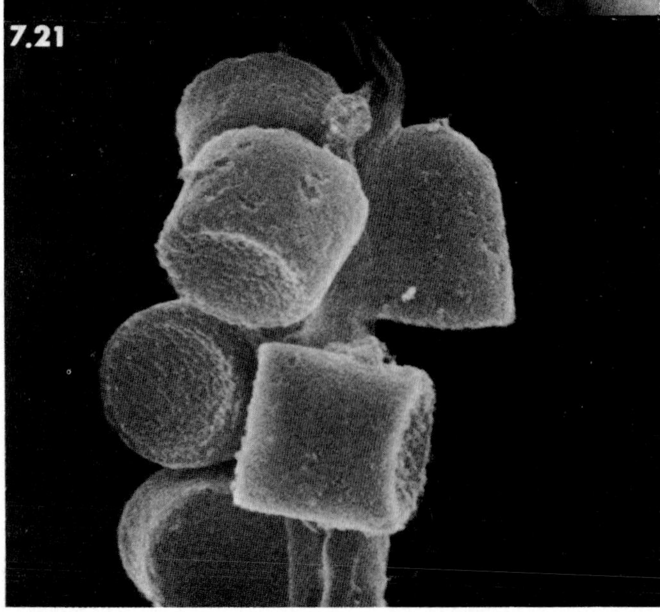

7.22

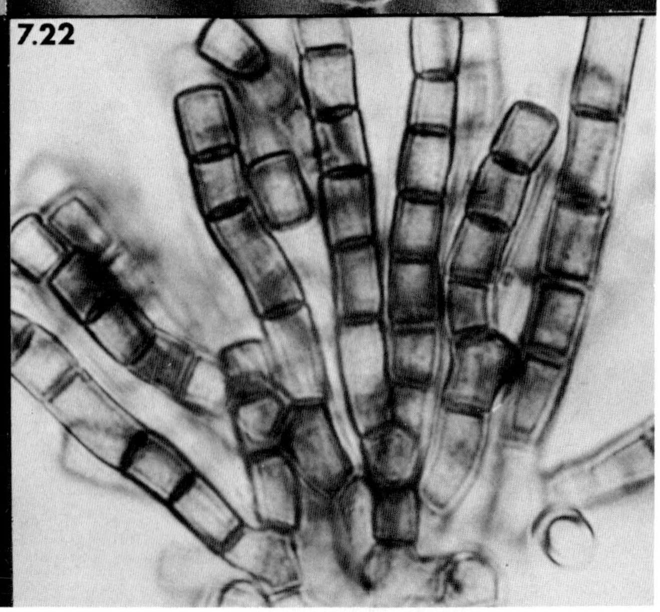

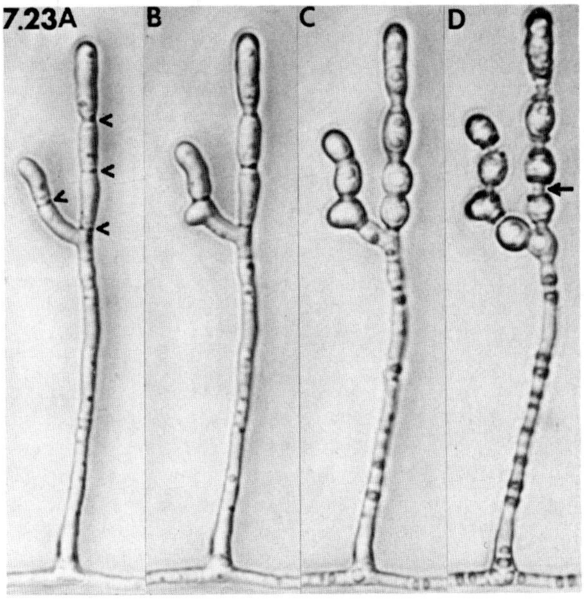

Fig. 7.23A—D *Oidiodendron truncatum.* Time-lapse sequence of holoarthric development. Arrowheads in (A) indicate septa. Arrow in (D) locates circumscissile separation of connective. T = 0, 4, 22, 69½ hours, respectively. ×1,500.

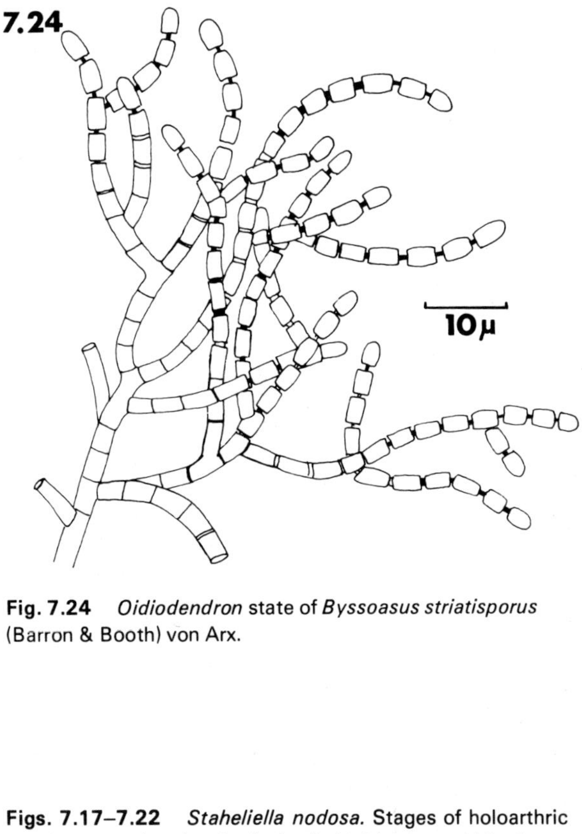

Fig. 7.24 *Oidiodendron* state of *Byssoasus striatisporus* (Barron & Booth) von Arx.

Figs. 7.17–7.22 *Staheliella nodosa.* Stages of holoarthric development showing fertile hyphal initials at conidiophore apex (7.17), branched, elongated and septate fertile hyphae (7.18–7.20) and holoarthric conidia (7.21) after disarticulation of the chains of cells (7.22). ×5,200, 3,100, 1,400, 1,800, 7,100 and 800, respectively.

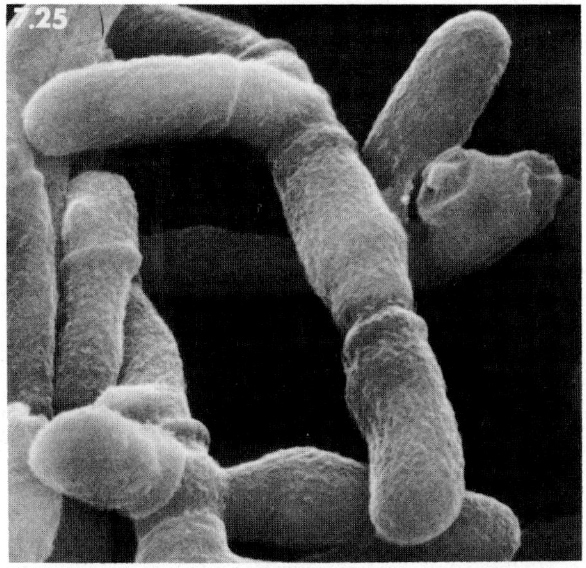

Fig. 7.25 *Oidiodendron truncatum.* Early stage of holoarthric ontogeny. Prior to differentiation of connectives. ×8,700.

connectives is actually due to intussusception of new wall material into the existing septum (Fig. 7.28A) causing it to thicken (Fig. 7.28B). A lytic zone differentiates across the middle of the thickened septum (arrows in Fig. 7.28B) which eventually results in secession of adjacent conidia. Although conidiogenesis in *O. truncatum* is holoarthric, the bipolar growth of the maturing conidia distinguishes it from development in *Geotrichum*.

The prominent conidiophores of *Botryomonilia scheeleae* Goos & Pirozynski (Fig. 7.24) (up to 7 mm high) branch sympodially and produce a penicillate head of fertile hyphae which later become septate. The cells then round off in basipetal succession and differentiate into conidia. Goos and Pirozynski (1975) recognized that during maturation, the conidia are 'held together in branched chains only by a thin strand of protoplasm connecting the septal pores' and the terminal cells are 'more or less hemispherical'. An obvious developmental similarity exists between species of *Botryomonilia* and *Oidiodendron.*

Conidium ontogeny in *Amblyosporium spongiosum* (Pers.) Hughes (Figs. 7.30–7.32) demonstrates yet another variation of holoarthric development. Basipetal chains of conidia form with alternate cells differentiating as *disjunctor cells* between adjacent conidia (Pirozynski, 1969). The clusters of fertile hyphae are supported by branches of the conidiophore (Figs. 7.30, 7.32). Early in conidiogenesis, the disjunctors are cylindrical and apparently turgid (arrows in Fig. 7.31), but as the conidia mature, they undergo autolysis and collapse. This eventually leads to disarticulation of the chain. Ontogeny of disjunctors is not entirely understood and requires thin-section investigation.

Conidiogenesis in species of *Oidiodendron, Botryomonilia* and *Amblyosporium* demonstrate the coexistence of *de novo* growth and conversion during conidiogenesis which is

7.26

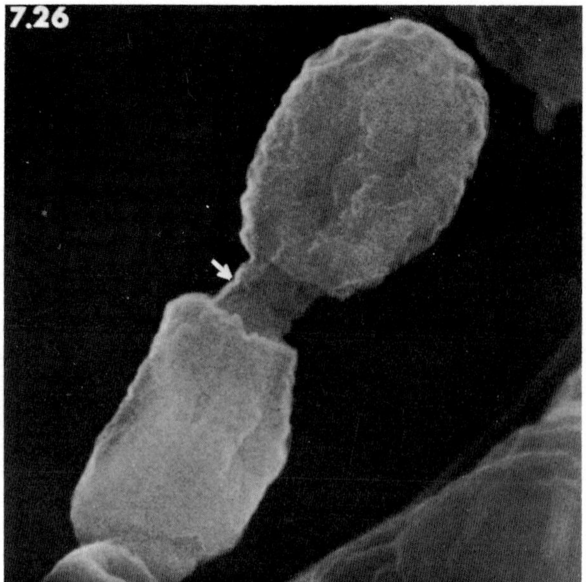

7.27

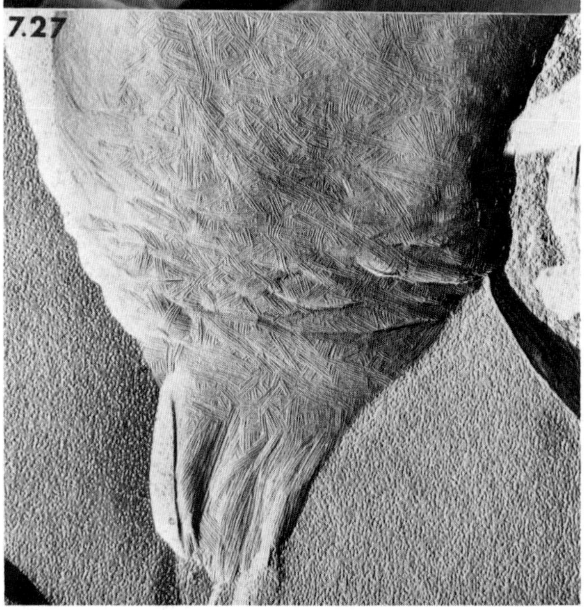

7.28

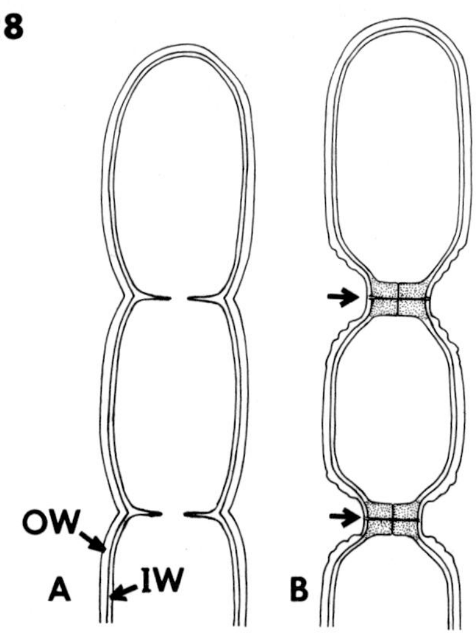

Figs. 7.26–7.27 *Oidiodendron truncatum.* Apical and subapical conidia of chain (7.26) showing connective and circumscissile separation (arrow). Freeze-fractured conidium (7.27) reveals longitudinally orientated rodlet fascicles on surface of connectives. These fascicles are continuous with those comprising lateral wall. ×13,200 and 33,000 respectively.

Fig. 7.28 Diagrammatic interpretation of holoarthric development and bipolar growth of conidia in *Oidiodendron truncatum.* Arrows indicate circumscissile separation of connectives where schizolysis occurs. IW, inner wall; OW, outer wall.

problematic in the formulation of definitions of blastic and thallic development. However, termination of polar growth of the fertile hyphae followed by conversion of hyphal segments into conidia, or disjunctor cells, are events in the development of these fungi which have convinced us to include them with other species demonstrating holoarthric ontogeny. The new growth which follows these initial stages of conidiogenesis represents a variation from the mechanism demonstrated by *Geotrichum.*

Conidiogenesis in the yeasts, *Trichosporon beigelii* (Küchenmeister & Rabenh.) Vuill. and *Trichosporonoides oedocephalis* Haskins & Spencer, is morphogenetically plastic. In *Trichosporon beigelii* conidia are holoarthric (Figs. 7.33, 7.34). The conidial walls are rough and fibrous, similar to species of *Cryptococcus* Kützing (Lurie *et al.*, 1971; Al-Doory, 1971). Later, these same conidia swell (arrows in Fig. 7.34) and form chains of secondary holoblastic conidia (Fig. 7.35). The latter may also arise directly from cylindrical, holoarthric conidia (Fig. 7.37A). Additional variability is shown in Fig. 7.36, which indicates that apical growth of the fertile hypha may occur (arrow) concomitantly with holoarthric and holoblastic development. Similar observations of conidiogenesis in *Trichosporon beigelii* were reported by King and Jong (1976). Haskins (1975) illustrated developmental plasticity in *Trichosporonoides oedocephalis* (Fig. 7.37B–E), a yeast with both basidiomycetous and ascomycetous affinities. Chains of holoarthric conidia and holoblastic conidia are produced (Fig. 7.37B–D). The latter may form directly from septate hyphae or from seceded arthric conidia (Fig. 7.37D), may arise synchronously from multiple growing points of an ampulla at the apex of a conidiogenous cell (Fig. 7.37E), or by simple budding. Species of *Trichosporon* and *Trichosporonoides* are typical of many imperfect yeasts in their ability to alternate between mycelial and budding states under certain environmental conditions (Garrison *et al.*, 1970; Lane and Garrison 1970; Garrison *et al.*, 1971; Oujedsky *et al.*, 1973). This dimorphic character is related to the variation between holoarthric and holoblastic modes of conidiogenesis.

A few monilioid Hyphomycetes have also been reported to demonstrate comparable morphogenetic plasticity (Carmichael, 1971). In *Moniliella acetoabutens* Stolk & Dakin (Fig. 7.37F) acropetal chains of blastic conidia arise directly from disarticulated fertile hyphae. The anamorphs of several

7.29

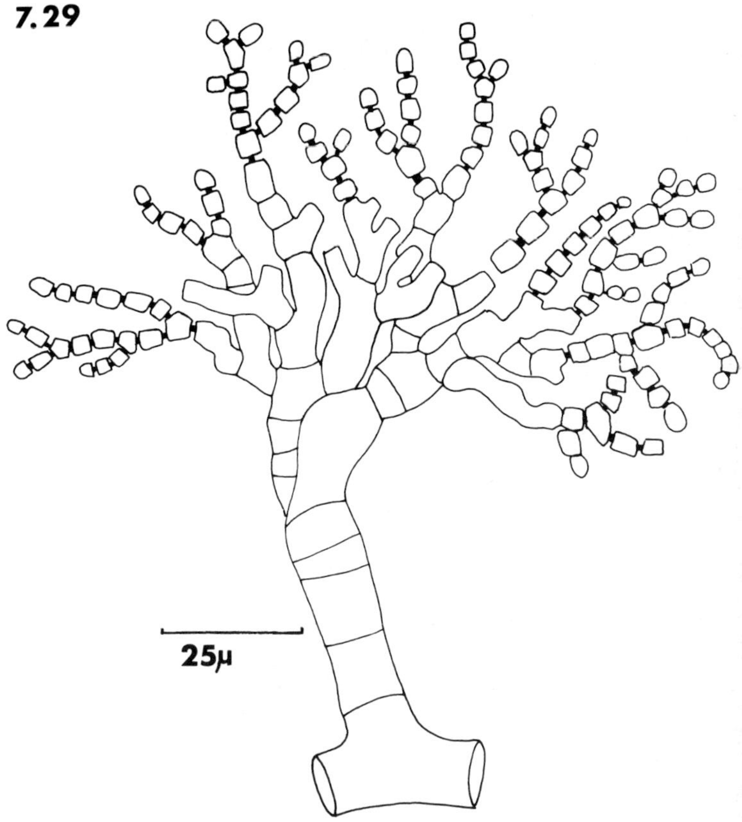

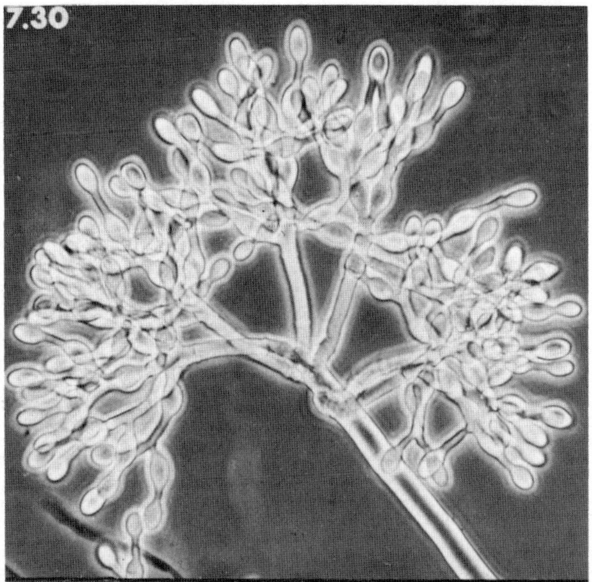

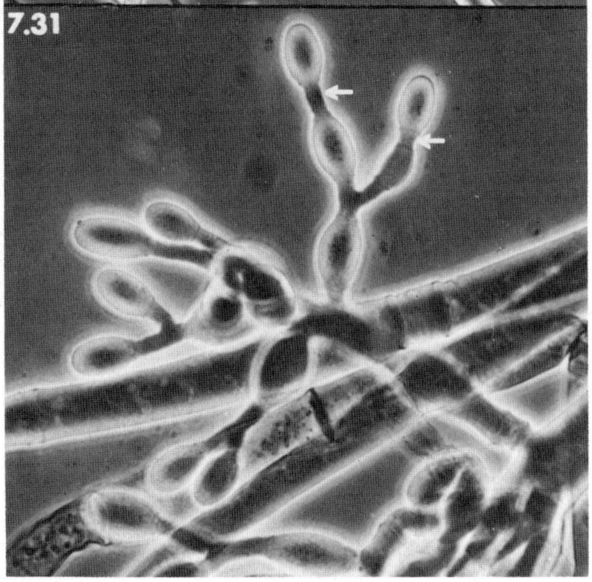

7.32

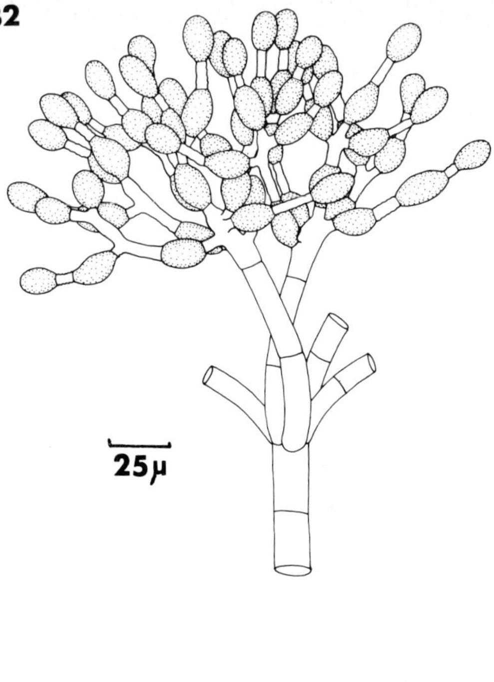

Fig. 7.29 *Botryomonilla scheeleae* Goos and Pirozynski.

Figs. 7.30–7.32 *Amblyosporium spongiosum.* Chains of holoarthric conidia produced by fertile branches of conidiophore (7.30, 7.32) and disjunctor cells (arrows in 7.31) between adjacent conidia. 7.30, 7.31; ×512 and 1,024, respectively.

genera of the Gymnoascaceae (e.g., *Arthroderma* Currey, Fig. 7.37G; *Gymnoascus* Baran.) often referred to *Chrysosporium* Corda, demonstrate both fragmentation of fertile hyphae and blastic development of microconidia. Thallic-arthric and phialidic development occur together in *Chalara elegans* Nag Raj & Kendrick (Fig. 7.37H). The chains of holoarthric conidia, in early development, resemble large multiseptate conidia (M. B. Ellis, 1971b) but later fragment. Some authors have referred to these as chlamydospores (Clough and Patrick, 1972; Nag Raj and Kendrick, 1975). In addition, prominent, cylindrical phialides up to 100 μm long are simultaneously produced (arrows in Fig. 7.37H).

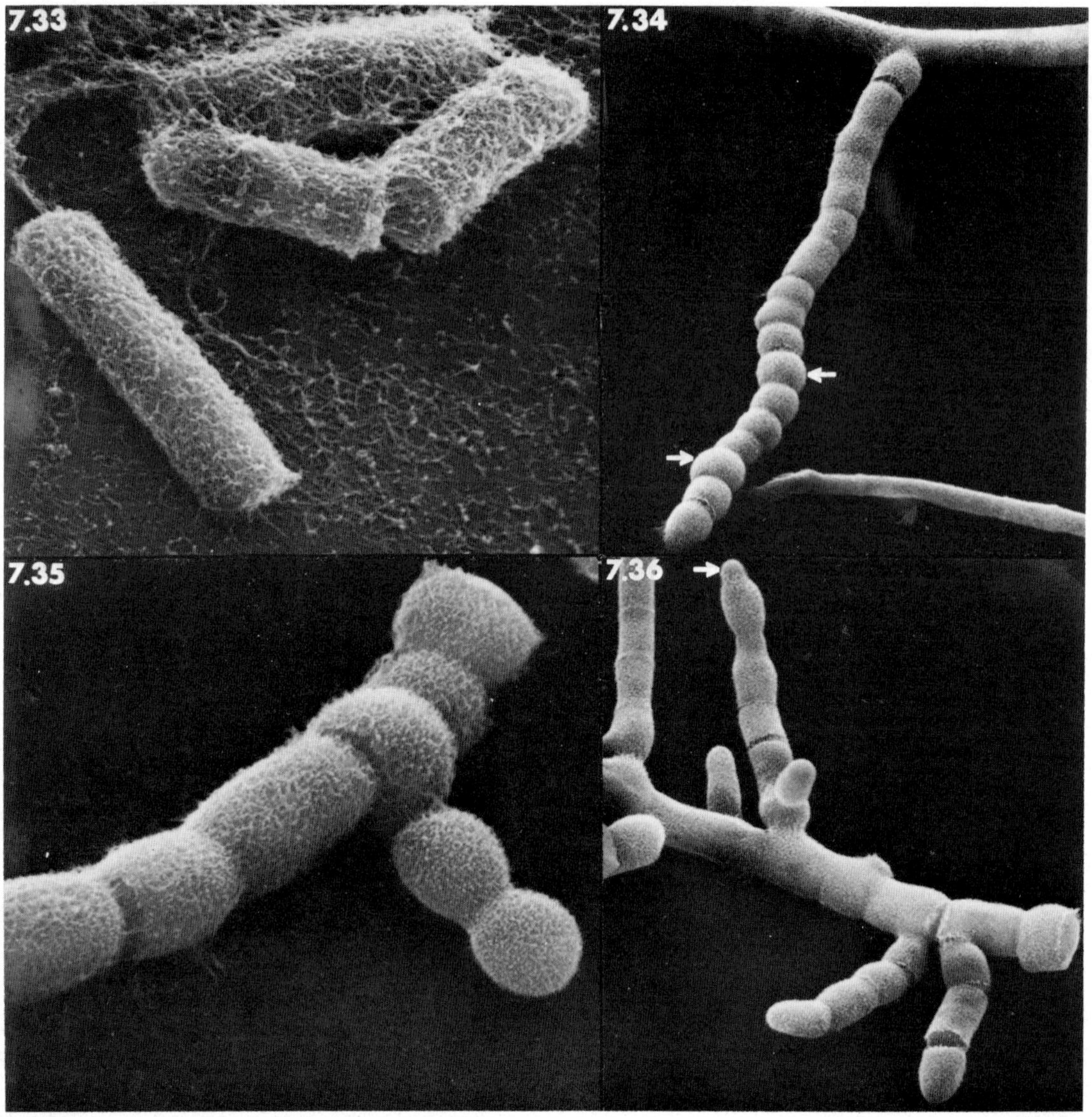

Figs. 7.33–7.36 *Trichosporon beigelii.* Stages of holoarthric (7.33, 7.34) and holoblastic development (7.35, 7.36). Arrows in 7.34 indicate holoarthric conidia which have swollen and later produced acropetal chains of holoblastic conidia (7.35). Arrow in 7.36 locates apical initial of acropetal chain. ×5,800, 2,000, 5,000 and 1,800, respectively.

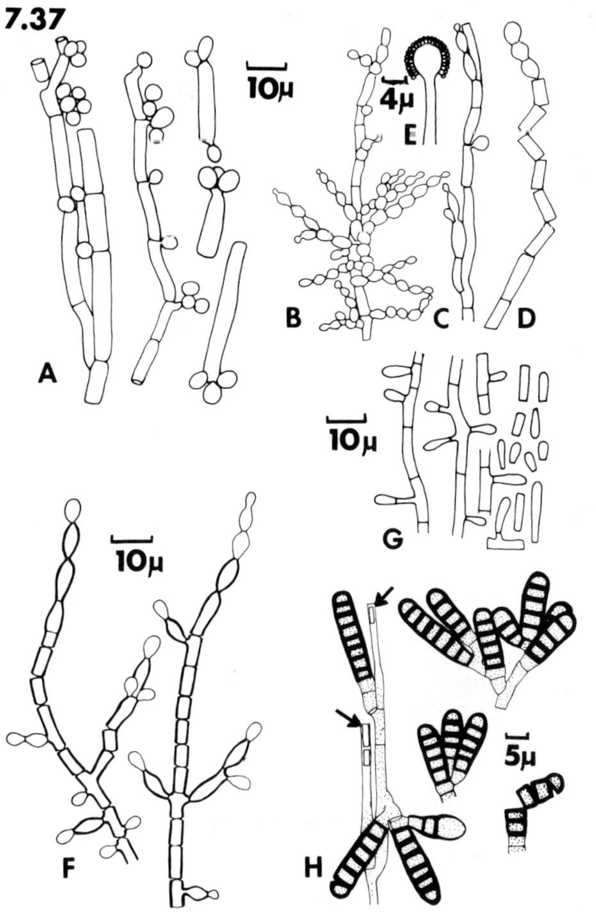

Fig. 7.37A *Trichosporon beigelii.*

Fig. 7.37B–E *Trichosporonoides oedocephalis.* Formation of acropetal chains of holoblastic conidia (B, C), holoarthric conidia (D) and polyblastic ampullae (E).

Fig. 7.37F *Moniliella acetoabutens.*

Fig. 7.37G *Chrysosporium* state of *Arthroderma curreyi* Berk.

Fig. 7.37H *Chalara elegans.* Arrows indicate phialides and endogenous conidia.

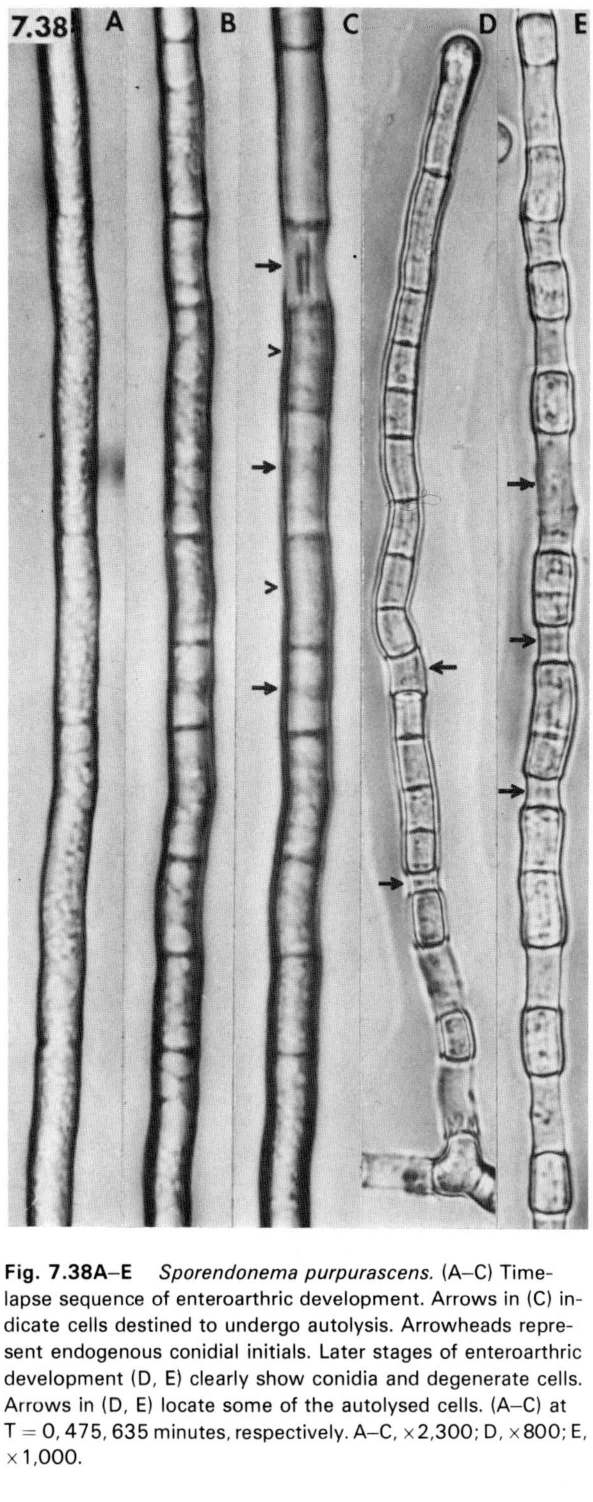

Fig. 7.38A–E *Sporendonema purpurascens.* (A–C) Time-lapse sequence of enteroarthric development. Arrows in (C) indicate cells destined to undergo autolysis. Arrowheads represent endogenous conidial initials. Later stages of enteroarthric development (D, E) clearly show conidia and degenerate cells. Arrows in (D, E) locate some of the autolysed cells. (A–C) at T = 0, 475, 635 minutes, respectively. A–C, ×2,300; D, ×800; E, ×1,000.

Enteroarthric conidia

Sporendonema purpurascens (Bon.) Mason & Hughes exemplifies enteroarthric development (Figs. 7.38–7.45). As in holoarthric ontogeny, conidiogenesis is initiated by cessation of hyphal tip growth and formation of septa, at first acropetally, but later randomly along the length of the hypha (Fig. 7.38A–C). However, a distinguishing characteristic of enteroarthric development is the conversion of certain hyphal compartments into conidia while other cells undergo autolysis. The cells indicated by arrows in Fig. 7.38C are less dense and have begun autolysis. The arrowheads locate cells which have begun to differentiate into conidia. Later developmental stages are shown in Fig. 7.38D, E (arrows indicate degenerate cells). In Fig. 7.38E it is apparent that conidia have differentiated within the wall of the fertile hypha. The wall of the degenerating cells eventually breaks down and releases the endogenous propagules. Ultrastructural features of this mode of conidiogenesis are revealed first by the SEM. In Fig. 7.39 several conidia are visible within the wall of a

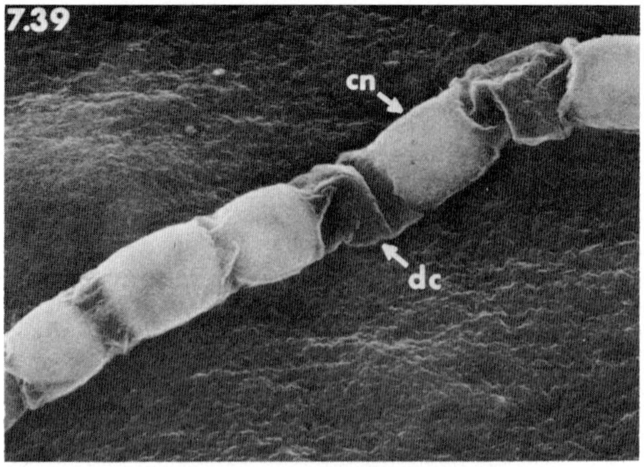

Fig. 7.39 *Sporendonema purpurascens.* Fertile hypha showing endogenous conidia (cn) and collapsed, degenerate cells (dc). The autolysed, thin-walled cells permit 1° electrons of the SEM to pass through them reducing 2° electron emission, which results in a dark image. The endogenous conidia are visible because the 1° electrons readily pass through the thin hyphal wall but generate 2° electrons as they strike the thickened wall of the conidium. The material was examined using an accelerating voltage of 30 KV. ×2,500.

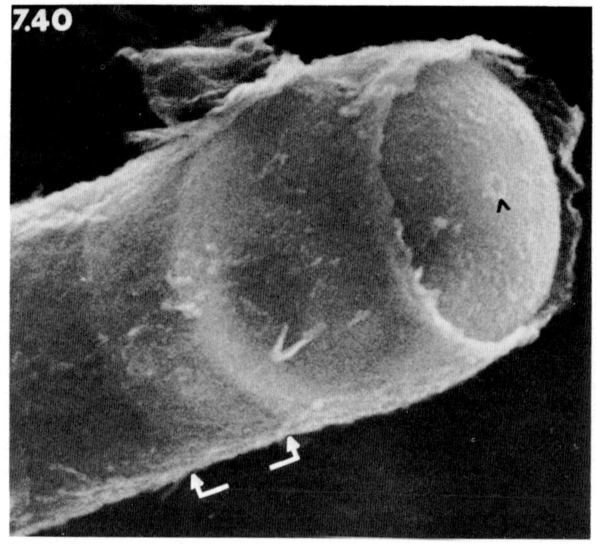

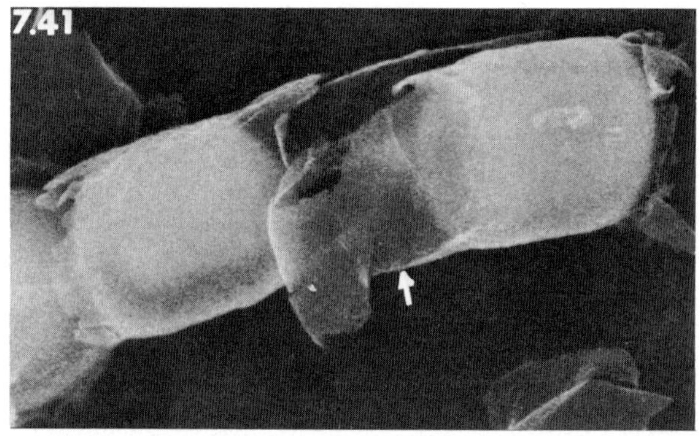

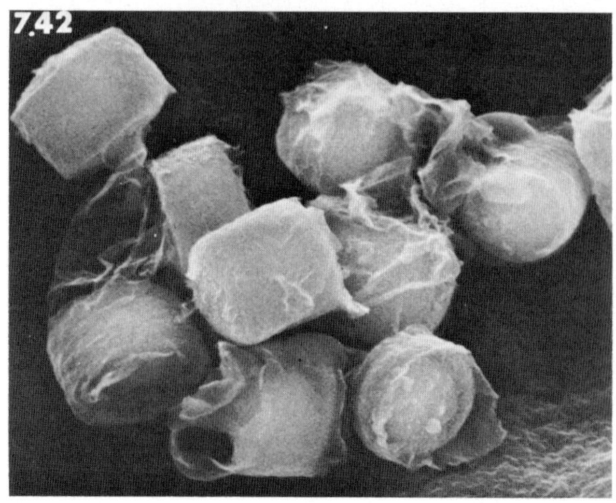

Figs. 7.40–7.42 *Sporendonema purpurascens.* Stages of enteroarthric development. Outline of endogenous conidial initial in 7.40 reveals septal pore (arrowhead) and outer wall of fertile hypha. Bracket represents cell adjacent to conidial initial. Late developmental stages demonstrate autolysed cell (arrow in 7.41) and seceded conidia (7.42). ×6,600, 4,100 and 3,400, respectively.

fertile hypha while adjacent regions are empty and show collapse. These latter compartments have degenerated and do not have a thickened inner wall layer. An early stage of conidiogenesis (Fig. 7.40) reveals the outline of a conidium within the torn outer wall of the fertile hypha. The septal wall and pore (arrowhead) are exposed at one end of the conidium. Adjacent to the conidium, the outline of another cell is distinguished (bracketed by arrows) which is suggested to represent a thin-walled compartment of the fertile hypha whose cytoplasm has begun to autolyse. A septum separates the base of this cell from the rest of the hypha which has not differentiated further into conidia. In Fig. 7.41 the hyphal wall between two conidia has broken down (arrow) resulting in disarticulation of the chain of cells. Short, cylindrical conidia, each enveloped by the folded and partly torn wall of the fertile hypha, remain (Fig. 7.42).

Thin-sections of *S. purpurascens* have revealed aspects of wall differentiation and autolysis during enteroarthric development (Fig. 7.43). An endogenously formed conidium and adjacent degenerate cells are shown, enclosed by a thin, electron-opaque layer (OW) in various degrees of separation from the underlying wall. Between this outer layer and the thicker inner wall (IW) is an amorphous layer (A). The cytoplasm of the degenerate cells demonstrates an apparent breakdown of plasmalemma suggested by the presence of short fragments of membrane adjacent to the inner wall (arrowheads in Fig. 7.43). Swollen mitochondria (m) and diffuse nuclei (N) are also present and the cytoplasm is much less dense than that of the conidium, which contains typical vacuoles, nuclei, mitochondria, and profuse, often coalesced lipid droplets. The inner wall layer (IW) encompasses the conidium and forms its end walls. The developing septum is

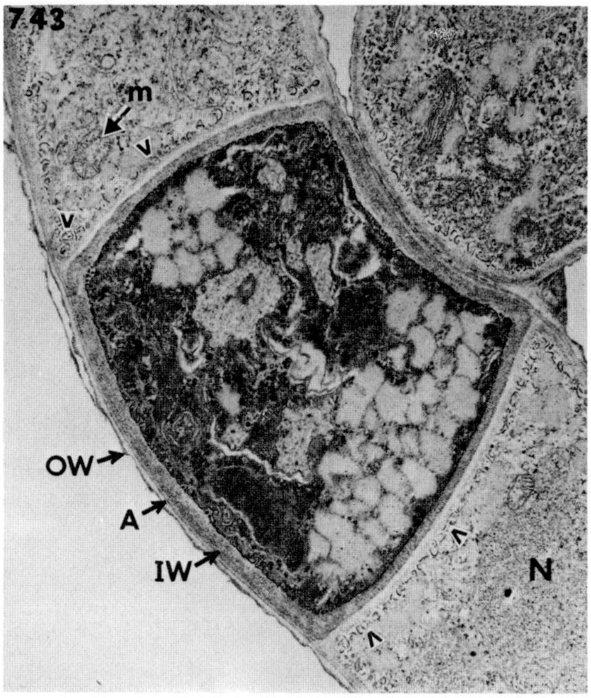

Figs. 7.43–7.45 *Sporendonema purpurascens.* Thin-section of conidium and adjacent, degenerate cells (7.43) showing differentiated wall layers. Arrowheads indicate fragments of cell membrane in degenerate cells. Freeze-fracture of cytoplasm of degenerate cell (7.44) shows hypertrophied mitochondrion (m) and irregular surface of inner nuclear membrane (arrows). Surface-fractured conidium (7.45) revealing randomly arranged rodlet fascicles. A, amorphous wall layer of fertile hypha; rodlets located beneath amorphous layer; IW, inner wall of conidium; np, nuclear pore; OW, outer wall layer of fertile hypha. ×13,200, 30,600 and 54,900, respectively.

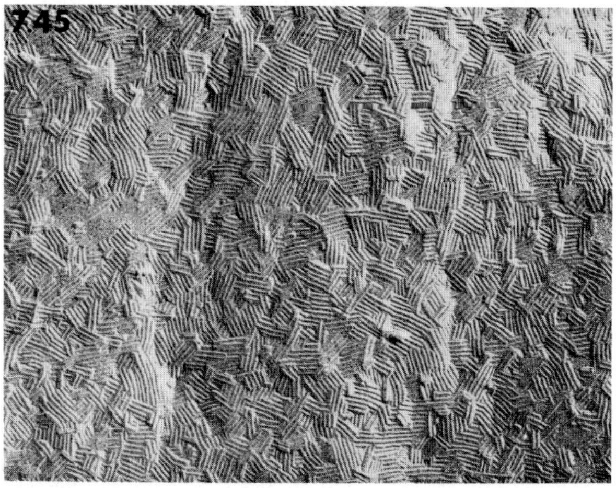

originally double-layered but the wall layer adjacent to the cytoplasm of the degenerate cells autolyses. As the inner conidial wall thickens, the wall of the degenerate cells continues to autolyse. This process occurs slowly from the cell membrane outwards so that the outer hyphal wall (OW) breaks down only after the endogenous conidium is mature. This aspect of the timing of conidial maturation and wall autolysis during enteroarthric development is probably critical. The mechanism of selective wall and cytoplasmic digestion in *S. purpurascens* is unknown and provides a challenging problem for future cytochemical and ultrastructural investigations.

The hypertrophied mitochondria with radially arranged cristae shown in thin-sections of degenerate cells (Fig. 7.43) are structurally similar to those revealed by the freeze-fractures (m, Fig. 7.44). Of particular interest is the structural detail of the fractured inner nuclear membrane which demonstrates linear ridges of particles extending randomly over its surface (arrows in Fig. 7.44). Surface-fractures of the outer membrane show bumps dispersed randomly between the nuclear pores. These structural irregularities may represent a stage of nuclear degeneration. Hemmes and Hohl (1973) pointed out in their study of nuclear degeneration in secondary sporangia of *Phytophthora palmivora* (Butler) Butler that the first evidence of this autolytic process 'occurred in the form of splitting and *un-zippering* of the two membranes of the nuclear envelope'. Complementary fractures of the nuclear envelope are necessary for further investigation of this phenomenon.

Surface-fractures of the conidial wall of *S. purpurascens* (Fig. 7.45) show a random pattern of short rodlet fascicles, comparable to the rodlet patterns on the surface of fertile hyphae. Cole (1973c) demonstrated similar arrangements of rodlet fascicles on the surface of conidia and fertile hyphae of *Geotrichum candidum.* In Chapter 1 it was proposed that changes in rodlet orientation may be correlated with changes in cell shape which occur during growth and differentiation of conidia. In the case of holo- and enteroarthric development, conidiogenesis occurs essentially by conversion of pre-existing elements which does not involve extensive *de novo* growth and, therefore, similar rodlet fascicle orientation persists on the surface of fertile hyphae and conidia.

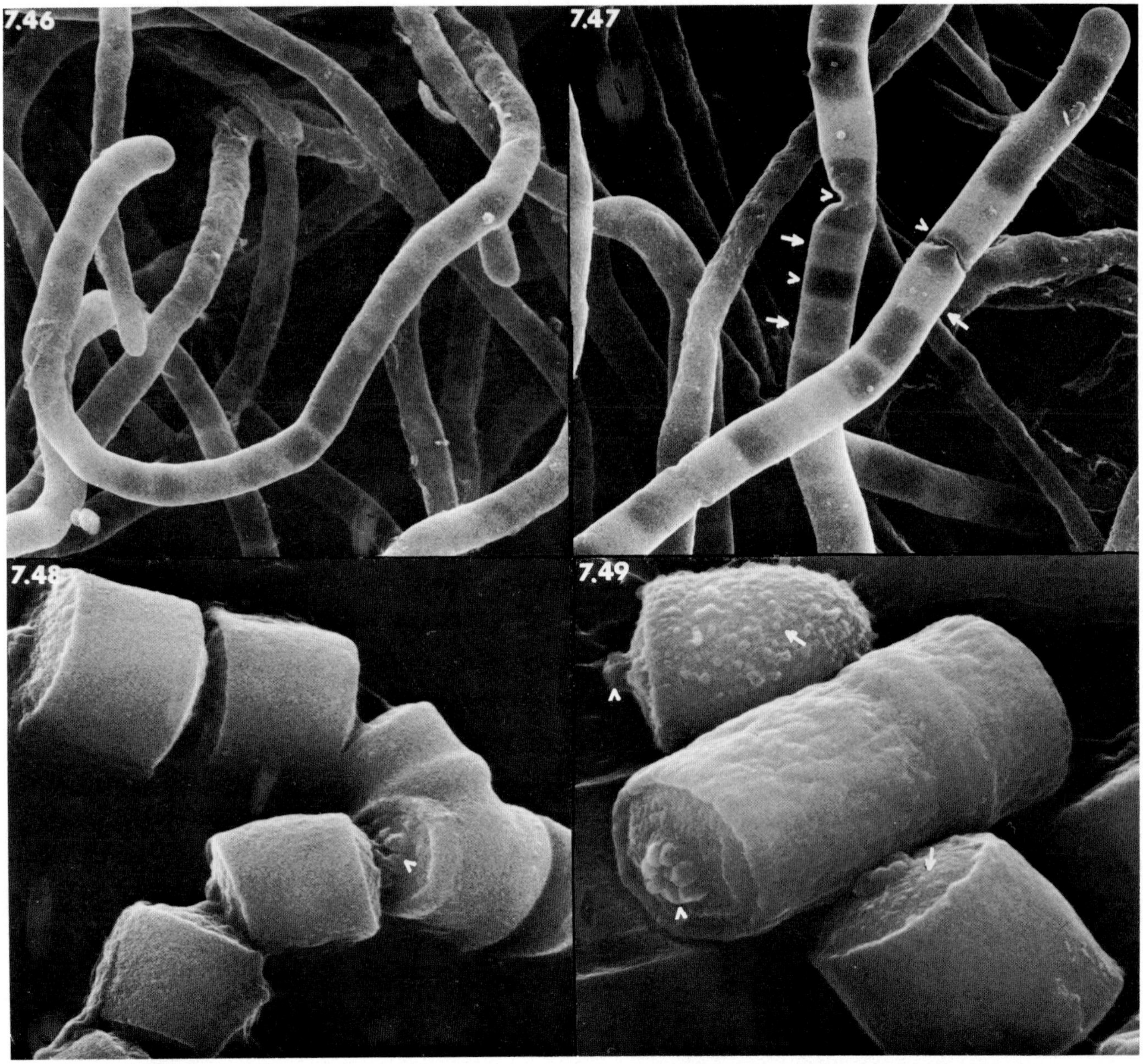

Figs. 7.46, 7.47 *Malbranchea sulphurea.* Typical, recurved hypha in (7.46) shows light and dark bands which later in development are identified (7.47) as endogenous conidia (arrows) and degenerate cells (arrowheads), respectively. ×2,400 and 3,200, respectively.

Figs. 7.48, 7.49 *Coremiella cubispora.* Enteroarthric conidia showing smooth, hyphal wall and inner, verrucose wall of seceded conidia (arrows in 7.49). Arrowheads (7.48, 7.49) indicate septal wall complexes. ×7,700 and 8,200, respectively.

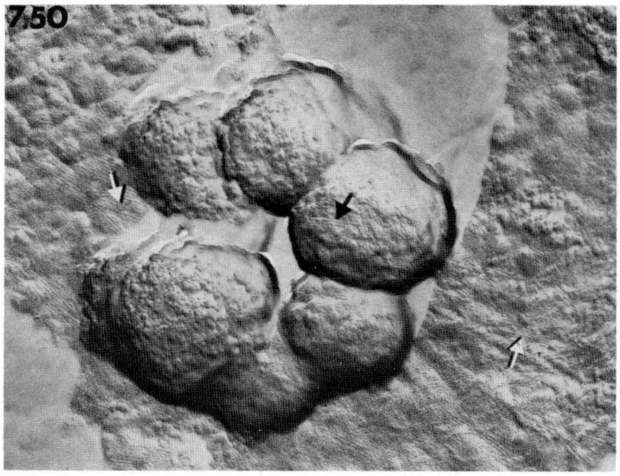

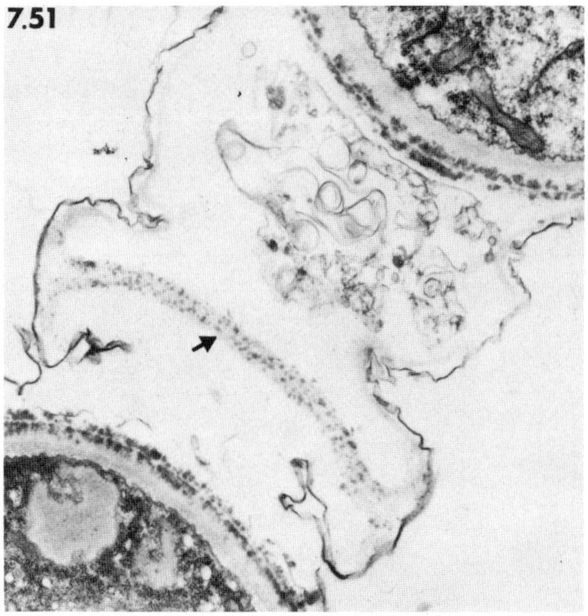

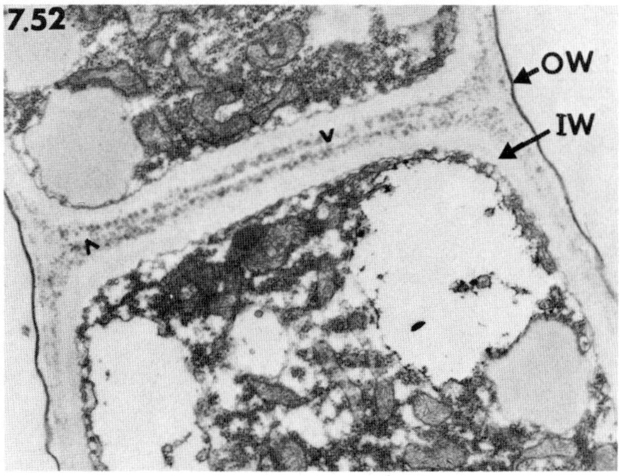

Malbranchea sulphurea (Miehe) Sigler & Carmichael also has enteroarthric development. The typical recurved hypha in Fig. 7.46 shows light and dark regions in the SEM, more clearly shown at a later stage in Fig. 7.47. The light bands (arrows in Fig. 7.47) represent endogenous conidia which have formed a thickened inner wall layer, while the dark bands (arrowheads) are cells destined to degenerate—some have already collapsed.

Coremiella cubispora (Berk. & Curt.) M. B. Ellis also produces enteroarthric conidia (Figs. 7.48–7.56) but demonstrates a slight variation from the previously described mechanisms. Branched (Fig. 7.48) and unbranched chains of conidia are formed, with and without intervening degenerate cells. The smooth outer wall of the fertile hypha (Figs. 7.48, 7.49) separates from the inner, verrucose, conidial wall to release the endogenous propagules (Fig. 7.49). A ring of globular wall excrescences surrounds the septal pore of seceded conidia (Figs. 7.49, 7.50). Under the light microscope, these septal wall complexes appear after the conidium has differentiated within the fertile hypha and separation of the double septum has occurred (arrows in Fig. 7.54). The replica of the septal wall complex in Fig. 7.50 shows six, fused globular excrescences of wall material in a circular arrangement with the septal pore at the centre. The outer wall layer of the septum is continuous with that of the excrescences. Thin-sections have revealed that the complex is composed of darkly pigmented deposits on the outer wall surface with no structural counterparts on the inner, cytoplasmic surface of the septum.

Thin-sections have also demonstrated that degenerate cells of the fertile hypha occur between differentiating conidia in *C. cubispora*, an event which characterizes enteroarthric development. Two adjacent degenerate cells are shown in Fig. 7.51, separated by remnants of a transverse septum. Parts of two endogenous conidia are visible. Conidia are also produced without intervening degenerate cells (Fig. 7.52). In this case, conidial secession occurs by schizolytic separation of the septal wall layers rather than the process described for *Sporendonema purpurascens* (Fig. 7.43).

Flora Pollack (personal communication) has suggested that *C. cubispora* does not demonstrate enteroarthric development, and considered the occurrence of degenerate cells as fatal accidents in the process of cell formation rather than a programmed event. However, our examination of nuclear-stained cells of *C. cubispora* (Figs. 7.53–7.55) shows typical features of enteroarthric ontogeny.

Figs. 7.50–7.52 *Coremiella cubispora.* Carbon-platinum surface replica of septal wall complex (7.50) showing rodlet fascicles (arrows). Thin-section of enteroarthric conidia separated by two degenerate cells (7.51). Arrow indicates remnants of septal wall. Thin-section of two adjacent conidia (7.52) showing lytic zone (arrowheads) between septal wall layers. IW, inner conidial wall; OW, outer wall of fertile hypha. ×47,600, 18,200 and 14,000 respectively.

Karyology of enteroarthric and holoarthric conidia

The vegetative hypha of *C. cubispora*, which was undergoing tip growth at the time of fixation, shows the typical absence of nuclei at the apex where secretory vesicles have accumulated and wall deposition and intussusception occurs (Bartnicki-Garcia, 1973). However, a large number of nuclei aggregate approximately 3–4 μm back from the tip and nuclear divisions have been observed on this subapical zone. When septation and conidial differentiation occurs in this region, each compartment contains 1–4 nuclei (Fig. 7.54). Although septation is random, there are enough nuclei distributed along the subdividing hypha for each cell to receive at least one nucleus. As wall differentiation progresses, some nuclei are found in cells which will degenerate (arrowheads in Figs. 7.54, 7.55). When the wall of the fertile hypha eventually ruptures and conidia secede, partially decomposed cytoplasmic material is lost from the degenerate cell.

Holoarthric development in *Geotrichum candidum* (Figs. 7.56, 7.57) is a less wasteful process since cytoplasmic material is not lost during conidiogenesis. The initial arrangements of nuclei in the vegetative and fertile hyphae are similar to those of *Coremiella cubispora.* At the time of secession, each holoarthric conidium contains 1 or 2 nuclei (arrows in Fig. 7.56), with uninucleate conidia predominating (Kendrick and Chang, 1971). Large hyphal segments, which undergo additional fragmentation into shorter cylindrical cells (arrowhead in Fig. 7.57), often show division and/or migration of nuclei to opposite poles of the cell before septation occurs.

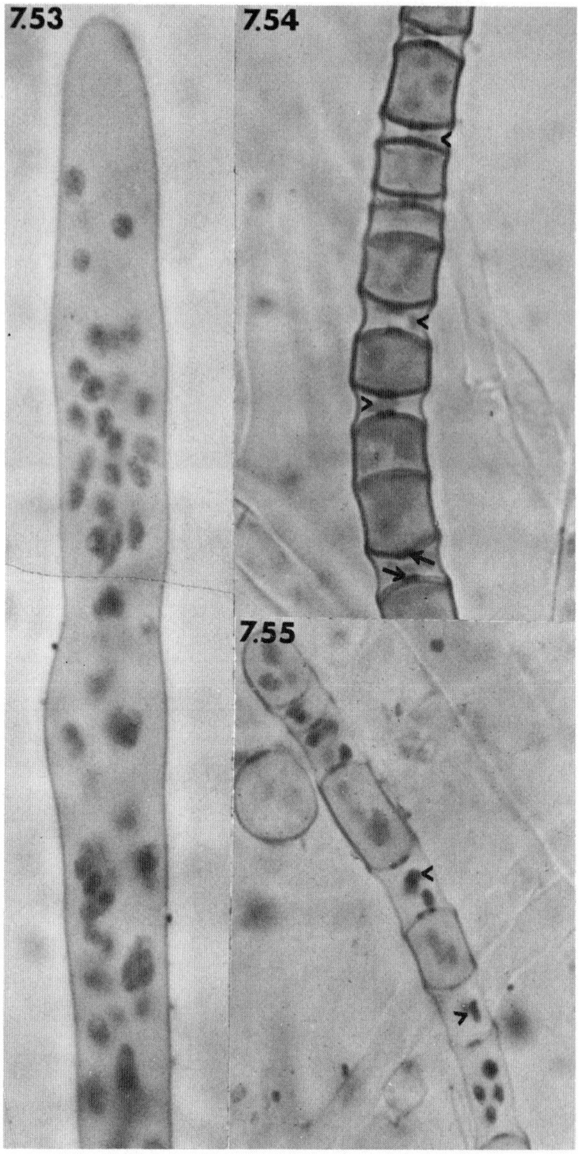

Figs. 7.53–7.55 *Coremiella cubispora.* Giemsa and aceto-orcein stained preparation of vegetative (7.53) and fertile hyphae (7.54, 7.55). Arrows in 7.54 locate septal complexes of two adjacent conidia. Arrowheads in 7.54 and 7.55 represent nuclei undergoing autolysis in degenerate cells. ×3,120, 1,248 and 1,000, respectively.

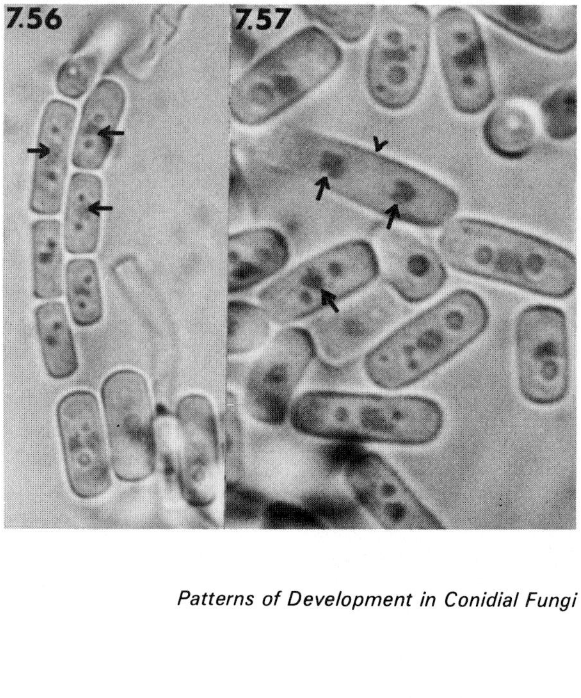

Figs. 7.56, 7.57 *Geotrichum candidum.* Nuclear stained conidia. Arrows in 7.56 and 7.57 locate nuclei. Arrowhead in 7.57 indicates large cell which would have later divided into two conidia. ×1,250 and 3,120, respectively.

Comparison of holoarthric and enteroarthric ontogeny

In the diagrammatic interpretation of these two modes of development (Fig. 7.58), the acropetal and then random order of septation and initial karyological events are similar (Fig. 7.58A, B). However, differences between holoarthric (Fig. 7.58A, B, F) and enteroarthric (Fig. 7.58A–E) development are soon manifested as wall differentiation occurs. During the formation of holoarthric conidia in *G. candidum,* the fertile hypha is divided up into progressively smaller compartments, each cell containing 1 or 2 nuclei. The final wall relations of mature conidia are shown in Fig. 7.58F.

During enteroarthric development (Fig. 7.58C–E), septation produces cells containing 1–4 nuclei (Fig. 7.58C). Differential deposition of new wall material signals which cells will become conidia and which will degenerate (Fig. 7.58D). The inner walls of those cells which will become conidia thicken, while the walls of adjacent cells programmed to degenerate do not thicken (Fig. 7.58D). Actually, even at this stage, the inner wall of the degenerate cells shows signs of autolysis. Selective breakdown of the inner wall layer provides the mechanism for eventual release of the endogenously formed conidia from the hyphal envelope (Fig. 7.58E). Separation of the septal wall layers of adjacent conidia (Fig. 7.58D) is essentially the same mechanism that occurs during holoarthric development (Fig. 7.58F).

The examples of thallic-arthric development illustrated in this chapter demonstrate the important characteristics of this mode of conidiogenesis, but certainly do not account for the full range of morphogenetic plasticity. Developmental and ultrastructural investigations of *Antromycopsis broussonetiae, Scytalidium lignicola* Pesante, *Oncocladium flavum* Wallr., *Arthrographis langeroni* Cochet and other species would be useful.

Fig. 7.58A–F Diagrammatic interpretation of holoarthric (A, B, F) and enteroarthric (A–E) development emphasizing karyological events and aspects of wall differentiation. Arrows in (A, B) point to dividing nuclei. Arrows in (F) locate micropores through septa. Arrowheads in D indicate degenerating nuclei. A, amorphous wall layer of fertile hypha; IW, inner wall layer; OW, outer wall layer.

7.58

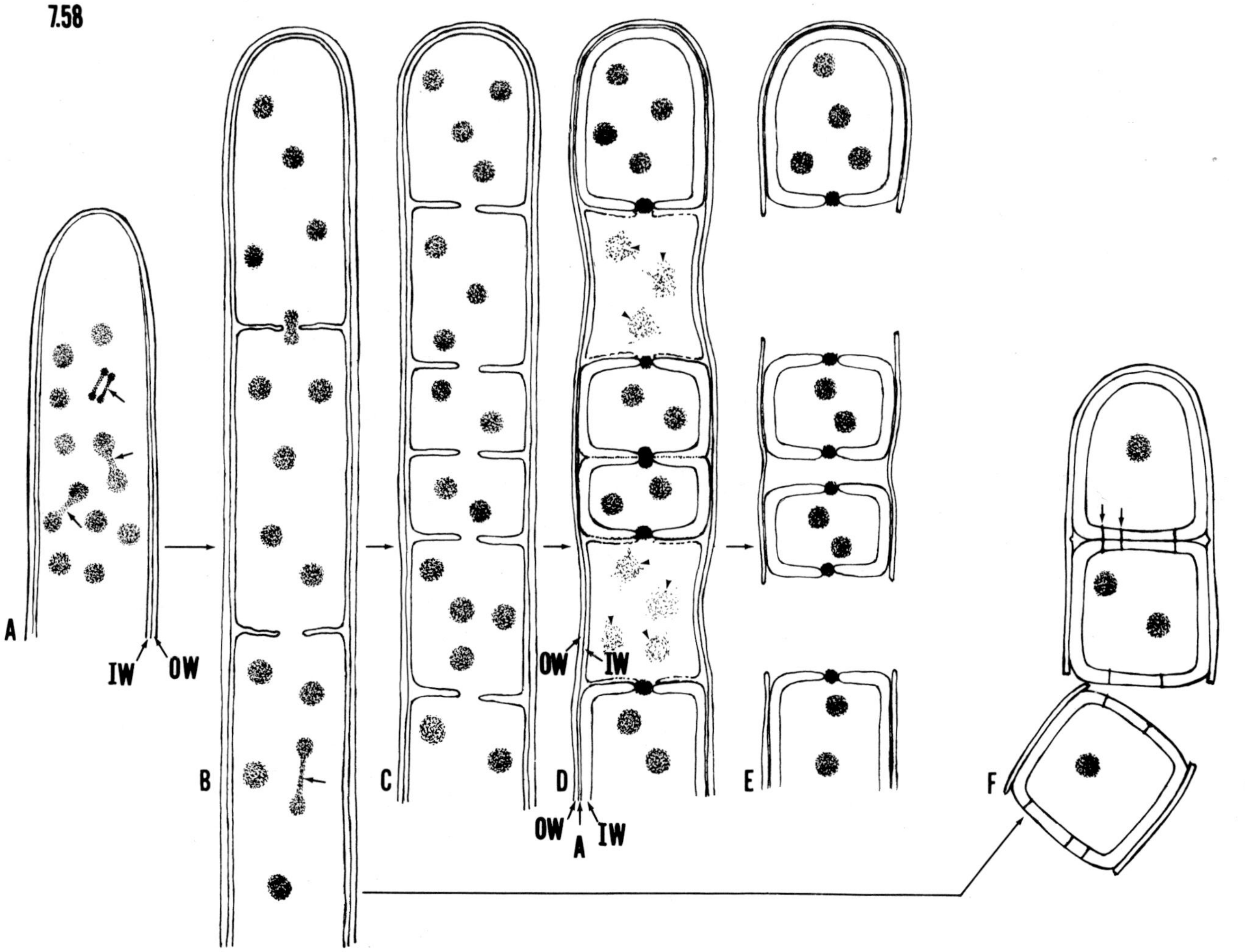

8 Classification of Developmental Processes

The two basic modes of conidiogenesis in the Deuteromycetes are referred to as blastic and thallic. The conidium differentiates from a fertile hypha either by the 'blowing-out' and *de novo* growth of part of the hyphal element (i.e., blastic development), or by the conversion of a segment of a hypha which may involve some enlargement and secondary wall growth (i.e., thallic development). The major differences between these two developmental processes are summarized in Table I.

Thallic development is subdivided into two processes distinguished by the arrangement and number of conidia produced. Terminal and intercalary, solitary, thallic conidia constitute one category, exemplified by conidiogenesis in species of *Microsporum, Epidermophyton* and *Trichophyton.* This is referred to as holothallic development and involves conversion of an entire hyphal segment into a single aseptate or multiseptate conidium. The second kind of thallic development involves conversion and fragmentation of determinate hyphae into chains of conidia and is identified as the thallic-arthric mode.

Three principal kinds of arrangement of blastic conidia on conidiogenous cells are identified: solitary, botryose and catenulate. Solitary blastic conidia are produced singly or successively from the conidiogenous locus of the fertile cell. A succession of solitary (i.e., individual) conidia is produced basipetally from terminal or acropleurogenous loci of the conidiogenous cell. Conidia formed in basipetal succession either adhere in chains or accumulate in droplets. We have here restricted use of the term *catenulate* to blastic conidia produced in acropetal chains. Development of such chains is usually dependent upon the maintenance of cytoplasmic continuity between adjacent cells while differentiation of basipetal chains is not dependent upon this condition and adjacent cells adhere simply because of wall connectives. Individual conidia may secede without interrupting development of the chain. The only known exceptions to this developmental concept of basipetal chains of conidia are demonstrated by *Wallemia sebi*, anamorphs of *Erysiphe* and other species discussed under *Variations in Basauxic Development* in Chapter 6. Botryose, blastic conidia develop synchronously or asynchronously from the usually swollen apices of conidiogenous cells. A summary of these subdivisions is presented in Table II.

Additional categorization of developmental patterns is

Table I. Comparison of developmental features of blastic and thallic conidia

	Origin from conidiogenous cell	Growth during early stage of conidial initiation	Protoplasmic differentiation during early stage of conidial initiation	Growth during conidial maturation
Thallic	*Blowing-out* of part of fertile hypha (conidiogenous cell)	Polarized; restricted to the *apical dome* of conidial initial	Cluster of secretory vesicles localized at apex associated with polarized tip growth	Polarized at apical dome, depolarized involving the entire cell, or some combination of these two conditions; results in high degree of variation in conidial shape
Blastic	Conversion of terminal or intercalary, fertile hyphal element	Involves swelling of entire hyphal element	Secretory vesicles and lomasomes distributed randomly in hyphal element associated with *secondary intussusception*; no polar clusters of secretory vesicles	Depolarized; low degree of variation in conidial shape

Table II. Summary of the different arrangements of blastic and thallic conidia

Origin of conidia	Arrangement of conidia on conidiogenous cells	Developmental characteristics and variations
Blastic	Solitary	a. Single, terminal b. Successive, acro-pleurogenous c. Successive, basipetal, accumulate in chains or droplets
	Catenulate	Acropetal chains
	Botryose	a. Synchronous clusters b. Asynchronous clusters
Thallic	Solitary	a. Single, terminal b. Single, intercalary
	Catenulate	Determinate chains

Table III. Categories of development based on conidial arrangement and wall relations

Origin of conidia	Arrangement of conidia	Wall relations between conidia and conidiogenous cells
Blastic	a. Solitary b. Botryose c. Catenulate	a′. Holoblastic b′. Enteroblastic
Thallic	a. Solitary b. Catenulate (Thallic-arthric)	a′. Holothallic a′. Holoarthric b′. Enteroarthric

based on wall relations between conidia and conidiogenous cells. If the outer conidiogenous cell wall remains continuous and encompasses the mature conidium, the process of conidiogenesis is referred to as holoblastic, or holothallic. If, however, rupture of the outer wall occurs and it separates from the inner conidial wall, the process is called enteroblastic or enterothallic (= enteroarthric) development. Formation of chains of thallic conidia is widely referred to as *thallic-arthric* development (derived from *arthrospore*, Vuillemin, 1910a). With respect to wall differentiation, the conidia of these fungi form either by conversion of the entire, determinate fertile hypha (holoarthric), or by differentiation of cells within the hypha (enteroarthric). The distinction between holo- and enteroblastic development is sometimes rather subtle and controversies have arisen even after thin-section examinations of conidiogenesis have been performed (Carroll and Carroll, 1974a; Brotzman *et al.*, 1975). In spite of scepticism that a classification of deuteromycetous fungi based on developmental characters can make use of wall relations (Harvey, 1974b; D. H. Ellis and Griffiths, 1977), certain differences in conidiogenesis have been identified in the previous chapters. Table III is a summary of the developmental categories based on conidial arrangement and wall relations.

The classification of kinds of conidiogenous cell development is straightforward. A conidiogenous cell produces one conidium or a plurality of conidia. During the latter process, the fertile cell may terminate extension growth when the first conidium is initiated (i.e., determinate conidiogenous cell), continue to elongate (i.e., proliferous) or shorten (i.e., retrogressive). Certain unique features of conidiogenous cells are also used in the classification of their development. For example, formation of a distinct pore through the wall of the conidiogenous cell occurs during poroconidium ontogeny. Such conidiogenous cells may be either determinate or proliferous. The phialide can be identified by the presence of a collarette. Although a fixed conidiogenous locus has been considered characteristic of phialidic development, thin-sections have revealed that some extension growth may occur with the formation of each conidium (i.e., formation of endogenous annellations). Also, formation of successive phialoconidia from the fertile apex of a phialide may cease and be followed by vegetative proliferation of the conidiogenous cell (sympodial or percurrent), culminating in production of a new conidiogenous locus at a higher level. The annellide, another differentiated conidiogenous cell which can be identified by the presence of annellations along its neck, produces a basipetal succession of solitary, blastic conidia. Formation of each propagule is followed by a percurrent proliferation of the inner conidiogenous cell wall usually through the scar of the previously formed conidium. The formation of successive, secondary conidia from both phialides and annellides is enteroblastic. On the other hand, primary conidium ontogeny from both kinds of conidiogenous cell is holoblastic. These and other interrelationships between conidium and conidiogenous cell formation are summarized in Table IV.

As pointed out in the introductory chapter, considerable interest has been attracted by the potential of using developmental characters in Deuteromycete systematics (S. J. Hughes, 1953). This stems primarily from the frustrations which systematists have encountered when they attempt to use conventional Saccardoan characters (Saccardo, 1886), such as mycelial pigmentation, conidial septation and arrangement of conidiophores and conidiogenous cells for the identification of deuteromycetous fungi. The high degree of variability frequently demonstrated by these morphological characters places considerable limitations on their taxonomic value. The disclosure of aspects of conidial and conidiogenous cell development has provided the fungal systematist with new characters which, because of their biological importance, are considered to be more genetically stable and, therefore, more taxonomically important than many Saccardoan characters. However, some of the original enthusiasm for use of these developmental features to solve taxonomic problems in the conidial fungi has waned during the past few years. In part, this has resulted from confusion generated by the publication of a surge of new terminology

Table IV. Categories of conidial and conidiogenous cell development

Origin of conidia	Arrangement of conidia	Wall relations between conidia and conidiogenous cells	Conidiogenous cell growth during formation of successive conidia	Special modes of conidiogenous cell development
Blastic	a. Solitary	a′. Holoblastic	a″. Determinate	a‴. Porogenous
	b. Botryose	b′. Enteroblastic	b″. Proliferous	b‴. Phialidic
	c. Catenulate		1. Sympodial	c‴. Annellidic
			2. Percurrent	
			3. Basauxic	
			c″. Retrogressive	
Thallic	a. Solitary	a′. Holothallic	a″. Determinate	
	b. Catenulate	a′. Holoarthric	a″. Determinate	
	(Thallic-arthric)	b′. Enteroarthric	b″. Proliferous	
			1. Sympodial	

(Cooke, 1974) and variations on Hughes's (1953) original, experimental classification. In addition, developmental concepts were initially based on light-microscopic examinations of fixed cells alone, and this often led to ambiguities, uncertainties, and misinterpretations in descriptive analyses. In many cases, the problems have been solved by time-lapse photomicrographic and electron-microscopic investigations. Unfortunately, these techniques are somewhat esoteric and the information is not readily accessible to many fungal taxonomists who still depend largely on light-microscopic characters for identification of Deuteromycetes. We have emphasized at several points in the previous chapters the importance of providing correlations between morphological characters revealed by special techniques, such as thin-sectioning, and characters which are visible with the light microscope. This process of correlation of data would, in turn, give rise to a more functional classification. It is clear, however, that there are some species whose subtle processes of conidiogenesis can only be defined by electron microscopy. On the other hand, many species demonstrate key morphological features which are recognizable with the light microscope and can be used, retrospectively, for identification of a particular mode of development. This was demonstrated by our examinations of retrogressive development in *Cladobotryum varium, Trichothecium roseum*, and *Basipetospora rubra* in Chapter 5. In addition, the presence of collarettes, annellations, and pores at the fertile apices of conidiogenous cells can usually be detected by light microscopy. Our interpretations of porogenous, phialidic and annellidic development, based on ultrastructural characters, provide the light microscopist with a better understanding of the differences between these ontogenetic processes. Similarly, the distinguishing features of basauxic development in species of *Spegazzinia, Arthrinium*, and *Dictyoarthrinium* were illustrated by the SEM (Chapter 6). The presence of cup-shaped *conidiophore mother-cells*, usually detected by light microscopy, is the key to identification of this mode of development.

Most of the developmental characters listed in Table IV and illustrated in this book, can be recognized in retrospect by careful examination of fixed cells under the light microscope. We hope that our enthusiasm in the application of developmental and ultrastructural features of conidia and conidiogenous cells to the systematics of the Deuteromycetes will rekindle the interest of taxonomists in these characters and motivate attempts to establish a functional classification.

9 Techniques for Developmental and Ultrastructural Studies of Conidial Fungi

Culturing techniques

Most of the fungal strains examined were obtained from the Centraalbureau voor Schimmelcultures (Baarn, Netherlands). Several subcultures were also obtained from the University of Waterloo, Center for Disease Control (Atlanta, Georgia) and private collections. All cultures have been maintained in the fungal collection of the Department of Botany, University of Texas at Austin, under the supervision of the senior author.

Except for *Erysiphe graminis*, all fungi were grown in pure culture on prepared media. The anamorph of *E. graminis* was parasitizing *Stenotaphrum secundatum* (Walter) O. Kuntze. A fresh collection of the fungus and its host were prepared for scanning electron microscope (SEM) study. The growth media used included potato dextrose, malt extract, corn meal, Sabouraud's glucose-peptone-yeast extract, and rabbit food agar (Stevens, 1974). The cultures were incubated at 22–31°C in continuous darkness. Thermophilic fungi (Cooney and Emerson, 1964) were grown at their optimal temperature (e.g., 45°C). Occasionally it was necessary to use a Blak-Lamp to stimulate sporulation.

Light microscopy

Preparations of fixed material for light microscopy were made by carefully excising cells from sporulating cultures using thin glass needles. The material was mounted on a slide in distilled water, lactophenol mounting medium (Stevens, 1974) or Shear's mounting fluid (Gams *et al.*, 1975) for immediate photomicrography. Line drawings were prepared with a camera lucida apparatus.

The time-lapse photomicrographic analyses of living cultures were performed using two previously described techniques. A thin glass culture chamber (Fig. 9.1), designed by Cole and Kendrick (1968b), proved satisfactory for examination of most fungi. The lower coverslip was sealed permanently to the slide with a waterproof cement. The vertical face of the hemispherical slab of agar was inoculated and then the upper coverslip was sealed to the slide with molten paraffin wax. A sterile Petri dish was lined with filter paper soaked in 50% glycerol, to maintain high humidity, and a V-shaped glass rod was laid horizontally on top of the wet paper to support the slide chamber. The entire apparatus was then incubated for several days in conditions optimal for conidiogenesis before examining the chamber under the light

9.1

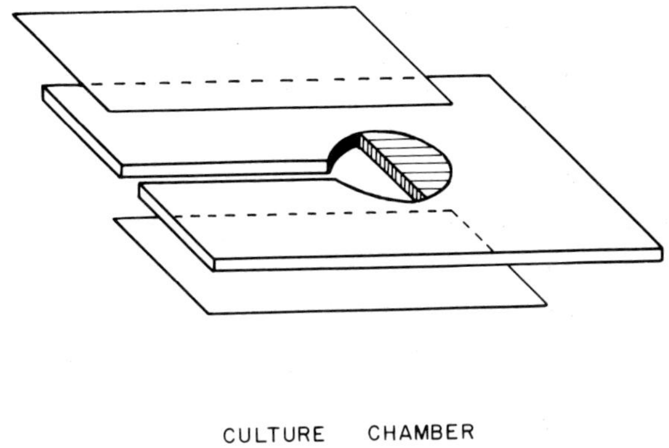

9.2

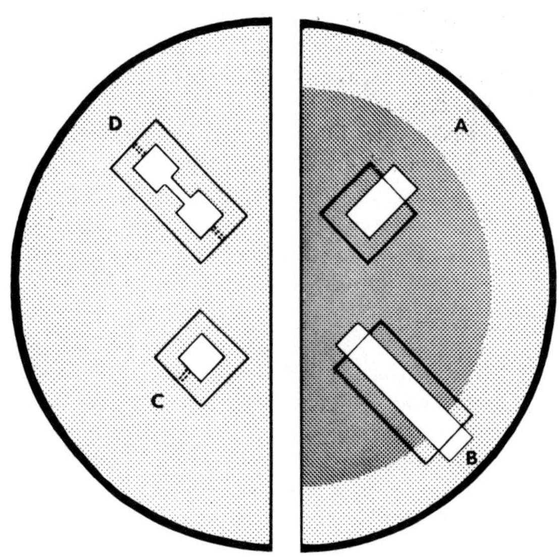

Fig. 9.1 An 'exploded' view of the thin glass culture chamber, showing the drilled and slotted slide, agar medium, and two coverslips which seal the upper and lower surfaces (after Cole and Kendrick, 1968b).

Fig. 9.2A–D Composite diagram of modified plate cultures as seen from above, showing agar (light stippling), fungal colony (dark stippling), channels (blank), coverslips (outlined), and pierced air passages (dotted lines) (after Cole *et al.*, 1969).

microscope. This glass culture chamber had several advantages, including (1) the fact that it was sufficiently thin to permit use of phase and Nomarski interference contrast illumination with a 100 X oil immersion objective, (2) its ability to support growth for several days without desiccation or staling, (3) its lack of contamination even after exposure on the microscope stage for several days and (4) its simple and inexpensive construction. Most chambers designed previously did not meet the condition of being thin enough to permit high magnification photomicrography. However, Heunert (1972) described several elegant thin culture chambers which will support growth of fungi both on solid agar media and in liquid culture. These chambers have been used for preparation of a 16 mm time-lapse film on conidiogenesis at the Institut für den Wissenschaftlichen Film in Göttingen (Cole, 1977).

A few species failed to produce conidia in the glass culture chamber which was probably due to limited aeration. An alternative technique was described by Cole *et al.* (1969) in which sporulating Petri dish cultures were modified to facilitate time-lapse photomicrography (Fig. 9.2). A thin layer of agar was poured into a standard plastic Petri dish. After the fungus began to produce conidia, a rectangular piece of agar and mycelium was removed with a sterile scalpel, exposing the base of the Petri dish (Fig. 9.2A, B). The radial channels were formed either at the margin (Fig. 9.2B) or close to the centre of the colony (Fig. 9.2A) depending on where most conidia were produced. The degree of aeration could be varied simply by extending one or both ends of the channel beyond the coverslip. Alternative procedures involved cutting shorter pieces of agar out of the freshly poured plate, inoculating the walls of the channel thus formed, and then immediately covering it with a glass coverslip which limited gaseous exchange to that occurring by diffusion through the agar (Fig. 9.2C, D). Aeration was sharply increased at a later stage of development by formation of an air passage under the coverslip with a sterile needle. This modified plate culture technique was used successfully for examination of conidiogenesis in several hyphomycetous fungi (e.g., *Scopulariopsis brevicaulis, Chalara paradoxa, Nodulisporium hinnuleum*).

Karyological studies of hyphomycetous fungi involved a nuclear stain procedure which was modified from Robinow and Caten's (1969) *direct aceto-orcein* technique. The fungi were grown on thin cellophane sheets (Kalle Aktiengesellschaft, Wiesbaden) overlaying the agar media in Petri dishes. The plate was first flooded with a modified Helly's fixative solution (5% $HgCl_2$ and 3% $K_2Cr_2O_7$ (mass/vol) in water with 6% (vol/vol) of a 3% solution of paraformaldehyde that was added immediately before use). Small squares of cellophane (about 3 mm) and sporulating mycelium were excised, placed in depression dishes that contained fresh fixative, and left at room temperature for 15 min. The rest of the procedure was the same as outlined by Robinow and Caten (1969). The mycelium separated easily from the cellophane during acid hydrolysis. At this stage, the mycelial mat could be moved from one solution to the next with a pair of fine-tipped forceps. The material was finally mounted on a glass slide in aceto-orcein stain, separated by gentle teasing, covered with a thin coverslip and sealed with paraffin. The cells were examined with an Olympus AH photomicroscope using a green filter. Photomicrographs were recorded on Kodak Panatomic-X 35 mm film.

Scanning electron microscopy (SEM)

Because of the delicate nature of most fungal material examined above, it was necessary to use a preparatory procedure for the SEM which preserved the original shape of the cells as well as the structural relationship between conidial chains or clusters, and conidiogenous cells. The most popular procedures for preparing soft biological tissue at present involves either the freeze-drying technique (Boyde and Echlin, 1973; Newman and Tailby, 1973), or the critical point drying method (Anderson, 1951; Horridge and Tamm, 1969; Fromme *et al.*, 1972; Porter *et al.*, 1972; Lewis and Nemanic, 1973; H. J. Marchant, 1973; Schramm and Petzold, 1973; Turner and Green, 1973; Kurtzman *et al.*, 1974; Lung, 1974; Tanaka and Iino, 1974). The latter technique has demonstrated very good preservation of conidia and conidiogenous cells of deuteromycetous fungi (Cole, 1973a, b, 1974, 1975b, 1976a, b, c) and has been used exclusively in this study.

The same chemical fixation procedure was used in preparation of all species for the SEM. A sporulating culture was first flooded with a mixture of equal volumes of 6% glutaraldehyde (Ladd Res. Ltd., Vermont) and 1% OsO_4 (Varlacoid Chem. Co., New Jersey), each prepared at 4°C in 0·1 M sodium cacodylate buffer (pH 7·1–7·4). Small blocks (approx. 5 cm square) of agar and mycelium were then excised from the flooded culture and placed in a vial containing fresh, cold fixative solution and left in the dark for 2 hours. The samples were then washed in buffer (10 times) and placed in 1% OsO_4 buffered as above for 2 hours. This is considered to be an important step in the preparation of soft biological material for the SEM. Kelley *et al.* (1973) have shown that osmium forms a thin coat (40–200Å in thickness) over the surface of cells which provides the specimen with some resistance to contamination from the electron beam during long term observation in the SEM. A heavier osmium coating can be achieved by use of thiocarbohydrazide because of its osmium-bridging properties. The specimens were subsequently washed in buffer (10 times), water (5 times) and then dehydrated in a graded ethanol series (15 minutes each), followed by washes in increasing concentrations of amyl acetate in absolute ethanol (15 minutes each) and finally in absolute amyl acetate. The blocks of mycelium were then placed in Beem capsules whose ends were cut and replaced with fine mesh screens permitting free flow of liquid through the capsules while retaining the fungal samples. The Beem capsules were loaded with the agar blocks in a small Petri dish filled with absolute amyl acetate inside the fume hood. Extreme care was taken to prevent desiccation of the specimens. The loaded capsules were then quickly placed in the bomb of the Denton DCP-1 critical point drying apparatus and the latter was filled immediately with liquid CO_2. The bomb was flushed with liquid

Table V. Schematic outline of preparatory procedures for scanning (SEM) and transmission (TEM) electron microscopy, and freeze-etching (FE)

Fungi grown in Petri dish cultures, or on agar-coated coverslips	6% glutaraldehyde + 1% OsO_4 (1:1 in buffer, 4°C)	1% OsO_4 (4°C)
		Buffer wash (10 times)
Glycerol		
Centrifuge cells; place pellet in depression of Balzers' support disc		Dehydration in graded ethanol series
Liquid freon-22	Block-stain (2% uranyl acetate in 70% EtOH)	Dehydration in graded amyl acetate series diluted with absolute EtOH
Liquid nitrogen	Wash and continue dehydration in ethanol series	
Place disc on cold stage of FE device		Critical-point drying (liquid CO_2)
	Wash in absolute acetone	
		Mount specimens
Fracture, etch, coat (carbon-platinum)	Embed in Spurr's low-viscosity plastic (20, 50, 80, 100%, continuous agitation)	
		Coat specimens (carbon-gold/palladium)
Clean replica (chromic acid)		
	Ultrathin sections	
Mount on copper grid		SEM
	Post-stain	
	TEM	

━━━━ SEM

──── TEM

- - - - FE

CO_2 for 2 minutes and then filled and sealed off for 2 minutes. It was then partially emptied (pressure in bomb reduced to approx. 250 psi) and slowly filled again (repressurized to 900 psi). This procedure was repeated 4 times. Finally, the bomb was sealed and its pressure increased to 1500 psi by heating with water at approximately 65°C. The CO_2 gas was subsequently released and the dried specimens were removed from the Beem capsules and mounted on aluminum stubs which were coated with conductive silver paint. The stubs were placed on an omnirotary table in a Denton DV 502 high vacuum evaporator and coated with about 100 Å (Russ and Kabaya, 1970) of gold-palladium (60:40). The material was examined in an AMR Model 1000 scanning electron microscope at 10–30 kV and 0–45° tilt of the specimen stage. The scanning electron micrographs were recorded with Polaroid Type 55 P/N film.

A variation of this procedure was designed to preserve the natural arrangement of conidia as they were formed from the fertile hyphae in culture. The fungus was grown on an agar-coated, thin glass coverslip permitting, first, examination with the light microscope, followed by fixation, critical point drying and subsequent examination in the SEM, with minimal disturbance to the mycelium (Cole, 1975b). Using this procedure, the intact chains of holoarthric conidia of *Geotrichum candidum* could be examined while the previous preparatory technique for the SEM resulted in disruption of the fragile chains. Successful preparation of coverslip cultures, however, depended on rapid growth and conidial development before desiccation of the thin layer of agar and was, therefore, limited to only a few species.

Transmission electron microscopy (TEM)

The fixation procedure used in preparing material for the TEM was the same as that for the SEM. The glutaraldehyde and OsO_4 treatments and ethanol dehydration were identical. Because of the integration of these early steps in the two preparatory techniques, it was possible to subdivide samples of the same culture for examination with each kind of electron microscope. The correlation of data from these two sources of ultrastructural information and light microscopy was both instructive and necessary (Bailey and Bisalputra, 1970; Begres *et al.*, 1970; Cole and Aldrich, 1971c; Sugimura, 1973; Winborn and Guerrero, 1974). In preparation for the TEM, the absolute ethanol wash was followed by two 15 minute washes in absolute acetone. The material was then placed in 25% Spurr's (1969) low-viscosity plastic, and prepared in absolute acetone, for 1 hour with continuous, gentle agitation on a rotating stage. The Spurr formula for medium hardness was used. The samples were then transferred to a 50% plastic solution for 4 hours, 80% plastic for 4 hours and 100% plastic overnight. The material was then removed from the vials, placed in fresh 100% plastic which contained 1 drop/ml of DMP-30 (polymerization accelerator) and flat embedded (1 in diam. pill bottle caps were used). A large number of blocks of mycelium excised from the same culture could be embedded in the same cap. The caps were placed in a vacuum oven at 70°C for 10 minutes to remove any air bubbles and residual acetone and then into a 70°C oven at atmospheric pressure overnight. Individual blocks of polymerized fungal material were then cut from the flat, plastic disc using a razor blade, mounted to the end of a short piece of plastic dowelling which could be clamped by the specimen holder of a Porter-Blum Mt II ultramicrotome. Thin-sections were cut with a diamond knife and picked up on formvar coated copper grids. The sections were poststained with 0·5% aqueous uranyl acetate (20 minutes) and Reynolds' (1963) lead citrate (7 minutes). The material was examined in a Hitachi HU-11E and Siemens Elmiskop 101.

Freeze-etching

Plate cultures producing large numbers of conidia were flooded with 10–30% glycerol and 10–20 drops of Teepol (a detergent; Shell product) and gently agitated until the mycelium was thoroughly wetted. The cultures were allowed to stand at room temperature from 4 to 12 hours. The surface of the agar was then carefully scraped with an acetone cleaned razor blade, which resulted in a suspension of conidia, conidiogenous cells and vegetative hyphae in the glycerol solution. Conidia were separated from the rest of the mycelium by passing the suspension through a fine-mesh nylon net. If conidia alone were to be examined, the filtrate was centrifuged at 1000 × g and a small droplet of the pellet was subsequently transferred to the depression of a Balzers' gold-nickel alloy disc and frozen in liquid Freon–22 followed by liquid nitrogen. High concentrations of fertile and vegetative hyphae could be alternatively collected from the nylon mesh, centrifuged in glycerol and frozen by the same procedure. The discs were placed on the precooled stage of a Balzers' BA 360M freeze-etch device and replicas were prepared following the procedure of Moor and Mühlethaler (1963). The freeze-etch device was equipped with a Balzers EVM 052 electron gun for carbon-platinum deposition. The replicas were cleaned in chromic acid, washed in distilled water (3 times) using acid-cleaned glassware, and picked up on formvar coated grids. The replicas were examined in the TEM.

Appendix
Patterns of Sporangiospore Development in the Zygomycetes

The purpose of this appendix is to demonstrate a few of the developmental similarities which exist between mechanisms of sporangiospore and sporangial formation, illustrated below, and processes of conidial and conidiogenous cell ontogeny already described. Patterns of development of deciduous sporangia in the Oomycetes have been reviewed by S. J. Hughes (1971a) and will not be discussed here. Instead, attention is directed to the Zygomycetes in which the lines of evolution of sporangial development are well illustrated (Hesseltine, 1952; Hesseltine and J. J. Ellis, 1973; Benjamin, 1959, 1960, 1961, 1963, 1965, 1966; Kreisel, 1969; Müller and Loeffler, 1976). The asexual states of many species are frequently confused with members of the Hyphomycetes (Boedijn, 1958; Clements and Shear, 1931; Zycha, 1935). The similarities between sporangio-sporogenesis and conidiogenesis represent examples of convergent evolution in the fungi and reinforce the concept that nonmotile, asexual propagules are produced by a limited number of ontogenetic processes which have evolved repeatedly in distantly related and unrelated groups of micro-organisms (S. J. Hughes, 1970, 1971a, c).

The Zygomycetes are now subdivided into four orders: the Mucorales, Kickxellales (Kreisel, 1969), Entomophthorales and Zoopagales. An additional order, the Endogonales (Moreau, 1953), is also recognized by a few workers. In recent classifications of the Mucorales (Zycha *et al.*, 1969; von Arx, 1970, 1974; Pidoplichko and Milko, 1971; Hesseltine and J. J. Ellis, 1973), morphological and developmental features of the asexual reproductive structures have been used as the most important taxonomic characters in separating families and genera. In a few of these fungi (e.g., species of *Mucor* Mich. ex Fr., *Zygorhynchus* Vuill., *Mortierella* Coemans), the zygosporic states have also facilitated the delimitation of species or genera (Gams, 1969; Gams *et al.*, 1972; Kuhlman, 1972, 1975; Schipper *et al.*, 1975), but in most Zygomycetes, the sexual spores are completely lacking, not reported, or are difficult to induce in culture and are rarely present on natural substrate.

Benjamin (1959) recognized three main patterns of sporangial development among the mucoraceous fungi: (1) sporangia relatively large, multispored, columellate; (2) sporangiola containing relatively small numbers of spores, columellae absent or rudimentary, and (3) sporangiola containing only one or two spores. The apparent evolutionary tendency in the Mucorales for reduction of the sporangium to a one-spored structure (Müller and Loeffler, 1976) has generated much of the confusion in differentiating the conidium, sporangiolum and sporangiospore. This aspect will be discussed later with a consideration of recent ultrastructural data derived from comparative, thin-sectioning examinations of conidial and sporangiospore formation.

In order to further categorize the modes of sporangial development, we present Table VI which separates the families of the Mucorales and Kickxellales into five groups. These subdivisions are not consistent for all species included in the two orders, but will suffice for the purposes of our discussion.

Table VI. Kinds of sporangial development in the Mucorales and Kickxellales[1]

Multispored sporangia	Mucoraceae (*Mucor, Rhizopus, Absidia*) Pilobolaceae (*Pilobolus*) Mortierellaceae (*Mortierella*) (also produces few-spored and monosporous sporangiola) Saksenaeaceae (*Saksenaea*) Endogonaceae (*Endogone*) Choanephoraceae (*Choanephora*) (also produces monosporous sporangiola)
Few-spored sporangiola	Radiomycetaceae (*Radiomyces, Hesseltinella*) Thamnidiaceae (*Cokeromyces*)
Monosporous sporangiola on vesicles	Choanephoraceae (*Choanephora*) Cunninghamellaceae (*Cunninghamella, Mycotypha*) Helicocephalidaceae (*Rhopalomyces*)
Few-spored merosporangia	Syncephalastraceae (*Syncephalastrum*) Piptocephalidaceae (*Syncephalis, Piptocephalis*) Dimargaritaceae (*Dimargaris*)
Monosporous merosporangia	Kickxellaceae (*Coemansia, Linderina*)

1. The Entomophthorales and Zoopagales are discussed separately at the end of this appendix.

Multispored sporangia

This is the most common and probably the most primitive mechanism of asexual reproduction in the Mucorales (Benjamin, 1959), and has no analogue in conidiogenesis among the Deuteromycetes. In spite of extensive light-microscopic investigations of this group of mucoraceous fungi, ultrastructural aspects of cleavage in a multispored sporangium have been reported so far for only a few species. Bracker's (1966, 1968) thin-section examinations of sporangial development in *Gilbertella persicaria* (Eddy) Hesseltine elucidate this process which is probably similar, at least in terms of the cytological events leading to spore differentiation, in all families of the Mucorales and Kickxellales. Figure A.1 summarizes the differentiation of a cytomembrane system which produces a ramifying, tubular, cleavage apparatus and, eventually, separation of spore initials (Bracker, 1968). The early pre-cleavage sporangium contains cisternal rings which are apparently capable of blebbing and forming vesicles. The latter differentiate into cleavage vesicles and later fuse with the plasma membrane. As pointed out in Chapter 1, Bracker (1968) interpreted the ring-like cisternal profiles as functional equivalents of the Golgi apparatus. Hammill (1976) described identical cisternal complexes in pre-cleavage sporangia of *Mucor mucedo* Link ex Fr. (Fig. A.2). Similar organelles were reported in the conidial initials of the Hyphomycete *Scopulariopsis brevicaulis* (Sacc.) Bain. (Cole and Aldrich, 1971a) and it was speculated that these cisternal rings are also Golgi equivalents (Chapter 1, Fig. 1.5). In both fungal groups, these presumably play important roles in wall growth and differentiation of reproductive cells. Dictyosomes of the kind found in typical plant and animal cells have been clearly shown only in the Oomycetes (S. N. Grove *et al.*, 1970; Haskins *et al.*, 1976) and in yeasts under certain growth conditions (Havelkova and Mensik, 1966). In the Basidiomycetes, Girbardt (1973) described the secretory organelles as 'Golgi-Sacculi'—the cisternae are not stacked,

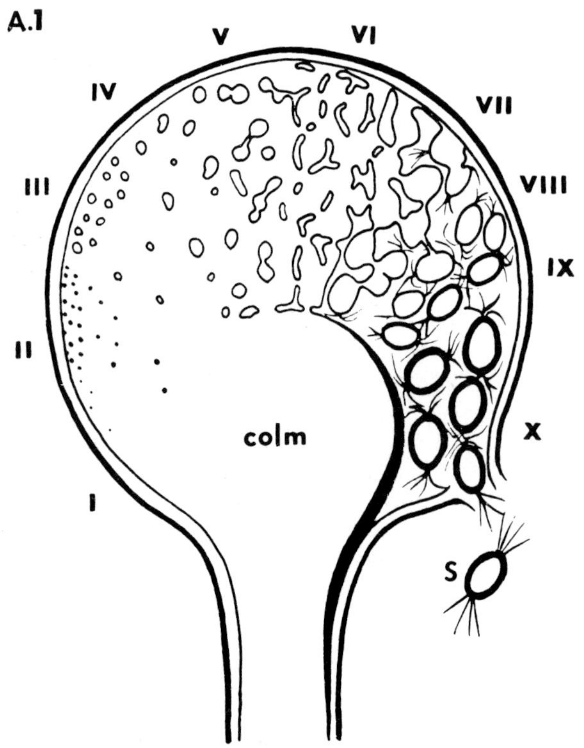

Fig. A.1 *Gilbertella persicaria.* Diagram summarizing sporangiospore formation. Only cleavage apparatus and walls are indicated. Spatial distribution of stages is for convenience in illustration only, and does not represent spatial progression within a single sporangium. Early pre-cleavage (I, II), mid-cleavage (III), late pre-cleavage (IV), early cleavage (V), mid-cleavage (VI), late cleavage (VII), early post-cleavage (VIII), mid-post-cleavage (IX), late post-cleavage or spore maturity (X). The discontinuity of the sporangial wall near X represents rupturing of the sporangium as it splits into halves at maturity. colm, columella; S, sporangiospore (after Bracker, 1968).

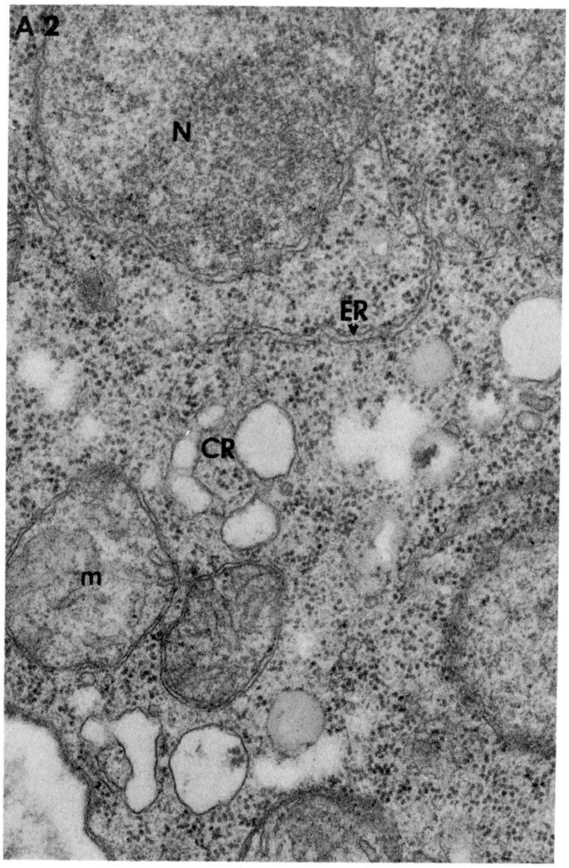

Fig. A.2 *Mucor mucedo.* Thin-section of pre-cleavage sporangium showing cisternal rings (CR). ER, endoplasmic reticulum; m, mitochondrion; N, nucleus. ×36,200 (from Hammill, 1976).

A.3

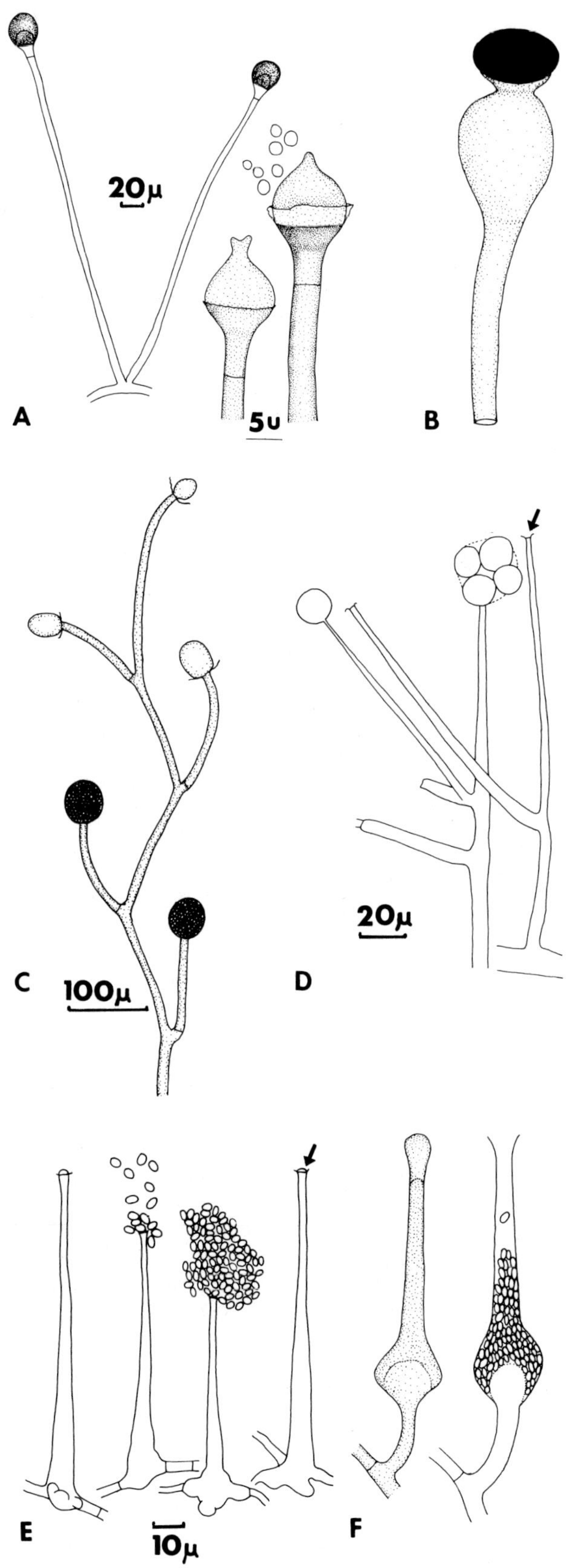

but appear often as solitary, swollen, blebbing sacs. The question of whether the cisternal rings found in the Zygomycetes and Deuteromycetes represent a reduced form of the Golgi complex may hold some phylogenetic significance.

After the spores mature, they are either released when the wall of the sporangium deliquesces (e.g., species of *Phycomyces* Kunze, *Absidia* van Tieghem; Fig. A.3A), or remain within the sporangium if its wall is persistent (e.g., certain species of *Mucor*). In the Thamnidiaceae, large, terminal multispored sporangia with deliquescent walls, and sporangiola with persistent walls, may occur together. In *Pilobolus* Tode ex Fr. (Fig. A.3B), the entire sporangium may be violently discharged from the sporangiophore apex and is sometimes mistaken for a seceded conidium. Forcible discharge of conidia is demonstrated by a few Hyphomycetes, such as species of *Alternaria* Nees ex Wallr., *Curvularia* Boedijn, *Zygophiala* Mason, *Pyricularia* Sacc. and others (Webster, 1952; Meredith, 1961, 1962, 1963; Hashioka, 1975), especially when humidity is reduced. Although the details of the spore dispersal mechanisms in these two groups of fungi may not be comparable, similar selection pressures may have been applied.

The development of successive sporangia from a Zygomycete sporangiophore may involve sympodial proliferation of the latter (e.g., *Mucor circinelloides* van Tieghem, Fig. A.3C). This is reminiscent of sympodial conidiogenous cells in the Deuteromycetes. S. J. Hughes (1971b) pointed out that; 'Successive percurrent proliferations of the cell subtending a sporangium (or conidium) are common in many but not all Phycomycetes' and presented examples in the Chytridiomycetes and Oomycetes. Percurrent proliferation is unknown in the Zygomycetes.

Most mucoraceous fungi with multispored sporangia possess a distinct columella at the sporangiophore apex. However, in the genus *Mortierella* (Fig. A.3D, E), the columella is indistinct and often obscured by the surrounding spores (e.g., *M. alpina* Peyronel, Fig. A.3E). When the wall ruptures and the spores are released, the fracture usually occurs close to the apex of the sporangium, leaving the impression of a *collarette* (arrows in Fig. A.3D, E). This feature, in

Fig. A.3A *Absidia glauca* Hagem. Deliquescent sporangia (after Domsch and Gams, 1972).

Fig. A.3B *Pilobolus crystallinus* (Wiggers) Tode. Terminal sporangium is violently discharged.

Fig. A.3C. *Mucor circinelloides.* Sympodial arrangement of sporangia on sporangiophore.

Fig. A.3D, E *Mortierella* spp. Sporangia have indistinct columellae which are often obscured by spores (D, *M. hyalina* (Harz) W. Gams; E, *M. alpina*). After spores are released, frill of ruptured wall at sporangial apex has appearance of collarette (arrows) (after Domsch and Gams, 1972).

Fig. A.3F *Saksenaea vasiformis.* Immature and mature sporangia (after Saksena, 1953).

addition to the elongate and tapered form of the sporangia of *Mortierella*, has occasionally led to their being misinterpreted as *phialides*. Of course, the similarities of these two kinds of reproductive structures are based only on superficial morphology and not development. Similarly, the mature sporangium of *Saksenaea vasiformis* Saksena, after rupture of its apex and release of sporangiospores (Fig. A.3F), somewhat resembles the mature phialide of *Chalara* (Corda) Rabenh. (Nag Raj and Kendrick, 1975). In contrast to *Mortierella*, the sporangium of *S. vasiformis* has a well-defined columella.

The formation of terminal and intercalary *chlamydospores* has been reported in the Zygomycetes. Some species of *Endogone* Link ex Fr. apparently produce both chlamydospores (Fig. A.4A) and zygospores in the same fruit body (Thaxter, 1922; Zycha *et al.*, 1969; Webster, 1970; Gerdemann and Trappe, 1974). The *chlamydospores* are thick-walled and tolerate broad ranges of pH, since they have been found in the guts of rodents, especially mice, rabbits and voles (Dowding, 1955; Webster, 1970). However, there is some controversy over whether these structures are actually *azygospores* (i.e., parthenospores), rather than *chlamydospores* (Gerdemann and Trappe, 1974). One of the problems of clearly defining development in the Endogonaceae is that, as yet, no successful, pure culture studies have been made on this group (Hesseltine and J. J. Ellis, 1973). On the other hand, thick-walled, pigmented structures, which resemble chlamydospores, or holothallic conidia produced by deuteromycetous fungi, have been reported by Benjamin and Mehrotra (1963) in *Mucor bainieri* Mehrotra & Baijal (Fig. A.4B–E). These spores may form in chains attached to substrate hyphae (Fig. A.4C), terminally at the apices of aerial hyphae (Fig. A.4D) or endogenously within a sporangiophore (Fig. A.4E). Similar chlamydospores were described in *Amylomyces rouxii* Eykman by J. J. Ellis *et al.* (1976). Comparative examinations of chlamydospores in the Mucorales and holothallic conidia are necessary to determine whether developmental similarities exist between these structures.

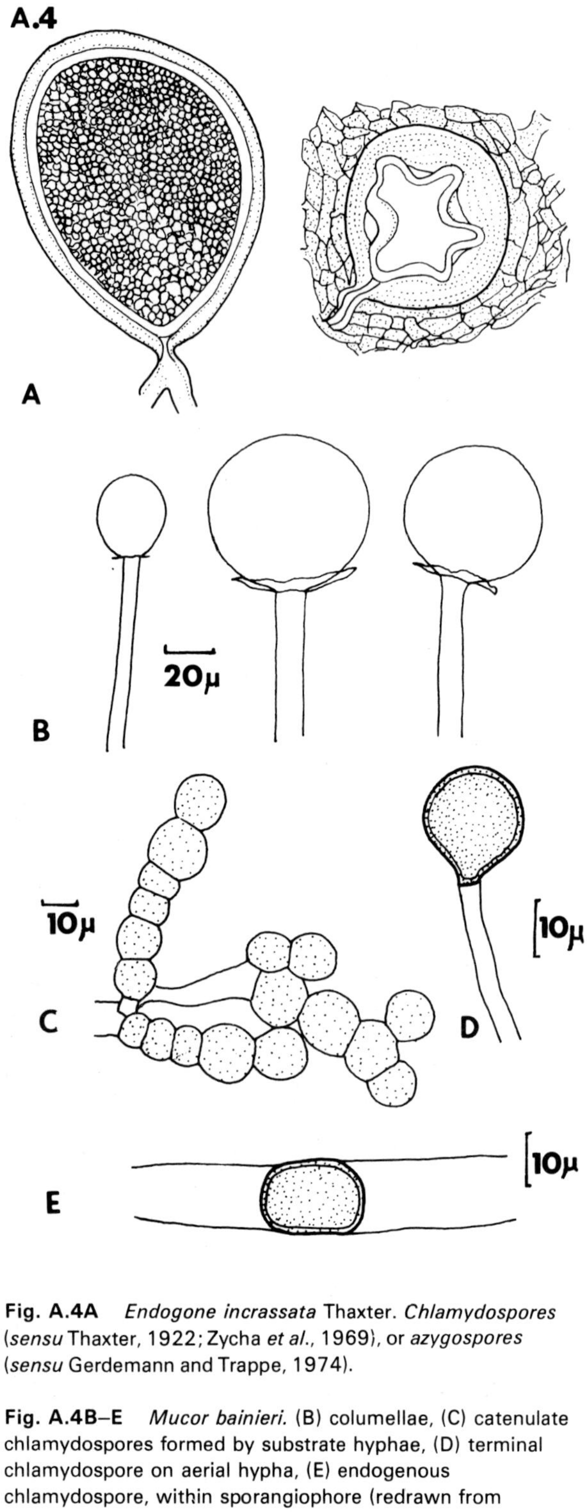

Fig. A.4A *Endogone incrassata* Thaxter. *Chlamydospores* (*sensu* Thaxter, 1922; Zycha *et al.*, 1969), or *azygospores* (*sensu* Gerdemann and Trappe, 1974).

Fig. A.4B–E *Mucor bainieri.* (B) columellae, (C) catenulate chlamydospores formed by substrate hyphae, (D) terminal chlamydospore on aerial hypha, (E) endogenous chlamydospore, within sporangiophore (redrawn from micrographs of Benjamin and Mehrotra, 1963).

Few-spored sporangiola on stalks or vesicles

A sporangiolum was defined by Ainsworth (1971) as 'a small sporangium without a columella, generally having a small number of spores'. However, Benny and Benjamin (1975, 1976) have shown that a distinct columella is often produced by members of this group of mucoraceous fungi and that the sporangiolar wall is persistent. Sporangiola are found mainly in the families Radiomycetaceae and Thamnidiaceae, but sometimes occur in combination with multispored sporangia, as in the Mortierellaceae and Thamnidiaceae.

Radiomyces spectabilis Embree gives rise to few-spored sporangiola on elaborate sporangiophores (Figs. A.5A–D, A.6). Initially, the sporangiophore apex swells to form a terminal vesicle which gives rise synchronously to a series of fertile branches, each terminating in smaller vesicles (arrows in Fig. A.5A). Over the surface of these vesicles, again in synchronous fashion, a large number of papillae develop

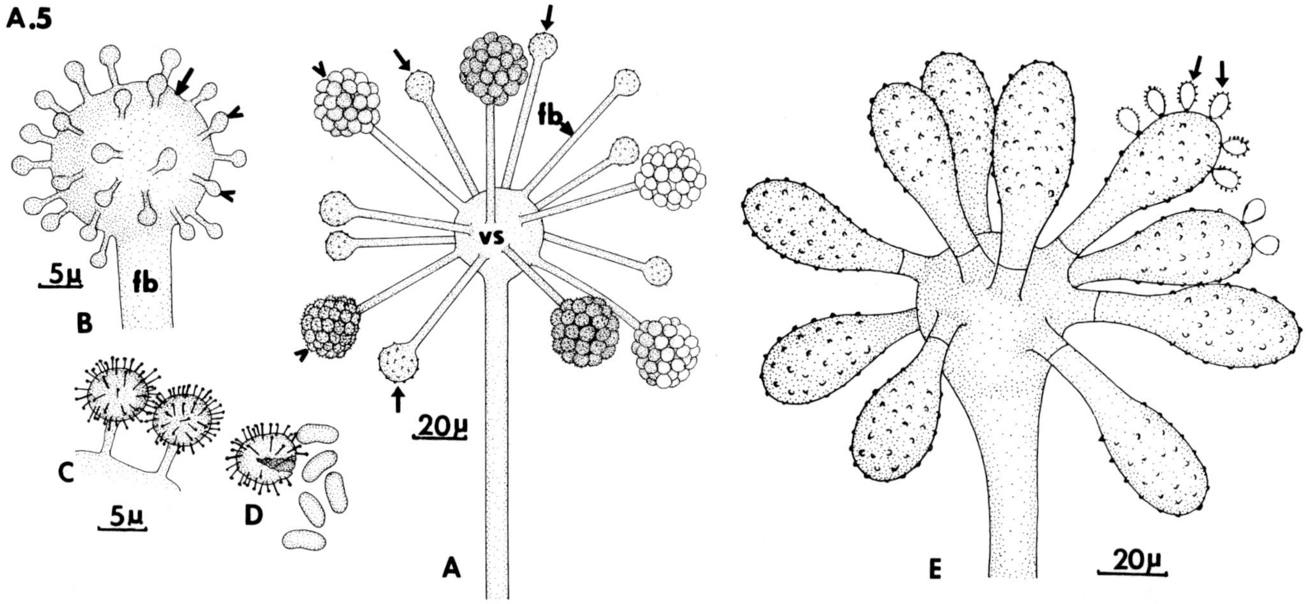

Fig. A.5A–D *Radiomyces spectabilis.* Formation of few-spored sporangiola. The vesicle (vs) of the sporangiophore gives rise to fertile branches which terminate in small vesicles (arrows in A, B). The latter produce sporangiola (arrowheads in A, B) which become appendaged (C) and eventually break open to release the sporangiospores (D). fb, fertile branch arising from vesicle of sporangiophore.

Fig. A.5E *Radiomyces embreei.* Arrows indicate monosporous sporangiola.

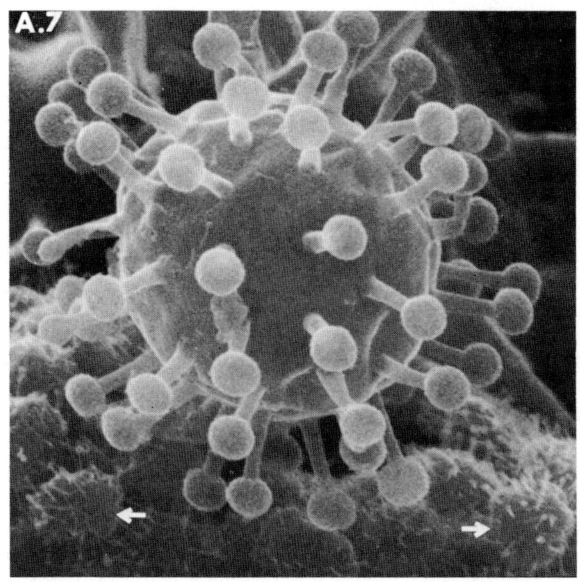

Figs. A.6, A.7 *Radiomyces spectabilis.* Sporangiophore bearing clusters of sporangiola at ends of fertile branches (A.6). Early stage of sporangiolar development (A.7) from small vesicle at tip of fertile branch (cf. A.5B). Arrows (A.7) indicate seceded, appendaged, mature sporangiola. ×600 and 3,200, respectively.

initially as conical outgrowths, but later become swollen apically (arrowheads in Figs. A.5A, B, A.7). These clustered, globose structures differentiate into sporangiola covered with short capitate appendages (Fig. A.5C, arrows in Fig. A.7) while still attached to the fertile vesicles by cylindrical denticles. They later secede from the denticles and eventually, by breakdown of the outer sporangiolar wall, release their spores (Fig. A.5D). This process of multiple branching from a terminal, fertile vesicle in *R. spectabilis* is reminiscent of the development of conidiophores of *Botrytis cinerea* Pers ex Fr. and *Botryosporium longibrachiatum* (Oudemans) Maire (Chapter 3). In the Hyphomycetes, lateral branches arise from the conidiophore axis and branch repeatedly before terminating in swollen ampullae which give rise to synchronously developing, holoblastic conidia. The sporangiophore of *Radiomyces embreei* Benjamin (Figs. A.5E, A.8) also forms

A.8

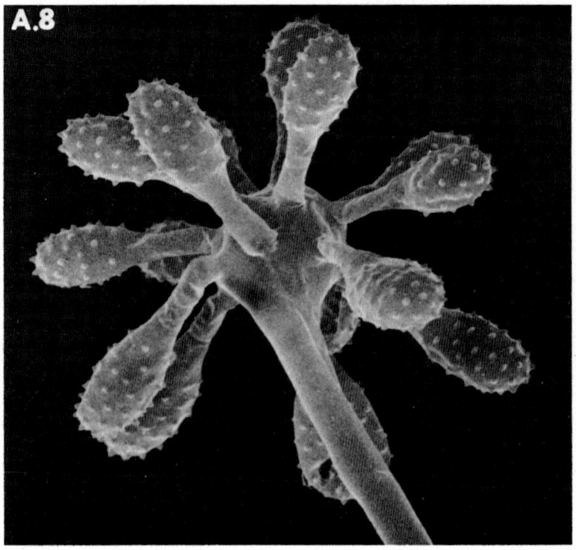

A.9

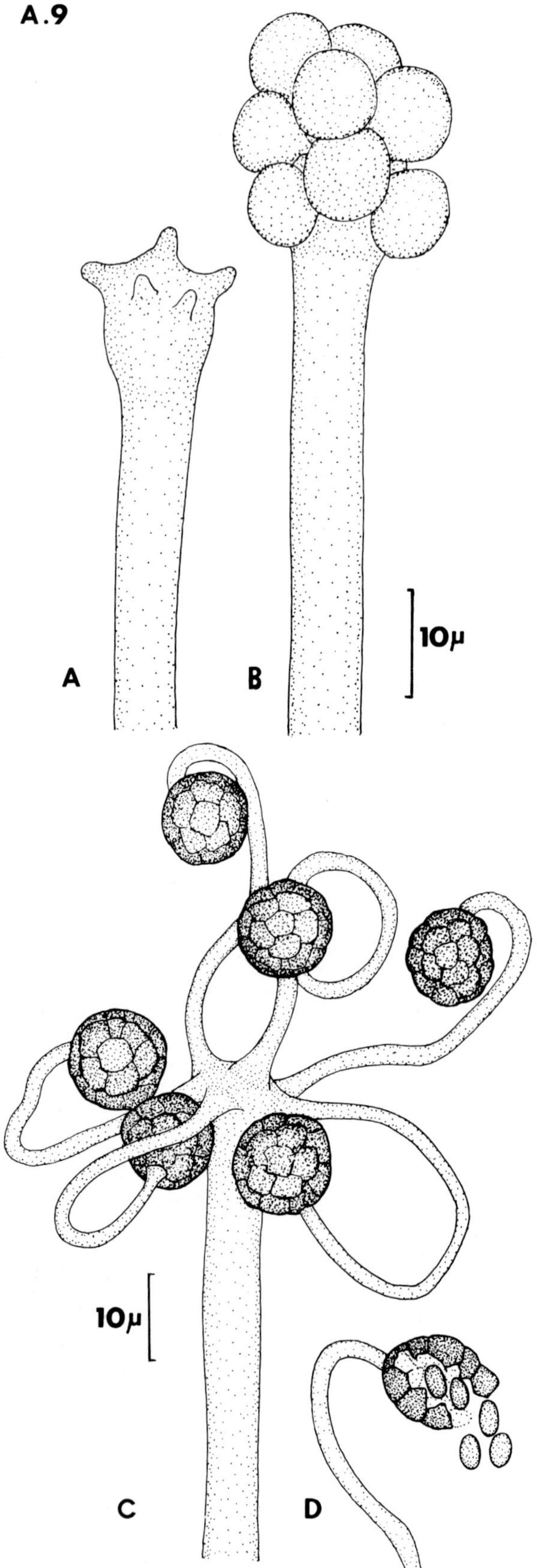

Fig. A.8 *Radiomyces embreei.* Mature sporangiophore from which the monosporous sporangiola have seceded. Denticles are visible on surface of swollen fertile branches. ×800.

Fig. A.9A–D *Cokeromyces recurvatus.* Stages of development of pedicellate sporangiola from sporangiophore vesicle. Thickened wall of sporangiolum eventually ruptures, releasing sporangiospores (D).

Figs. A.10–A.15 *Cokeromyces recurvatus.* Initial asynchronous proliferation of pedicels terminates in differentiation of apical sporangiola (A.10–A.12). Pedicels subsequently undergo intercalary growth (A.13) while thick-walled, pitted sporangiola remain attached to apices (A.14) but later secede passively (A.15). ×2,300, 4,100, 2,300, 1,300, 3,700 and 1,500, respectively.

a terminal vesicle from which several fertile branches arise, each bearing many ornamented sporangiola (arrows in Fig. A.5E). The sporangiola in Fig. A.8 have all seceded. In contrast to *R. spectabilis,* each sporangiolum of this species is monosporous and its developmental similarity to the synchronous, holoblastic, botryose conidium is even more striking.

Cokeromyces recurvatus Poitras (Figs. A.9–A.15) produces sporangiophores 0·5–1 mm high which form pedicellate sporangiola, each containing 10–25 spores. Early in sporangiolar development, tubular outgrowths of the terminal, sporangiophore vesicle arise synchronously (Fig. A.9A), but subsequently elongate at different rates. Extension growth then ceases and each outgrowth swells apically (Figs. A.9B, A.10, A.11), differentiating into a sporangiolum (Fig. A.12). The sporangiolar wall thickens and becomes pigmented and pitted. Normally, the sporangiola do not secede at this stage and those which have been detached in Fig. A.12 were probably removed during sample preparation. After the sporangiola have apparently matured, the pedicels undergo intercalary growth (Figs. A.9C, A.13). In several respects, this developmental process is comparable to

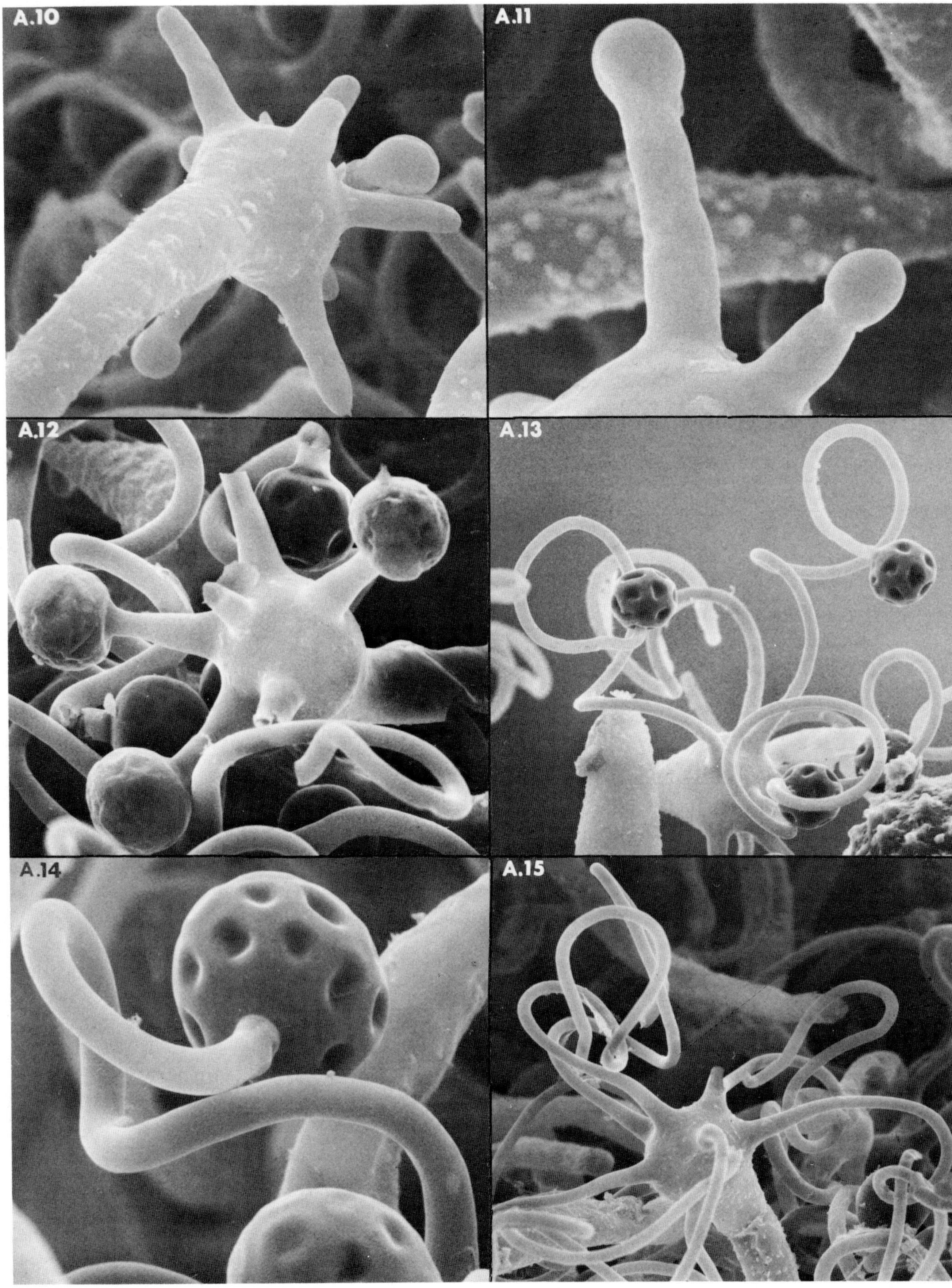
A.10
A.11
A.12
A.13
A.14
A.15

basauxic conidiogenous cell ontogeny in *Spegazzinia tessarthra* (Berk. & Curt.) Sacc. (Chapter 6). In each case, basauxic or intercalary growth of the filamentous structure (pedicel or conidiogenous cell) occurs only after the terminal propagule (sporangiolum or conidium) has matured. The thick, pitted wall of the sporangiola (Fig. A.14) eventually ruptures, releasing the sporangiospores (Fig. A.9D). Secession is apparently a passive process in *C. recurvatus*, since there is no recognizable weakened zone of dehiscence (i.e., region of schizolysis) at the base of the sporangiolum or pedicel (Figs. A.14, A.15). The processes of basauxic development in *Cokeromyces* Shanor and *Spegazzinia* Sacc. have evolved separately, but they demonstrate similar adaptations to insect dispersal. The pedicels of *C. recurvatus* easily become entangled with insect appendages, while the spiny conidia of *S. tessarthra,* precariously attached to their proliferated conidiogenous cells, are readily carried away and dispersed by passing foragers.

Monosporous sporangiola on vesicles

Choanephora cucurbitarum (Berk. & Rav.) Thaxter (Figs. A.16–A.19) produces both multispored sporangia and monosporous sporangiola. The columellate multispored sporangia rupture and release striate spores with spinelike, polar appendages (Fig. A16B, C). The monosporous sporangiola develop from swollen vesicles (Fig. A.16A). Hesseltine (1961) showed that *C. cucurbitarum* forms monosporous sporangiola at 25°C, while at 31°C multispored sporangia are produced instead. *Choanephora trispora* (Thaxt.) Sinha demonstrates similar dimorphism, except that here each sporangiolum releases 3 or 4 appendaged spores (Hesseltine and Benjamin, 1957). The sporangiolar initials of *C. cucurbitarum* develop synchronously from the fertile vesicle as conical projections (Fig. A.17), which later swell at their apices becoming ellipsoidal (Fig. A.18). The monosporous sporangiola remain attached to the vesicle by short pedicels (Fig. A.19). The wall of the sporangiolum, which is continuous with that of the fertile vesicle, is fused to the spore wall. The striations on the surface of mature spores (Hesseltine and Benjamin, 1957) may be the result of some separation of the two wall layers. Secession of the sporangiolum occurs by a fracture through its cylindrical base (arrow in Fig. A.18). The synchronous, holoblastic nature of sporangiolar development in *Choanephora* Currey is comparable to botryose conidial formation in *Spiniger* Stalpers and *Oedocephalum* Preuss. (Chapter 3).

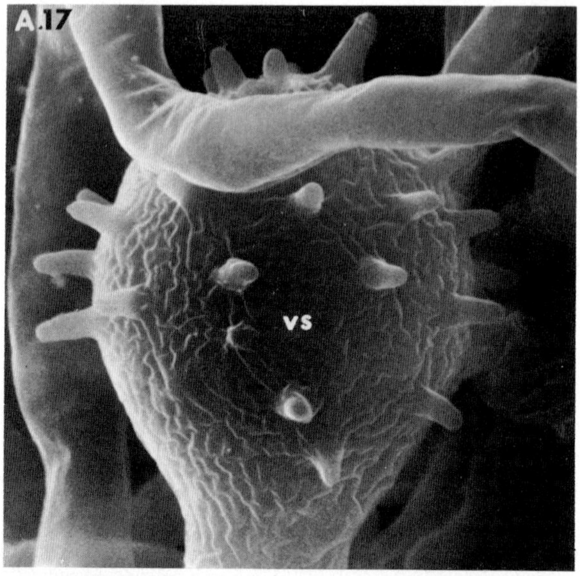

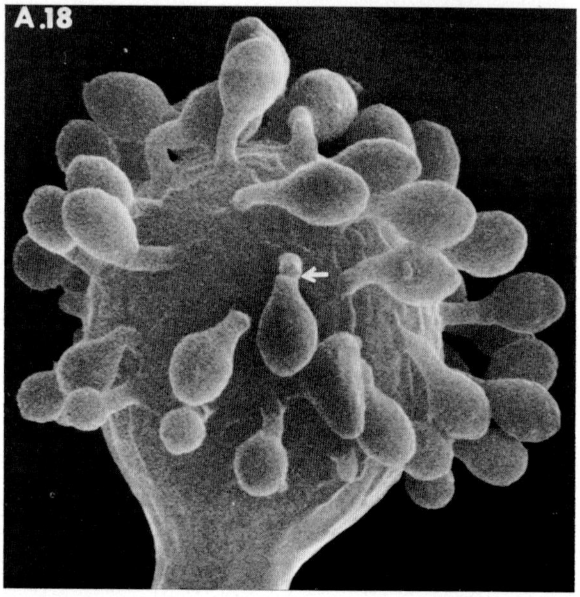

A.16

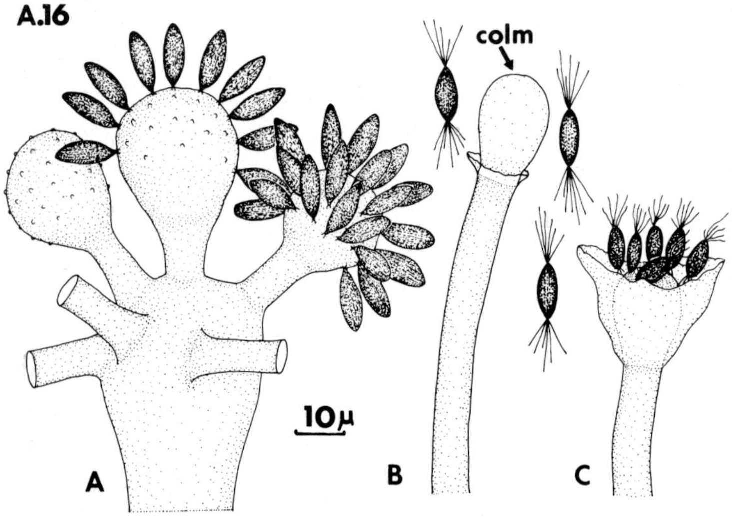

Fig. A.16A–C *Choanephora cucurbitarum.* (A) Monosporous sporangiola produced from cluster of fertile vesicles. (B, C) Multisporous sporangia showing distinct columellae (colm) and appendaged sporangiospores.

Figs. A.17–A.19 *Choanephora cucurbitarum.* Stages of synchronous development of holoblastic, monosporous sporangiola from sporangiophore vesicle (vs). Arrow in A.18 indicates fracture through pedicel. ×3,100, 2,300 and 2,900, respectively.

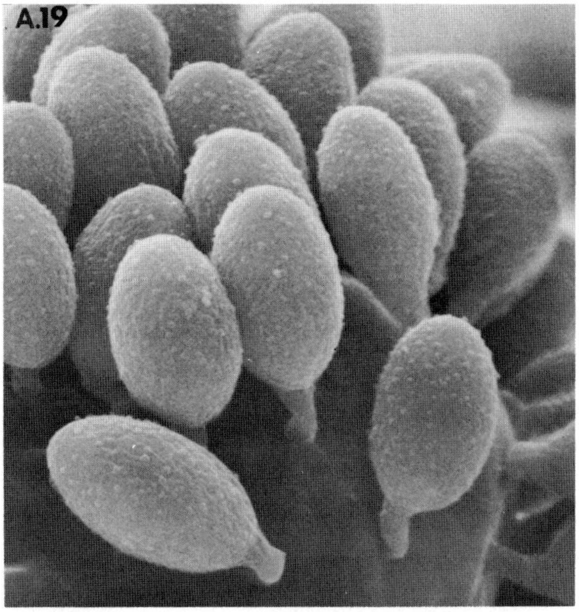

A.20

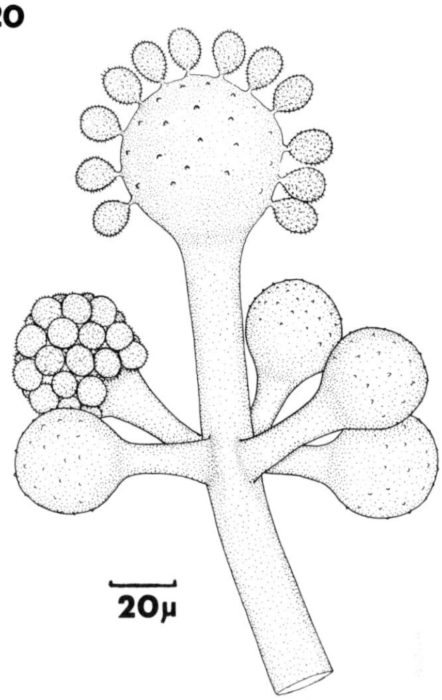

Fig. A.20 *Cunninghamella elegans.* Fertile vesicles which have formed at ends of branches of sporangiophore. Monosporous sporangiola arise from vesicles and eventually secede, leaving abstriction scars.

Figs. A.21, A.22 *Cunninghamella echinulata.* Young sporangiola which have formed holoblastically from the surface of the fertile vesicle. The spines first appear as small excrescences (arrow, A.21) and elongate synchronously (A.22). ×8,400 and 11,500, respectively.

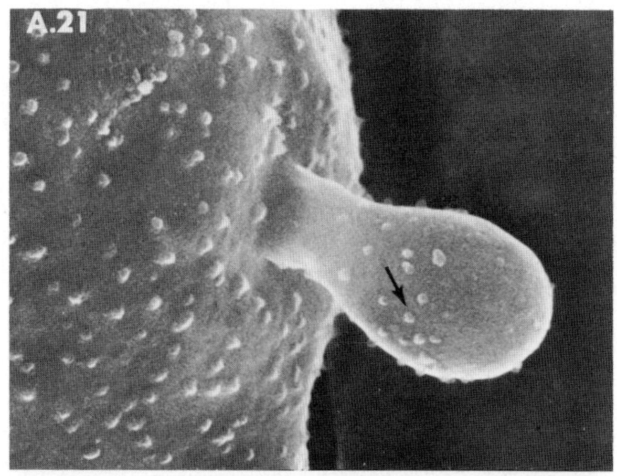

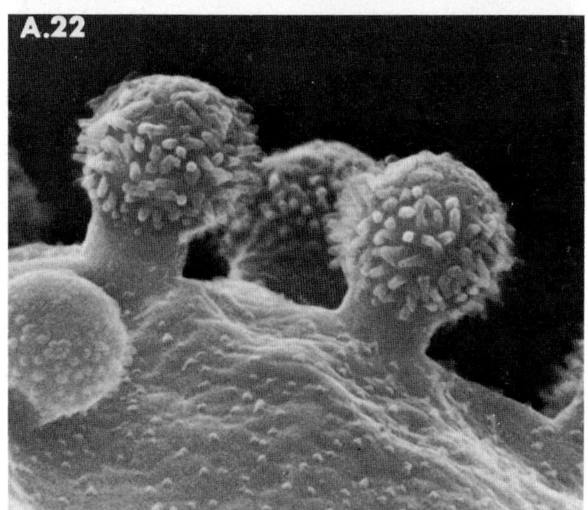

Similar monosporous sporangiolar ontogeny occurs in members of the Cunninghamellaceae (e.g., *Cunninghamella elegans* Lendner, Fig. A.20; *C. echinulata* (Thaxt.) Thaxt., Figs. A.21–A28), and Thamnidiaceae (e.g., *Mycotypha microspora* Fenner, Figs. A.29, A.31–A39). Again, morphological and apparent developmental similarities exist between these species and several hyphomycetous fungi. The elongated vesicles of *M. microspora* (Figs. A.31A, A.32) appear superficially similar to the fertile conidiogenous cells of the *Ostracoderma* state of *Peziza ostracoderma* Korf (cf. Chapter 3, Figs. 3.67, 3.68, 3.70C). The synchronous formation of monosporous sporangiola from vesicles of *Cunninghamella echinulata* is reminiscent of conidial development from the ampullae of *Oedocephalum.* Earlier misinterpretations of monosporous sporangiola as typical *conidia* have been generated primarily by the lack of a clear understanding of the differentiation of the sporangiolar and spore walls in Zygomycetes (S. J. Hughes, 1971a, c). This in turn, has led to erroneous conclusions that the development of monosporous sporangiola is unrelated to the endogenous cleavage in multisporous sporangia and sporangiola. However, mounting evidence is available, mainly from thin-section investigations, which indicates that the monosporous sporangiolum is simply a reduced sporangium: clear ontogenetic differences exist between the former and the conidium. The functional similarities between these two kinds of

Figs. A.23, A.24 *Cunninghamella echinulata.* Later stages of sporangiolar development. The mature sporangiola (A.24) are supported by short, cylindrical pedicels (arrows in A.23). ×3,500 and 2,300, respectively.

Figs. A.25–A.28 *Cunninghamella echinulata.* Thin-sections through pedicel (P, A.25), fertile vesicle (vs, A.25) and sporangiolum (S, A.26), showing structural relationships of the outer (OW) and inner (IW) wall layers (A.27, A.28). Note that the pedicel is almost completely occluded by inner wall material in A.25. L, lipid droplet; pm, plasma membrane. ×16,400, 8,500, 25,830 and 50,400, respectively (after Benny, in prep.).

Fig. A.29 *Mycotypha microspora.* Thin-section through dimorphic sporangiola (S_1, S_2) and supporting pedicels (P). At maturity the sporangiospore wall (IW) is complete, relatively thick and encompassed by a thin, sporangiolar wall (OW) which is visible in several places. Secession of the sporangiola occurs by a circumscissile rupture of the pedicels (arrows). vs, fertile vesicle of sporangiophore. ×24,000 (from Benny, in prep.).

asexual, caducous propagules are merely the result of convergent evolution.

Investigations of the differentiation of wall layers of ungerminated spores of *Cunninghamella elegans* (Hawker, 1966) have demonstrated the presence of 'a thin outer electron-dense layer corresponding to the sporangial wall, and a thick, less electron-dense, inner one, corresponding to the sporangiospore wall'. On the other hand, this distinction between two wall layers was not visible in ungerminated conidia of the Hyphomycete *Botrytis cinerea* (Hawker and Hendy, 1963; Buckley *et al.*, 1966). Although differences in wall formation during germination of sporangiospores and conidia have also been suggested (Bartnicki-Garcia, 1968; Hawker *et al.*, 1970; Dykstra, 1974a), the validity of these data have been challenged (Khan, 1975). Khan and Talbot (1975) pointed out that primary emphasis in the definition of a sporangiospore should be on the process of wall differentiation during its ontogeny. In their thin-section examination of *Cunninghamella echinulata*, Khan and Talbot (1975) illustrated early sporangiolar development, which is signalled by the appearance of 'localized weak spots . . . in the ampullar wall', which, 'yielding to internal pressure', first blow-out into pedicels and then swell more or less synchronously at their apices to become the spiny, monosporous sporangiola (Figs. A.21–A.24).

Benny (in prep.) also presented thin-section micrographs of this developmental process in *C. echinulata*, some of which are reproduced here (Figs. A.25–A.28). In Fig. A.25, the pedicel (P) is shown extending between the fertile vesicle (vs) and sporangiolum (S). An outer wall layer (OW) of the fertile vesicle extends over the surface of the pedicel and encompasses the young sporangiolum (Fig. A.26), and at higher magnification (Fig. A.27), appears as a thin, electron-opaque layer. An inner wall layer (IW, Fig. A.25), which is continuous between the fertile vesicle and sporangiolum, has also differentiated (Fig. A.27). The rate of thickening of this layer is greatest in the pedicel which later becomes almost completely occluded with wall material (Figs. A.25, A.26). The inner wall surrounding the fertile vesicle (IW, Fig. A.27) remains thin but thickens around the developing sporangiolum (IW, Fig. A.28). The outer wall at this stage (OW, Fig. A.28) appears to have partially degenerated. The inner wall layer (IW) becomes the wall of the sporangiospore,

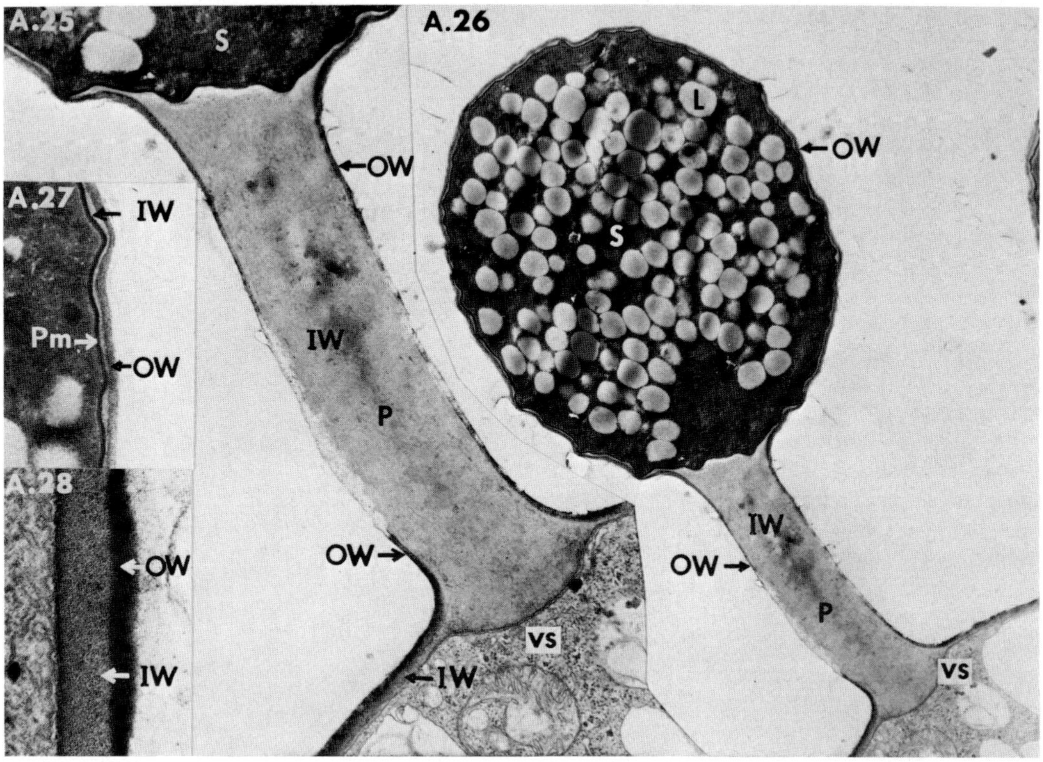
A.25
S
OW
IW
P
OW
IW
vs
A.26
L
OW
S
IW
OW
P
vs
A.27
IW
Pm
OW
A.28
OW
IW

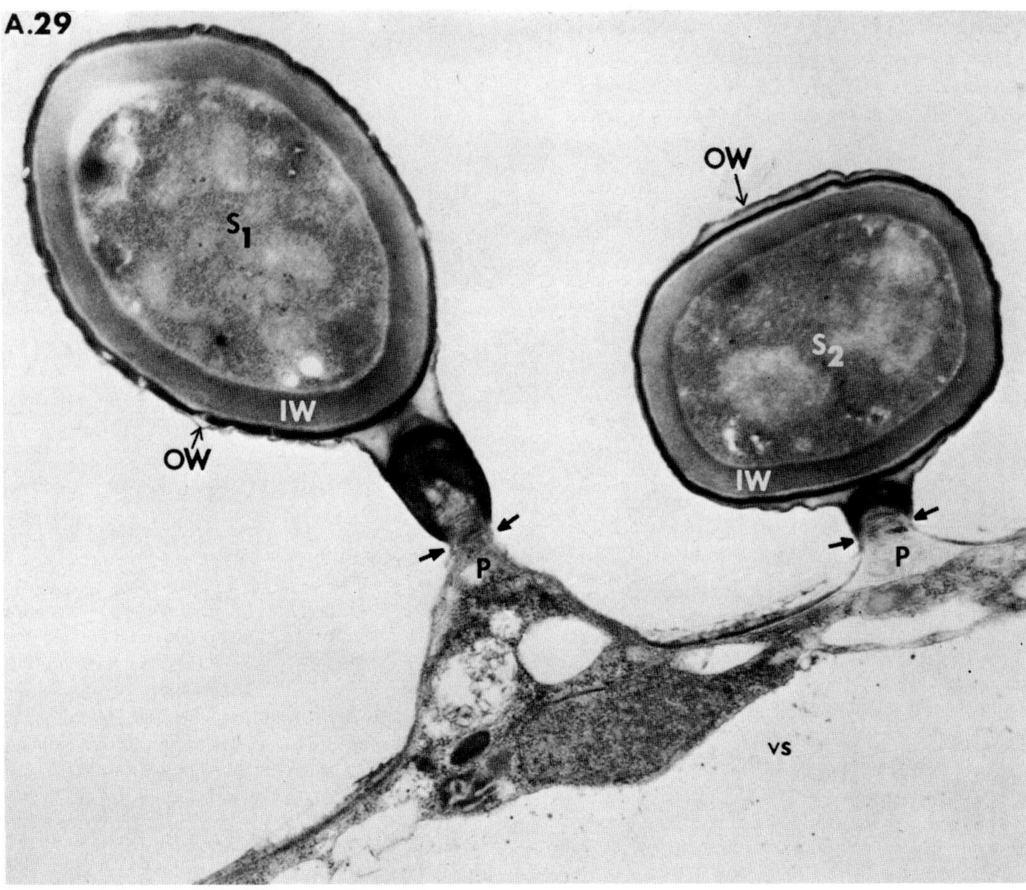
A.29
S1
IW
OW
OW
S2
IW
P
P
vs

while the outer wall layer (OW) is the sporangiolar wall and is suggested to be homologous with the outer wall of a multisporous sporangium. Secession occurs by a fracture through the pedicel and a short cylindrical denticle is left on the surface of the vesicle.

These ultrastructural investigations permit us to present a comparative, diagrammatic interpretation of wall differentiation during development of a monosporous sporangiolum (Fig. A.30A–D) and formation of holoblastic conidia in hyphomycetous fungi (Fig. A.30E–H). The major difference between these two schemes is evident during maturation of the two propagules. The inner wall of the sporangiospore is complete, surrounding the protoplasm contained within the sporangiolum (Fig. A.30C). Deposition of wall material in the pedicel and occlusion of the cytoplasmic canal occur prior to secession of the sporangiospore (Fig. A.30D). The inner wall layer of the holoblastic conidium is continuous with its basal septum which develops centripetally (Fig. A.30F). In many cases, the conidial septum is two-layered and differentiates a medial, electron-transparent zone along which schizolysis occurs (Fig. A.30G, H). The conidium, therefore, develops as a truly exogenous, blastic outgrowth of the conidiogenous cell. On the other hand, the sporangiospore differentiates endogenously within the evaginated sporangiolar wall. A closer structural association between the sporangiolar and spore walls presumably evolved concomitantly with the reduction from multispored sporangia to monosporous sporangiola.

In *Mycotypha microspora*, the slightly swollen, cylindrical sporangiophore (Figs. A.31A, A.32) produces two kinds of monosporous sporangiola (Khan and Talbot, 1975) which are arranged in two layers (Figs. A.31B, A.33). Some are elevated above the surface of the sporangiophore supported

A.30

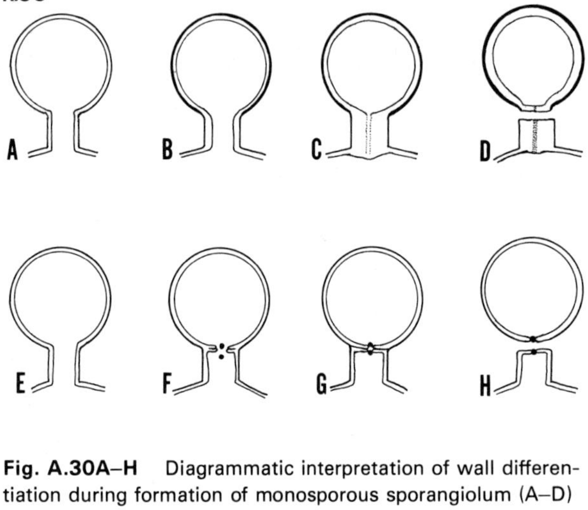

Fig. A.30A–H Diagrammatic interpretation of wall differentiation during formation of monosporous sporangiolum (A–D) and holoblastic conidium (E–H).

Fig. A.31A, B *Mycotypha microspora.* Dense cluster of monosporous sporangiola on elongated, cylindrical sporangiophore (A). Arrangement of dimorphic sporangiola (B).

A.31

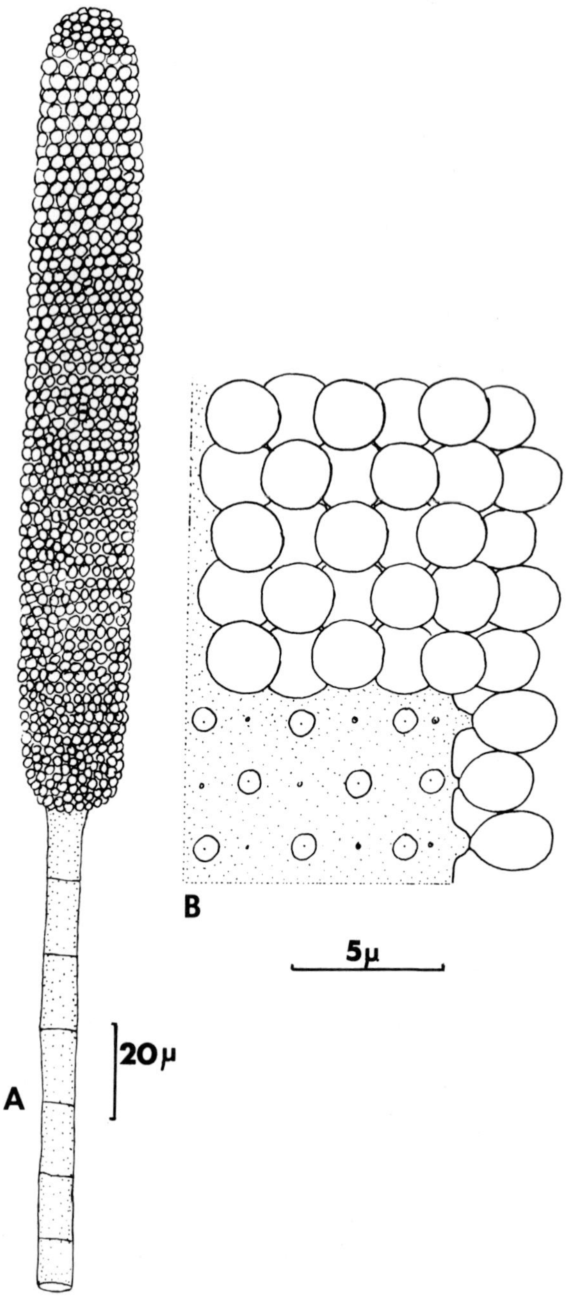

Figs. A.32–A.39 *Mycotypha microspora.* Elongated, cylindrical fertile vesicle (vs) at sporangiophore apex (A.32) which produces a dense cluster of dimorphic sporangiola (A.33–A.35). One kind of sporangiolum (S_1 in A.35, A.37) has an attenuated base and is supported by a short, rounded pedicel (P), while the other (S_2 in A.35, A.37) is attached to a flattened pedicel (arrows in A.36) close to the surface of the vesicle. Arrows in (A.35) indicate constricted region of pedicel where secession occurs. Pores are visible at the apices of pedicels after sporangiola have seceded (A.34, A.36). Freeze-fractured sporangiola (A.37–A.39) reveal distinct rodlet fascicles on their surface (A.38, A.39), morphologically similar to those on conidial surfaces. ×400, 1,200, 600, 5,500, 6,000, 4,900, 23,400 and 53,500, respectively.

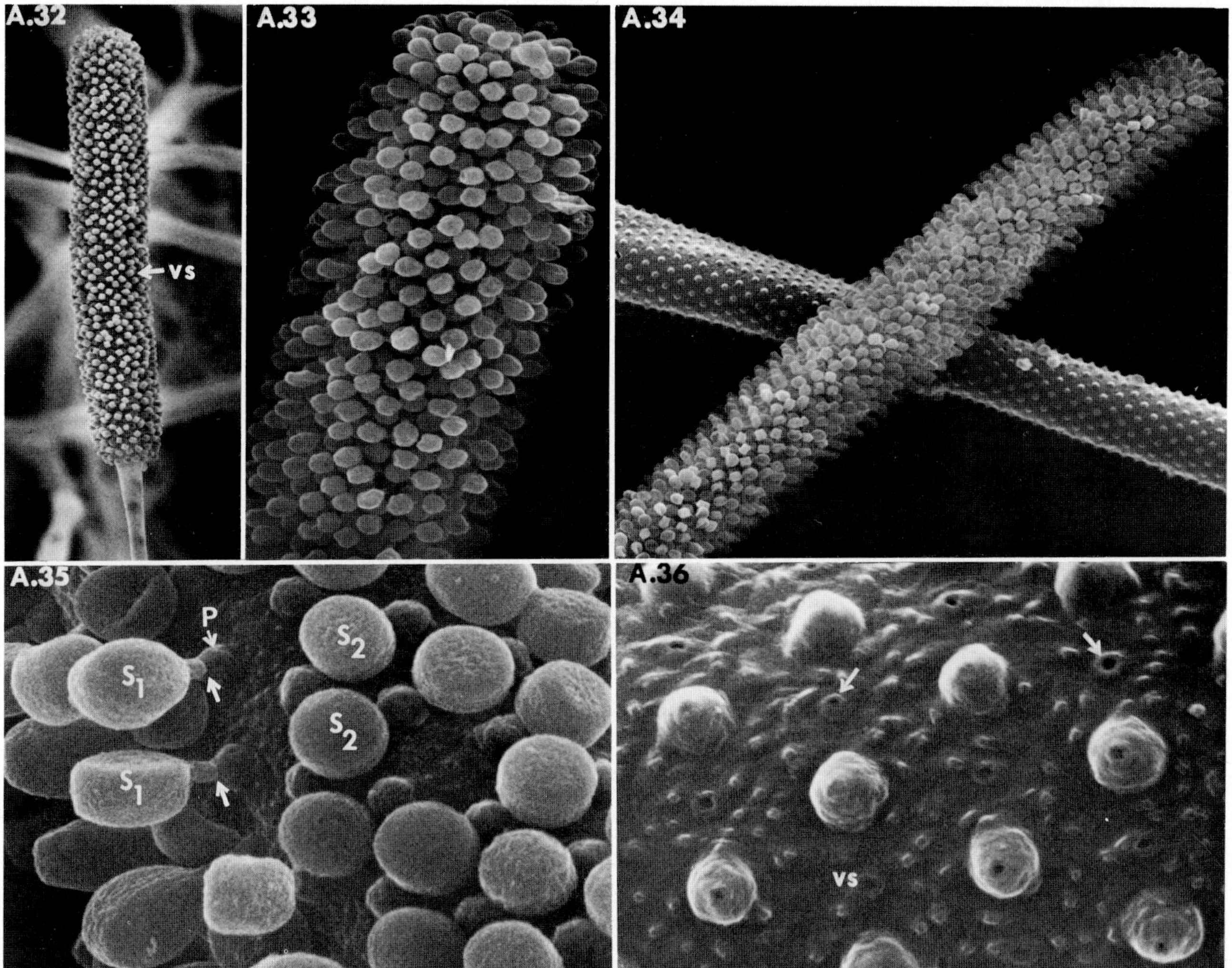
A.32
vs
A.33
A.34
A.35
P
S1
S2
A.36
vs
A.37
S1
S2
P
A.38
A.39

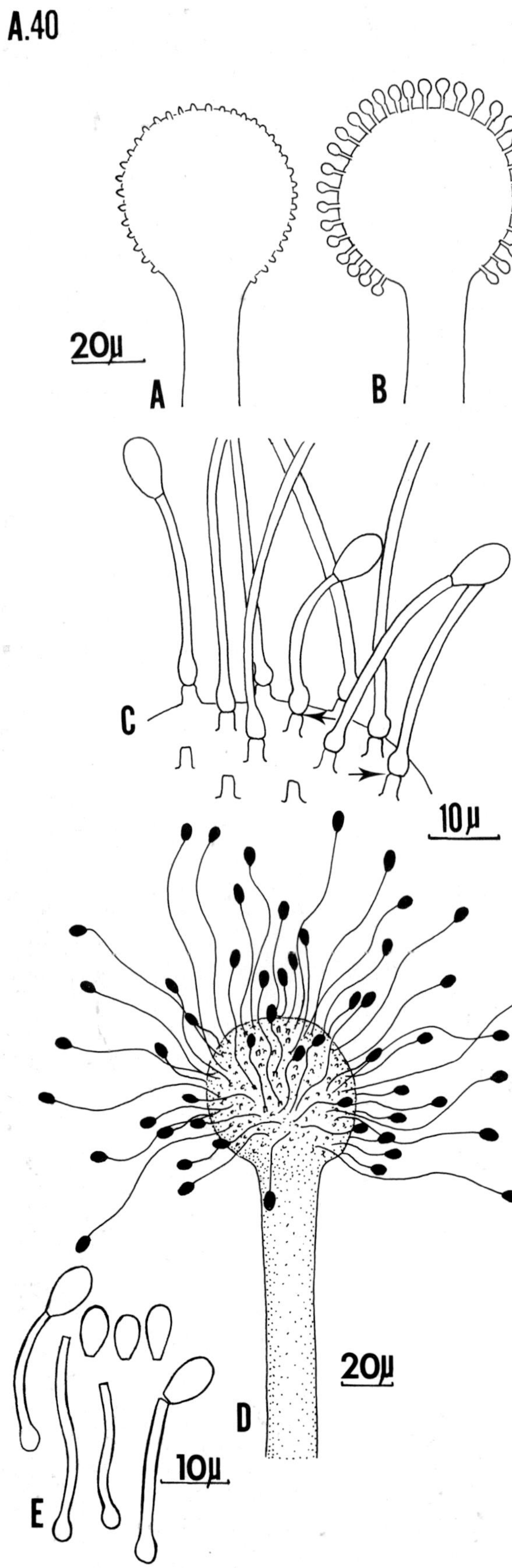

Figs. A.40A–E *Mycotypha poitrasii.* Synchronous development of monosporous sporangiola from fertile vesicle (A, B), followed by asynchronous, intercalary growth of cylindrical pedicels (B–D). Secession of sporangiola usually occurs at junction of swollen base of pedicel and cylindrical denticle (C), but also occurs at base of sporangiola (E).

Figs. A.41–A.46 *Mycotypha poitrasii.* Synchronous, holoblastic development of monosporous sporangiola (S) from surface of fertile vesicle (vs). Arrow in (A.44) indicates programmed zone of abstriction of pedicel from vesicle (cf., arrows in A.45). Unequal extension growth of pedicels, after differentiation of sporangiola (cf., A.44, A.45), results in complexity of intertwined filaments at sporangiophore apex (A.46). ×2,700, 2,000, 5,400, 6,300, 5,000 and 700, respectively.

by their attenuated base (Figs. A.35, A.37, A.38) and raised, papilla-like pedicels (P, Figs, A.35–A.37). The others are wedged between these pedicels and lie close to the sporangiophore surface (Fig. A.35). Wall differentiation during this developmental process is essentially the same as in *Cunninghamella echinulata* (Fig. A.29). The monosporous sporangiola readily secede (Fig. A.34) at a constricted region of their pedicels (arrows in Figs. A.29, A.35). Small pores are revealed at points of detachment (Fig. A.36), which may have provided for cytoplasmic continuity during sporangiolar development. This high degree of regularity in arrangement of asexual propagules has not been found in hyphomycetous fungi which produce synchronous, botryose conidia from ampullae (Kendrick *et al.*, 1968; Cole, 1973b).

Freeze-fractures of the surface of seceded sporangiola (Figs. A.37–A.39) have demonstrated rodlet fascicles which appear very similar to those in Hyphomycetes (Cole and Aldrich, 1971a; Cole, 1973b, c). Comparative investigations of the chemical composition of the rodlet layer in these two fungal groups has been initiated by the senior author; the results may prove interesting in light of earlier discussion of cell wall composition and fungal phylogeny (Bartnicki-Garcia, 1968, 1969; Gooday *et al.*, 1973).

Mycotypha poitrasii (Benjamin) Benny & Benjamin (= *Cokeromyces poitrasii* Benjamin) also produces monosporous sporangiola but at the apex of elongated pedicels (Figs. A.40–A.49). Initial stages of development are well synchronized (Figs. A.40A, B, A.41, A.42) and, as in *M. microspora,* the sporangiola are regularly arranged over the surface of the fertile vesicle (Figs. A.41–A.43). The tubular outgrowths swell apically (Figs. A.42, A.43) and differentiate into sporangiola (Fig. A.44). The pedicels subsequently elongate by intercalary growth (Fig. A.40B–D, A.45, A.46), an ontogenetic process described earlier for the mucoraceous fungus, *Cokeromyces recurvatus,* and some hyphomycetous fungi with basauxic development (Chapter 6). In *M. poitrasii,* an abstriction zone is visible at the base of the differentiated sporangiolum before pedicel elongation (arrow in Fig. A.44). Subsequent rupture of the pedicel at this point (arrows in Figs. A.40C, A.45) confirms that this weakened zone is programmed. The monosporous sporangiola may also

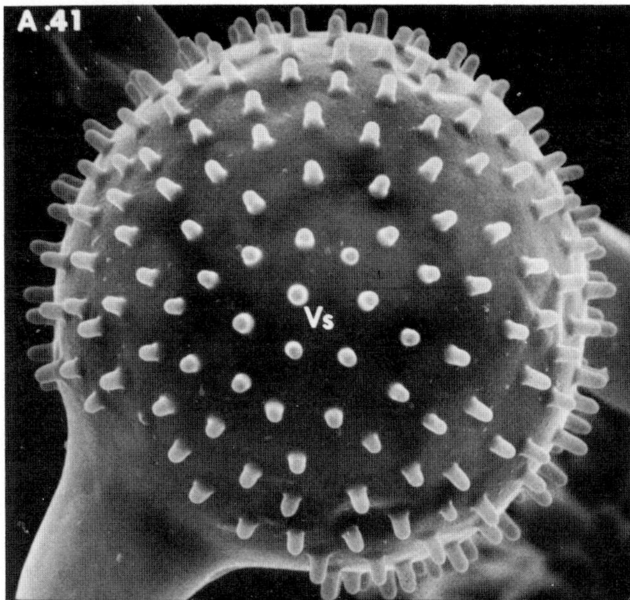
A.41
Vs

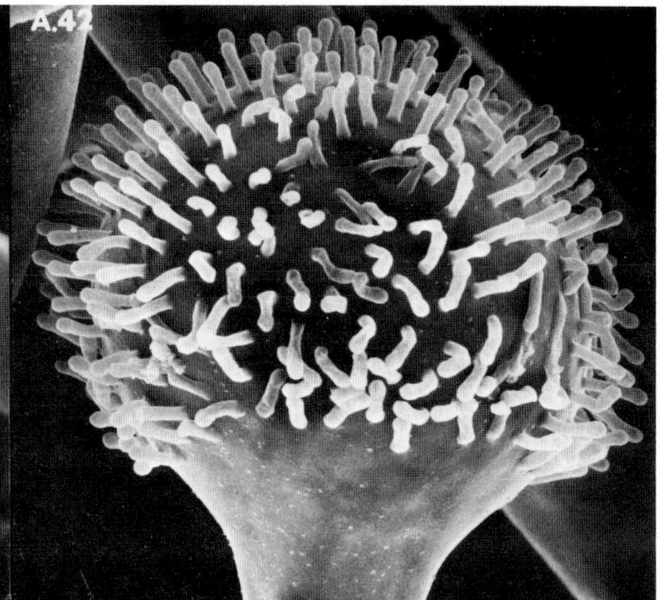
A.42

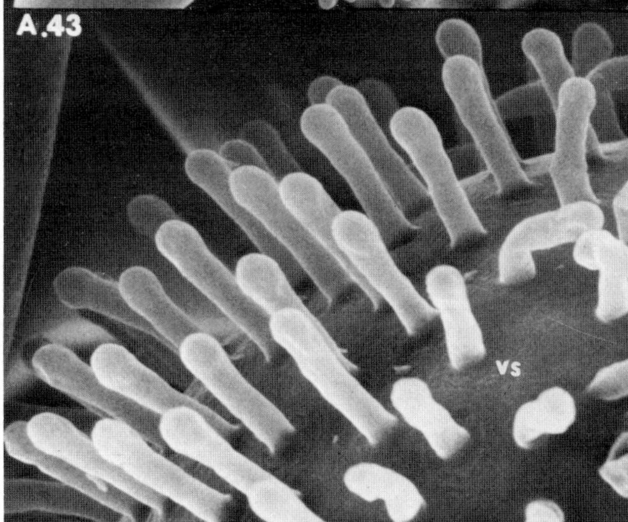
A.43
vs

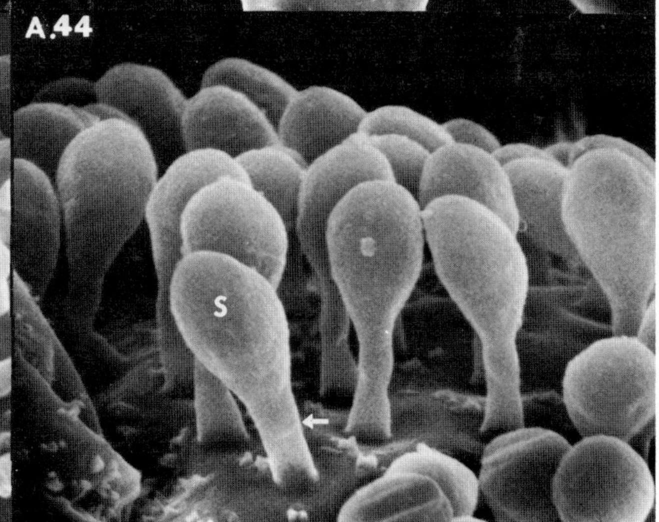
A.44
S

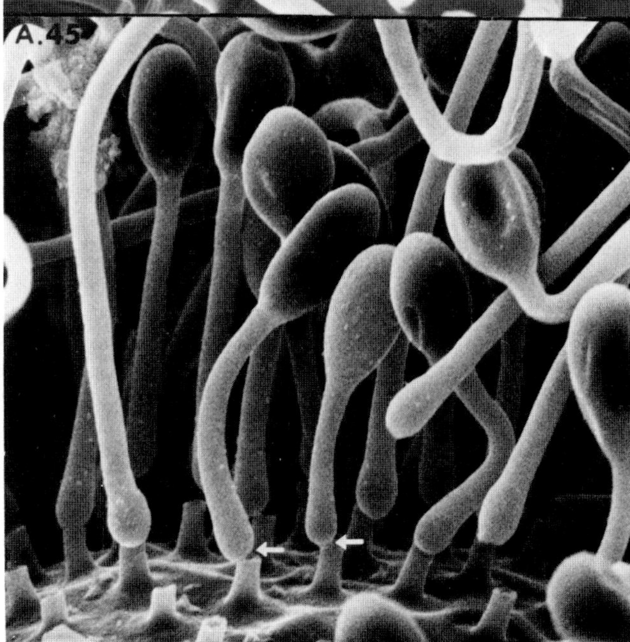
A.45

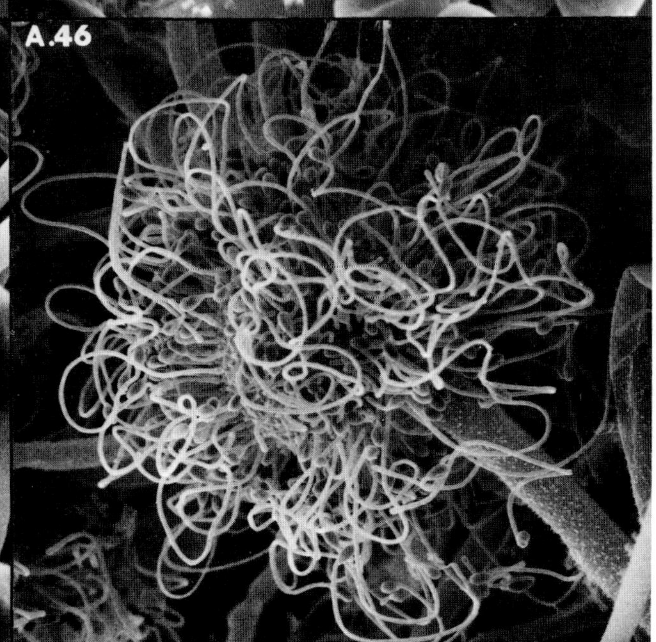
A.46

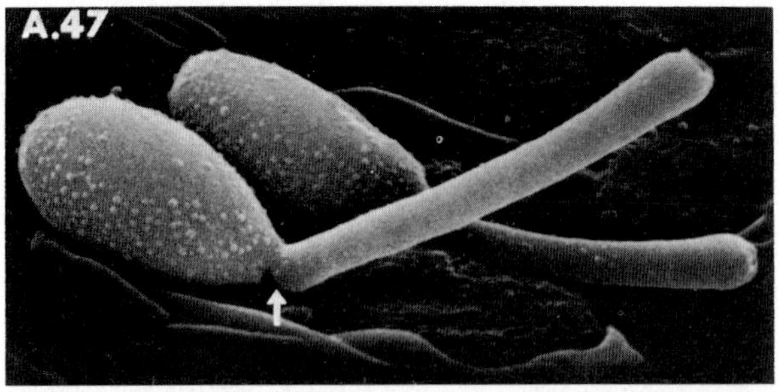

Figs. A.47–A.49 *Mycotypha poitrasii.* Seceded sporangiola and pedicels. Arrow (A.47) indicates rupture at base of sporangiolum. Short, hollow, cylindrical denticles are left on surface of sporangiophore vesicle (A.48, A.49) after pedicels have seceded. ×6,600, 1,100 and 10,200, respectively.

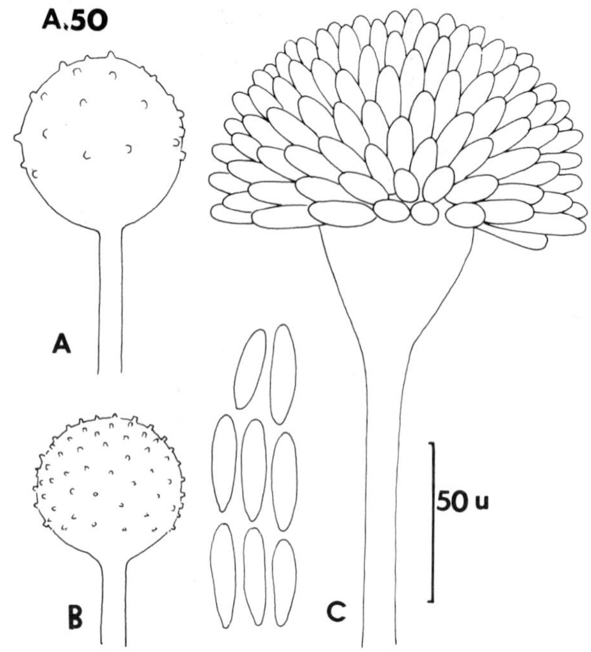

Fig. A.50 *Rhopalomyces* spp. Sporangiophores give rise to synchronously developing, monosporous sporangiola from their fertile vesicles (A,C). Denticles on surface of fertile vesicle after secession of sporangiola (B) (after Drechsler, 1955a).

secede from the apex of the pedicels (Fig. A.40E, arrow in Fig. A.47) although less easily. The pedicels have all seceded from the sporangiophore vesicle in Fig. A.48, leaving short, hollow, cylindrical denticles on its surface (Fig. A.49).

Other species forming monosporous sporangiola from vesicles are found in the Helicocephalidaceae, exemplified by the nematophagous fungus *Rhopalomyces* Corda (Fig. A.50).

Few-spored merosporangia

The Syncephalastraceae, Piptocephalidaceae, and Dimargaritaceae are characterized by production of several endogenous spores, arranged uniseriately within cylindrical outgrowths or *merosporangia* (Martin, 1940), which arise from the usually swollen, fertile apices of simple or branched sporangiophores. As in other groups of Zygomycetes already discussed, the propagules in these families have frequently been misinterpreted as exogenously produced conidia (Brefeld, 1872; de Bary, 1886, 1887; Fischer, 1892; Schroeter, 1897; Lendner, 1908; Migula, 1910). However, Benjamin (1959) was convinced of the mucoraceous affinities of these species and presented an elegant description of their morphology. More recently, Benjamin (personal communication) has agreed with Kreisel's (1969) separation of the Kickxellales from the Mucorales and his suggestion of affinities between the Piptocephalidaceae and Zoopagaceae (Zoopagales). This would leave only *Syncephalastrum*

A.51

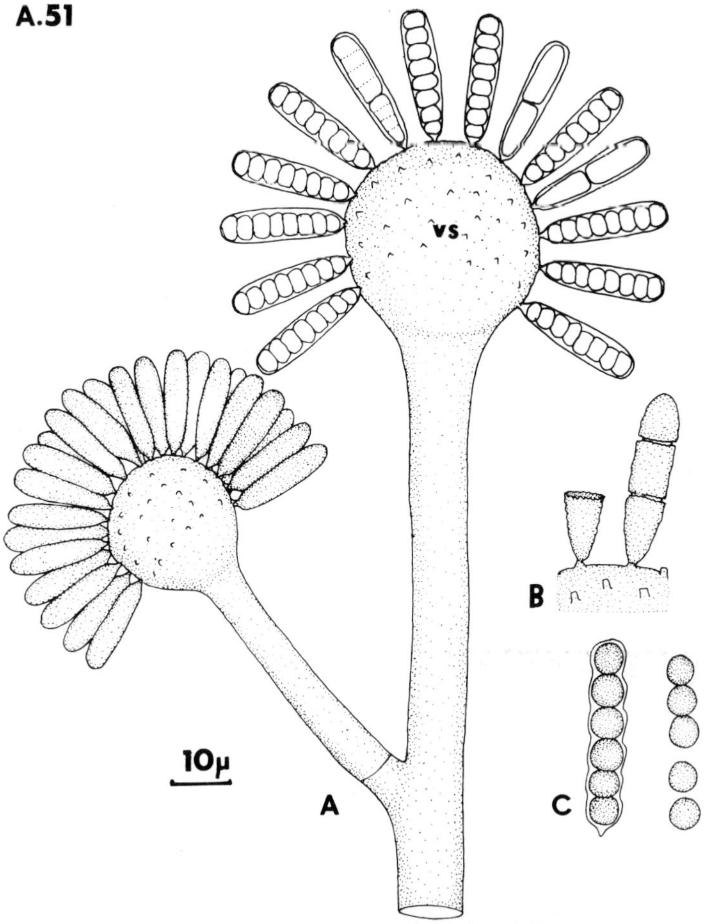

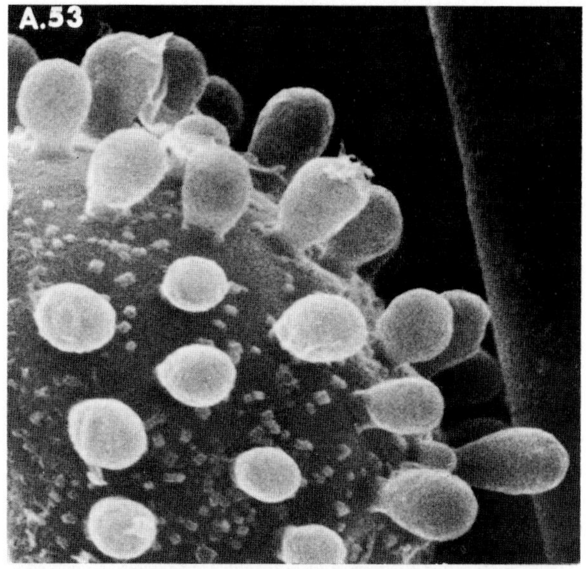

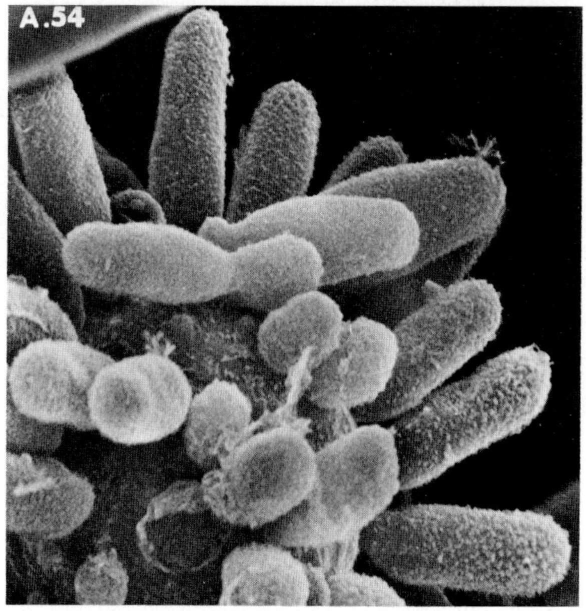

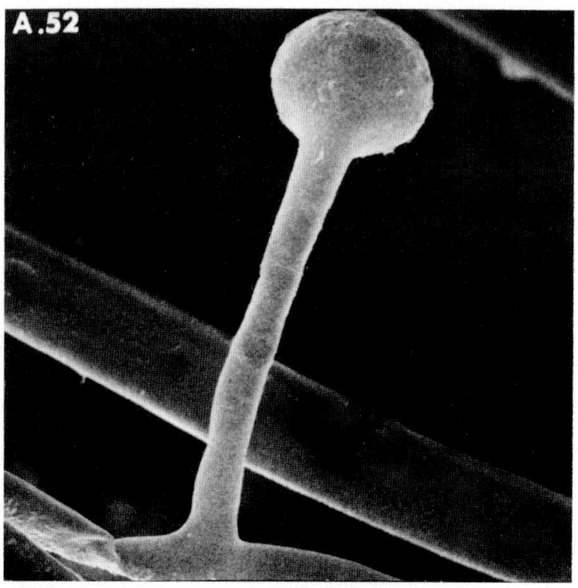

Figs. A.51–A.54 *Syncephalastrum racemosum.* Synchronous formation of merosporangia from fertile vesicle (vs) of sporangiophore (A.52–A.54). Sporangiospores are formed uniseriately within merosporangia (A.51A) and secession occurs by rupture of the merosporangial wall between adjacent spores (Fig. A.51B). A.52–A.54; ×2,100, 7,000 and 3,000, respectively.

Schroeter in what Benjamin (1959) called the merosporangiferous Mucorales. Recent ultrastructural examinations of *Syncephalastrum* (Fletcher, 1972), *Piptocephalis* de Bary (Jeffries and Young, 1975, 1976) and several members of the Kickxellaceae (Young, 1968a, b, 1971; Benny and Aldrich, 1975) have provided evidence that spore formation in merosporangia occurs by typical protoplasmic cleavage and wall differentiation. However, thin-section examinations of multispored species of *Piptocephalis* and *Syncephalis* van Tieghem & Le Monnier are needed for comparison with other multisporous sporangial forms. Baker *et al.* (1977) examined thin-sections of *Syncephalis sphaerica* van Tieghem and demonstrated cleavage of the merosporangial protoplasm into a uniseriate arrangement of endogenous sporangiospores.

In *Syncephalastrum racemosum* Cohn ex Schroeter (Figs.

A.51–A.58), merosporangia arise synchronously from the sporangiophore vesicle (Fig. A.53), elongating to form tubular structures (Fig. A.54). The sporangiophore in Fig. A.55 supports a dense cluster of merosporangia, each attached to the terminal vesicle by a narrow denticle at its base (d in Fig. A.56). In more or less synchronous fashion, the contents of each merosporangium differentiate into a single row of usually 5–10 sporangiospores (Fig. A.51A), although up to 18 have been recorded (Benjamin, 1959). When mature, the spores are free within the merosporangium and may be released by rupture of its outer wall, which commonly occurs between adjacent spores (arrows in Fig. A.56) and may be lytic.

The merosporangia also secede at their base leaving short, denticulate abstriction scars on the vesicles (arrowheads in Figs. A.56, A.57). Occasionally, the basal portion of the merosporangium remains attached to the vesicle after the sporangiospores above have been released (Fig. A.51B, A.57). The seceded spores are smooth-walled (Fig. A.51C), but remnants of the merosporangial wall may remain attached to their surface (Fig. A.58).

Endogenous sporangiospore differentiation is reminiscent of enteroarthric development described in Chapter 7, and illustrated by species of *Sporendonema* Desm. ex Fr., *Coremiella* Bubák & Krieger, and *Malbranchea* Sacc. In both cases, the asexual propagules are differentiated as chains within the cylindrical, fertile cells. The outer hyphal wall of the hyphomycetous fungi eventually breaks down and releases the endogenous conidia, a process which is similar to that just described. However, several aspects of enteroarthric conidiogenesis including sequential septation and degeneration of alternate cells in the fertile hypha (Chapter 7, Fig. 7.58A–E), are not found in the merosporangiferous fungi.

Benjamin (1959) pointed out that the sporangiophores and merosporangia of members of the genus *Syncephalis* superficially resemble the conidiophores of *Gliocephalis hyalina* Matruchot, as originally described by Matruchot (1899). The Hyphomycete, however, produces phialides and actually bears no relation to *Syncephalis* (Embree, 1963; Barron, 1968). The merosporangial initials of *Syncephalis nodosa* van Tieghem arise synchronously (Fig. A.59A) and soon branch dichotomously from a basal cell (Figs. A59B, C, A.60). The branches then become slightly constricted between the endogenously formed sporangiospores (Figs. A.59D, A.60). The merosporangium subsequently disarticulates (Figs. A.59E, A.61) and the spores accumulate in droplets at the sporangiophore apex. This latter

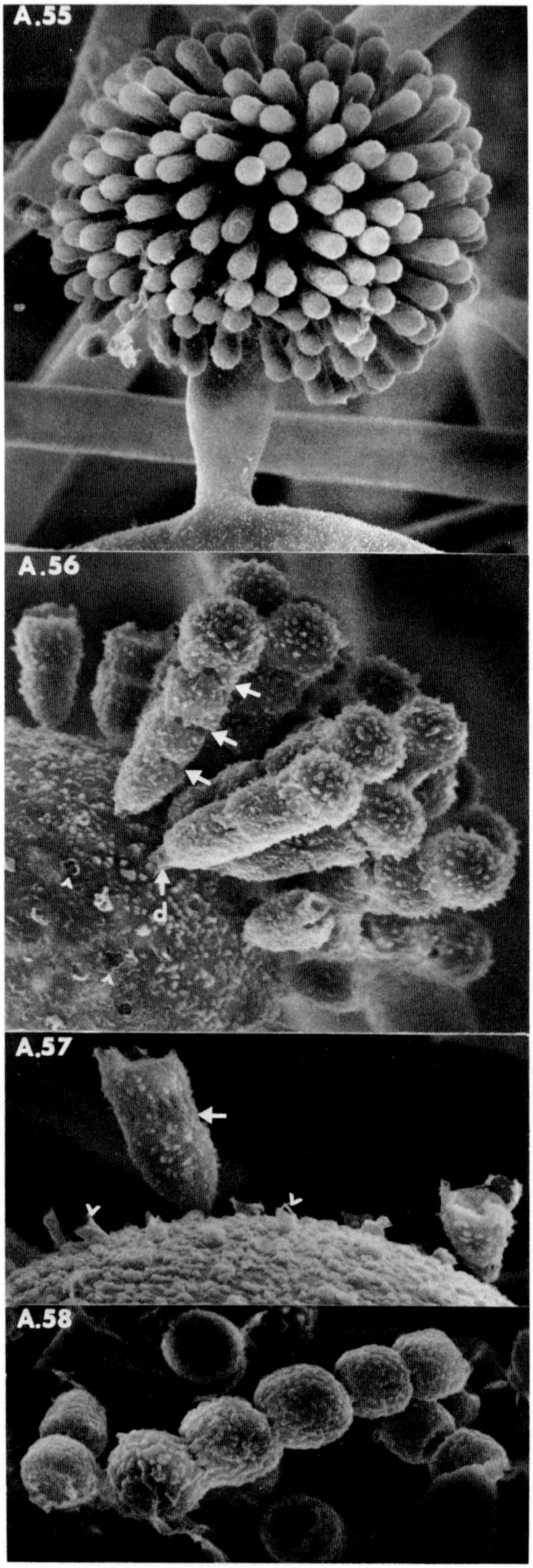

Figs. A.55–A.58 *Syncephalastrum racemosum.* Dense cluster of merosporangia attached to sporangiophore apex. Attachment is by narrow denticles (d in A.56) and secession leaves abstriction scars on surface of vesicles (arrowheads in A.56, A.57). Secession of sporangiospores may occur basipetally, leaving segments of merosporangia attached to vesicle (arrow in A.57). The outer merosporangial wall remains attached to seceded spores (A.58). Arrows in A.56 indicate ruptured merosporangial wall between adjacent spores. ×1,400, 3,600, 4,100 and 4,200, respectively.

A.59

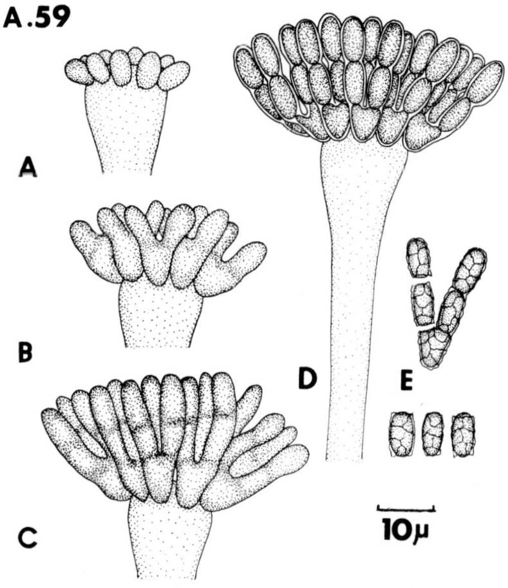

A.60

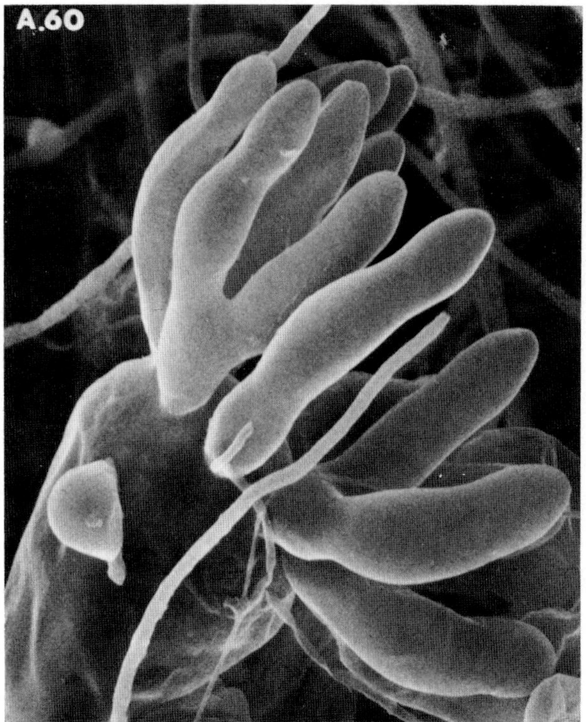

A.61

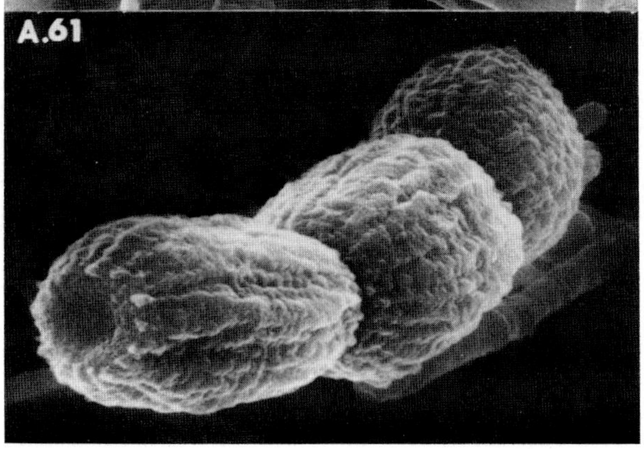

Figs. A.59–A.61 *Syncephalis nodosa.* Synchronous formation of dichotomously branching merosporangia from sporangiophore apex. The mature merosporangia disarticulate by rupture of the outer wall between adjacent spores (A.59E, A.61), which resembles arthric development in conidial fungi. A.60, A.61; ×2,800 and 3,500.

phenomenon, which may be due to exudation of mucilage by the fungus, commonly occurs in species of *Syncephalis, Dimargaris* van Tieghem, and several species of *Piptocephalis.* Comparable production of slimy heads of conidia at apices of conidiogenous cells is widespread in Deuteromycetous fungi (Wakefield and Bisby, 1941; Kendrick and Carmichael, 1973), and in both groups is probably an adaptation for insect dispersal. After disarticulation of the merosporangia in *S. nodosa*, their outer wall layer persists on the surface of the seceded spores and, possibly because of production of the mucilaginous exudate, the surface becomes characteristically wrinkled as the spores mature (Figs. A.59E, A.62). The row of abstriction scars on the surface of the sporangiophore wall in Fig. A.63 demonstrates the regular arrangement of merosporangia around the circumference of the fertile vesicle.

Evolutionary convergence is apparent between merosporangial and sporangiospore formation in *Syncephalis nodosa* and conidiogenesis in the Hyphomycete *Staheliella nodosa* van Emden (Chapter 7, Figs. 7.17–7.22). The latter forms holoarthric conidia by fragmentation of fertile hyphae which arise synchronously from the apex of the conidiophore.

The dichotomously branched sporangiophores of *Piptocephalis lepidula* (Marchal) Benjamin (Figs. A.64–A.70) give rise to a fertile vesicle, or *head-cell*, at the tip of each branchlet (Dobbs and English, 1954). The merosporangia at first develop synchronously as bulbous outgrowths from the surface of the head-cells (Figs. A.64A, B, A.65, A.66), then proliferate acropetally (Fig. A.64C) producing a new terminal cell (Fig. A.67). Two endogenous spores are then differentiated successively within the merosporangial wall (Fig. A.64D), which later ruptures at the medial constriction between them (Fig. A.68). The wall of the merosporangium remains fused to the surface of the spore (Fig. A.64E, arrow in Fig. A.68). A transverse septum forms at the base of the head-cell (arrows in Figs. A.64D, A.69) and the latter may secede after maturation of the merosporangia (Figs. A.64F, A.70).

Analogous ontogeny is found in conidial fungi, for example, the development of acropetal chains of holoblastic conidia from synchronously formed, primary conidia on the ampullae of *Gonatobotryum apiculatum* (Peck) Hughes and *Nematogonium ferrugineum* (Sow. ex Pers.) Hughes (Chapter 3). The branched chains of sporangiospores in *Piptocephalis arrhiza* van Tieghem & Le Monnier (Benjamin, 1959) are comparable to the chains of conidia in *G. apiculatum.* Benjamin's (1959) illustrations of the nuclear cycle in *P. lepidula* is also similar to that reported for *G. apiculatum* by Kendrick and Chang (1971). In each case, the nuclei which accumulate in the fertile vesicle, or ampulla, first

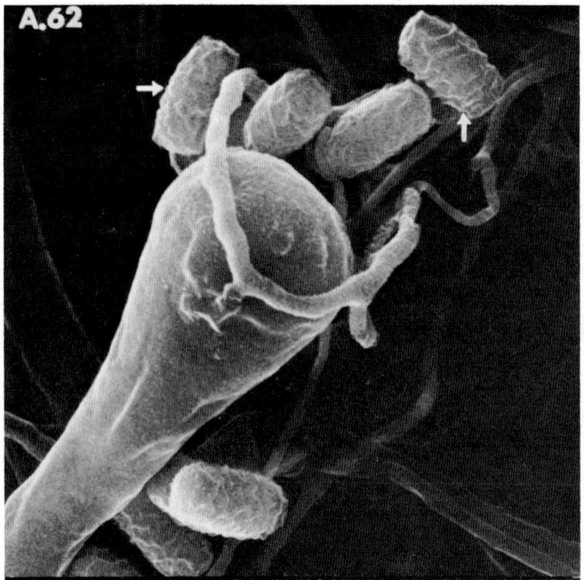

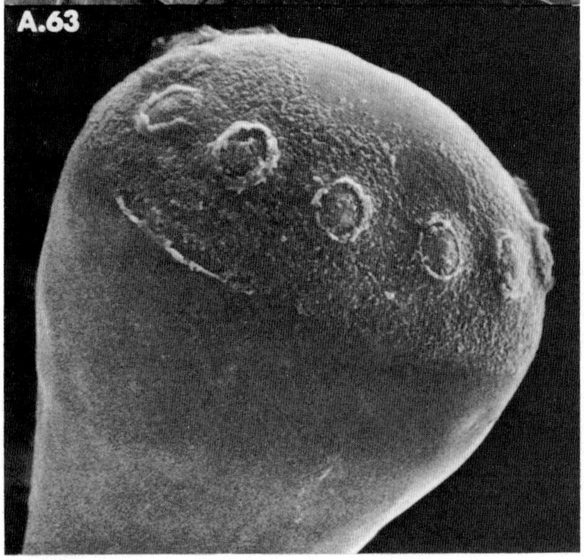

A.64

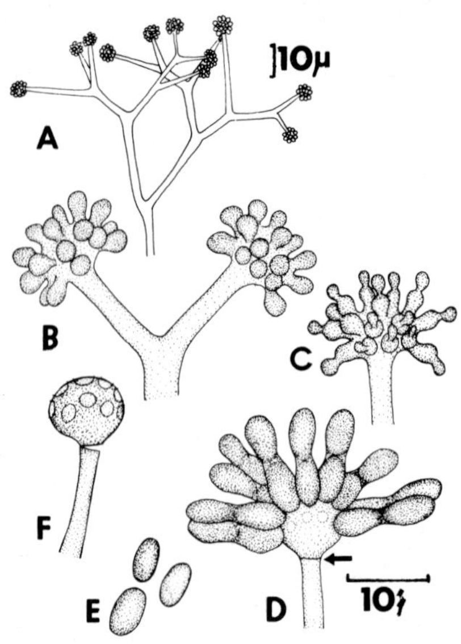

Figs. A.62, A.63 *Syncephalis nodosa.* Sporangiophore apices after merosporangia have seceded. The wrinkled remains of merosporangial wall are visible on the surface of sporangiospores (arrows in A.62). Abstriction scars on surface of fertile vesicle (A.63). ×1,500 and 3,100, respectively.

Fig. A.64 *Piptocephalis lepidula.* Stages of merosporangial development from fertile vesicles (*head-cells*) at the apex of sporangiophore branches. Merosporangia proliferate acropetally (A.64C). Arrow in A.64D locates septum at base of vesicle. The latter usually secedes (A.64F) after disarticulation of merosporangia.

Figs. A.65–A.70 *Piptocephalis lepidula.* Clusters of bulbous merosporangial initials produced from fertile vesicles (A.65, A.66), which later proliferate acropetally and give rise to two endogenous sporangiospores. Secession occurs at medial constriction and base of merosporangia (A.67). Merosporangial wall remains fused to surface of spore (arrow in A.68). Distinct abstriction scars are left on the surface of fertile vesicle (vs, A.69), which often secedes at this late developmental stage by a fracture through septum at its base (arrows in A.69, A.70). Arrowheads in (A.69) locate apices of sporangiophore branches from which fertile vesicles have seceded. ×1,300, 5,700, 5,400, 6,600, 3,800 and 9,500, respectively.

migrate into the initial series of outgrowths, divide again as development of the acropetal chain continues, and one of the daughter nuclei migrates into the newly formed merosporangial, or conidial, initial. This sequence of nuclear division and migration probably occurs in all similarly developing acropetal chains of asexual propagules in the fungi.

Dimargaris cristalligena van Tieghem (Figs. A.71–A.74) and *D. bacillispora* Benjamin (Figs. A.75–A.81) demonstrate blastic and arthric development, respectively. The sporangiophores of *D. cristalligena* produce a complexity of sporiferous branchlets (Figs. A.71A, B, A.73) which develop from a large terminal, fertile vesicle (vs, Fig. A.72). The primary, cylindrical outgrowths of this vesicle branch and produce swollen cells which in turn give rise to whorls of merosporangia (Figs. A.71B, A.73). The acropetal development of the merosporangia (Fig. A.71A), spore differentiation and secession (Figs. A.71C, D, A.73), and associated nuclear cycle in this species are comparable with those processes described above for *Piptocephalis lepidula.* In Fig. A.74, the scanning electron microscope reveals abstriction scars on the surface of the terminal vesicle. The outer sporangiophore wall, represented by the rings of torn wall material, surrounds the exposed, inner wall layer which has formed the *septa* at the base of the fertile branches. Each septum is perforated by a central pore (sp); all but one of these (arrow) are occluded. Benjamin (1959) carefully examined the morphology of these septa under the light microscope and speculated: 'It would seem that the very remarkable and highly complex septa of the Dimargaritaceae and the Kickxellaceae . . . may somehow be actively associated with the movement of protoplasm through the vegetative and

A.65
A.66
A.67
A.68
A.69
vs
A.70

A.71

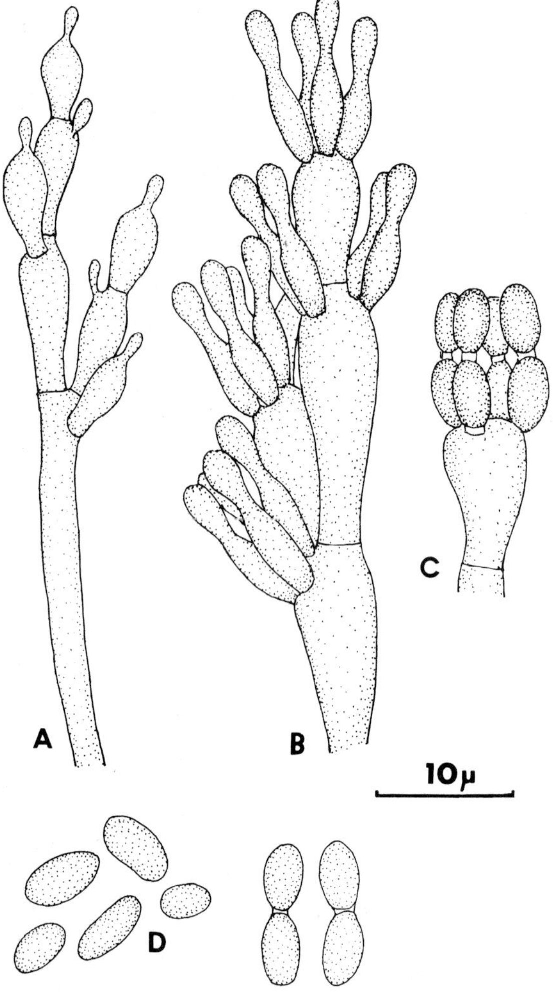

Fig. A.71A–D *Dimargaris cristalligena.* Sporiferous branches which have produced whorls of acropetally proliferating merosporangia (A, B). Secession occurs by rupture of merosporangial wall at medial constriction and base of chains (C, D).

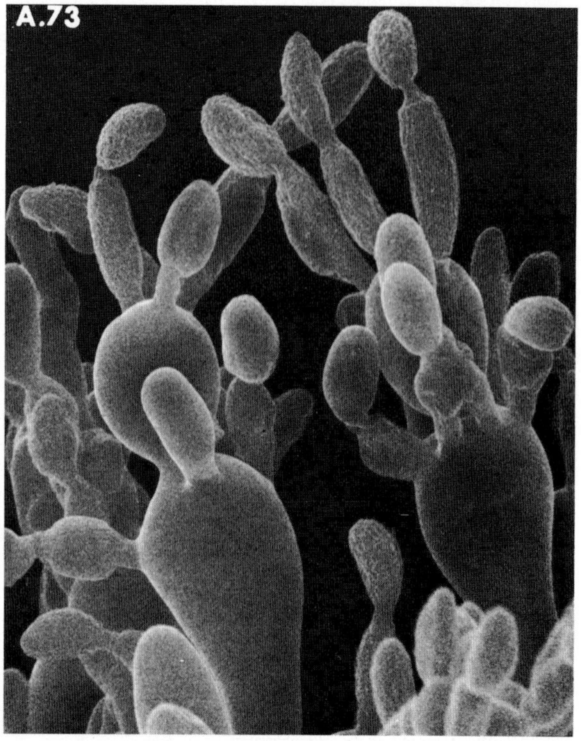

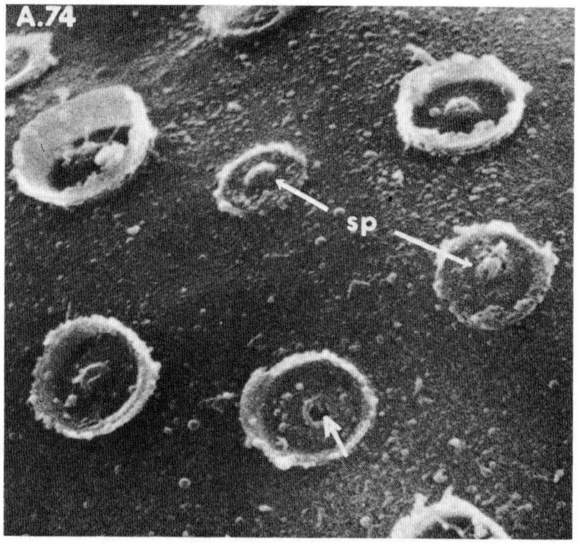

Figs. A.72–A.74 *Dimargaris cristalligena.* Sporiferous branches arising from fertile vesicle (vs, Fig. A.72). Whorls of merosporangia which have arisen from apices of fertile branches (A.73). Abstriction scars on surface of vesicle (A.74). Rings of outer wall of sporiferous branches surround central, septal pores (sp), all of which appear occluded except one (arrow). ×800, 3,200 and 6,900, respectively.

fruiting hyphae and that the plugs formed in the median concavities of these septa may function as valves which in some way regulate such movement'. Most representatives of the Dimargaritaceae and Kickxellaceae demonstrate regularly septate hyphae unlike most other *lower fungi.* A discussion of the fine structure of septa in the Deuteromycetes and their possible functions in cytoplasmic streaming and conidial secession was presented in Chapter 2. However, few ultrastructural and developmental data on septation in the Zygomycetes are yet available (Young, 1969a; Benny, 1972; Benny and Aldrich, 1975) and an extensive comparative study on this aspect of sporangiospore and conidial ontogeny would be most instructive.

The sporangiophores of *Dimargaris bacillispora* also produce a dense cluster of fertile branches which bear whorls of rod-shaped merosporangia (Figs. A.75A, B, A.78, A.79). The merosporangial initials arise as cylindrical growths from the apices of the swollen branchlet cells (Figs. A.75A, A.76), and continue to elongate without becoming constricted (Figs. A.75B, A.77). The merosporangia then cease growth and disarticulate (*arthric* development), each giving rise to a pair of bacilliform spores (Figs. A.75C, A.80). Disarticulation appears to be synchronous for a whole sporangiophore and is associated with some swelling of the spores (Fig. A.75C). Parts of the outer merosporangial wall remain fused to the surface of seceded spores (Figs. A.75D, A.81).

A.75

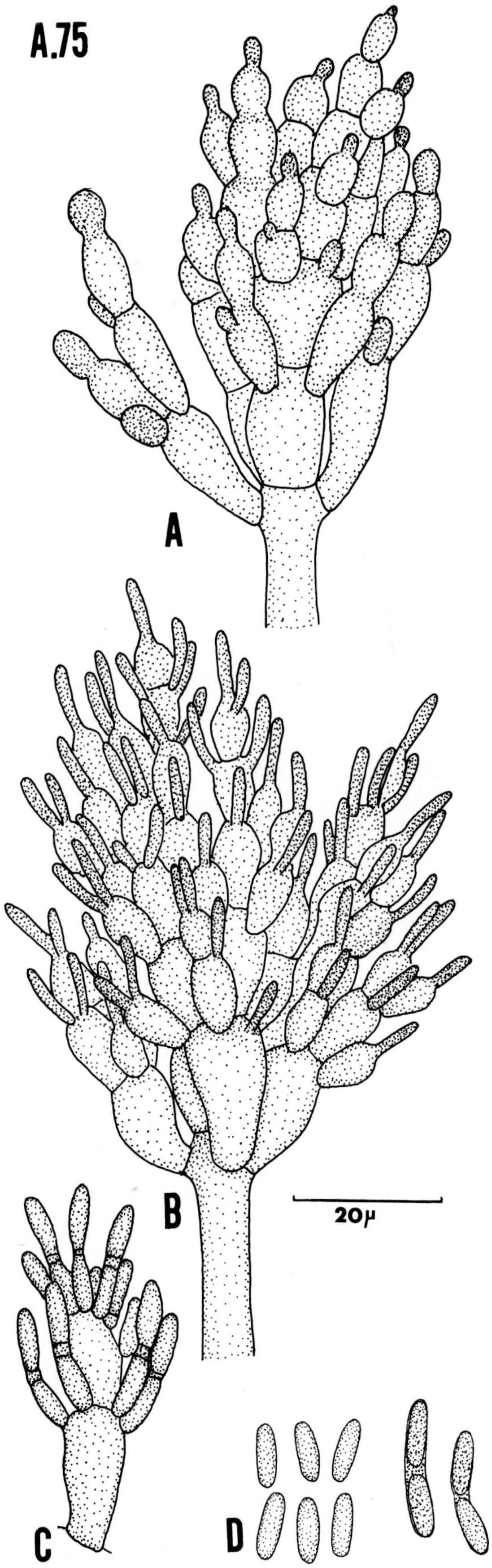

Fig. A.75A–D *Dimargaris bacillispora.* Sporiferous branches (A) which produce rod-shaped merosporangia at their apices (B). Merosporangia cease elongation, then sporangiospores differentiate and disarticulate (C, D).

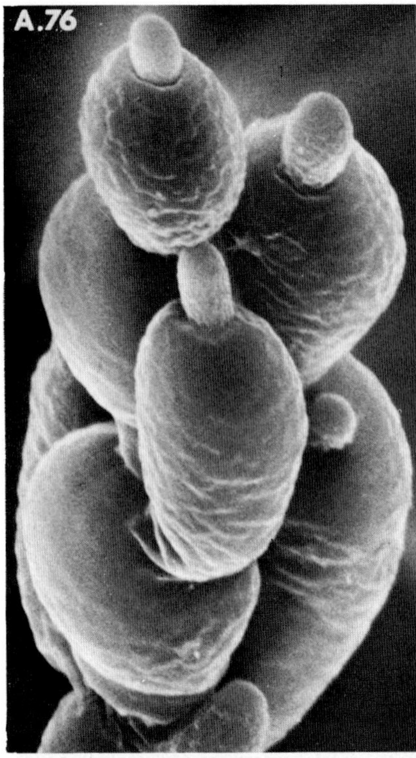
A.76
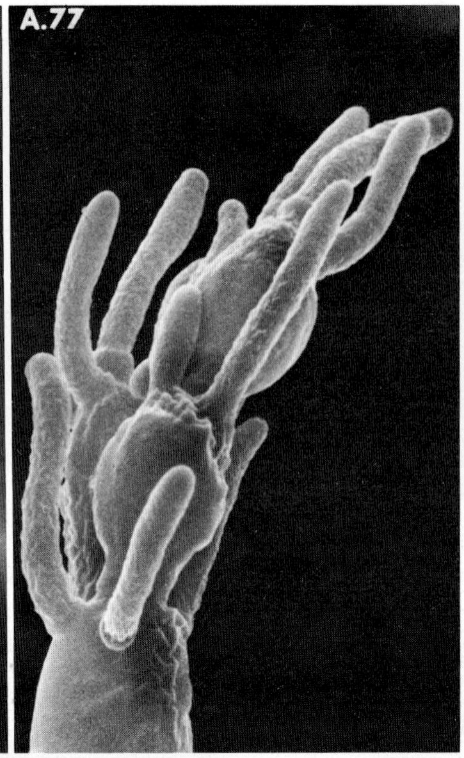
A.77
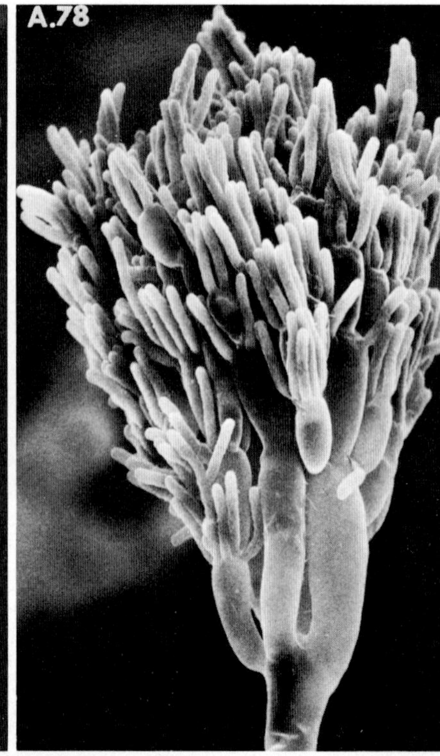
A.78

A.79
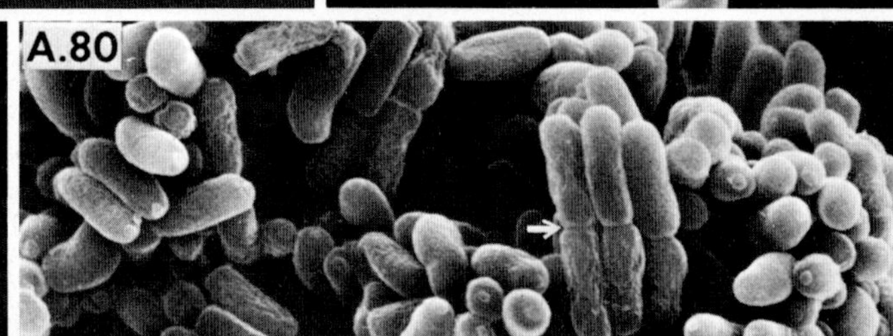
A.80
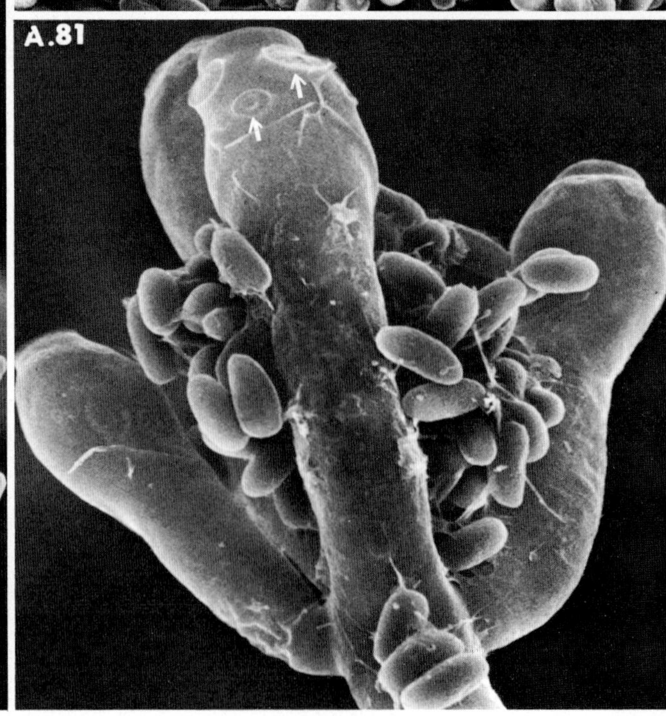
A.81

A.82

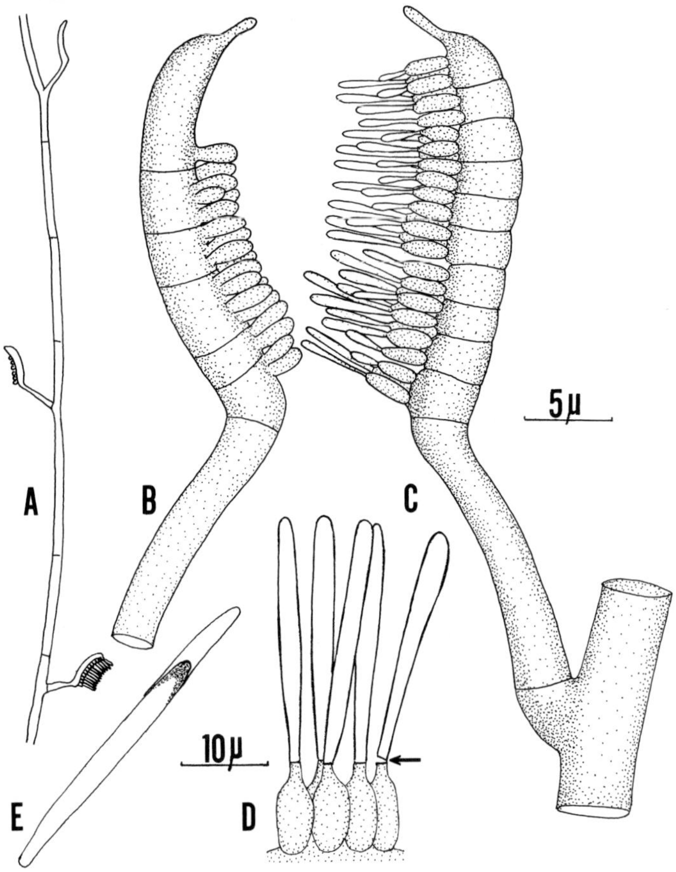

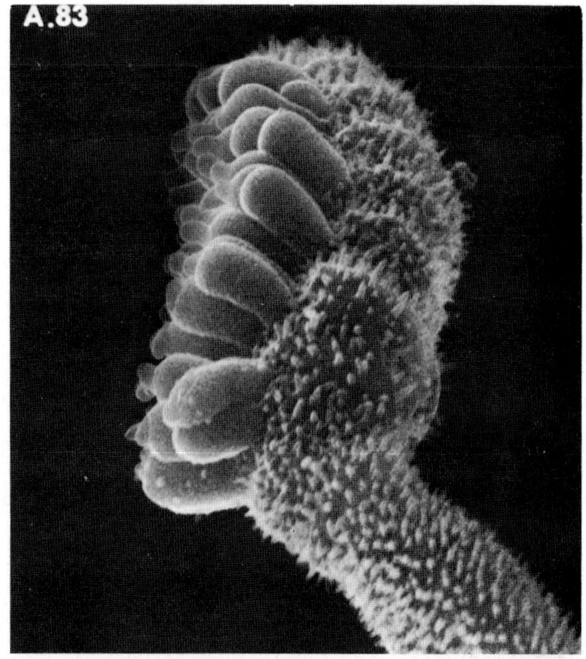

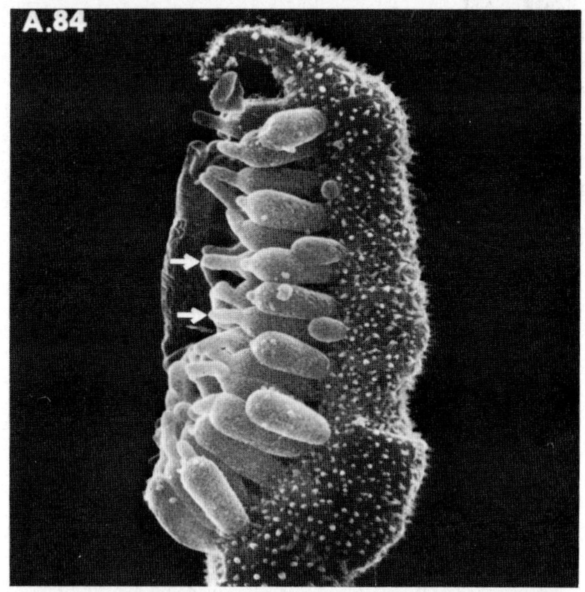

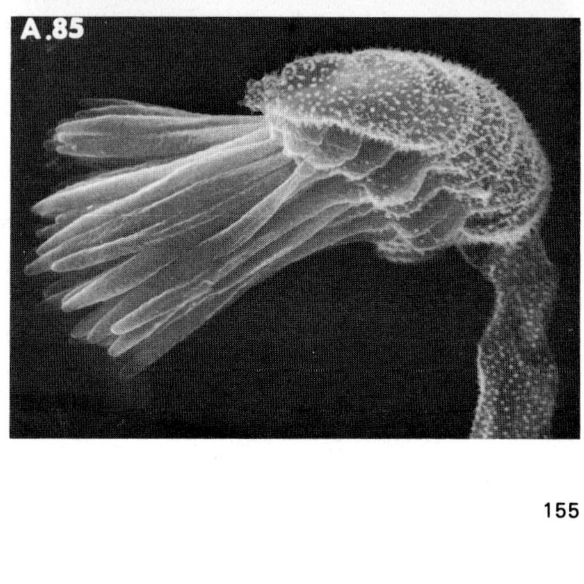

Fig. A.82–A.85 *Coemansia aciculifera.* Alternate arrangement of sporocladia on sporangiophore (A.82A). Septate sporocladia bear ovoid *pseudophialides* (A.82B, A.83) which produce special monosporous sporangiola (A.82C, A.84, A.85). These secede (A.82D) and the sporangiospores later emerge from the merosporangial wall (A.82E). Arrows in (A.84) indicate merosporangial initials. A.83–A.85; ×4,900, 3.400 and 2,500, respectively.

Figs. A.76–A.81 *Dimargaris bacillispora.* Merosporangial initials (A.76) arise in clusters from apices of cells of sporiferous branches (A.77–A.79). Mature merosporangia are rod-shaped (A.77) and give rise to two sporangiospores which secede by disarticulation (arrow, A.80). The sporangiospores swell during maturation (A.81) but remain partly enveloped in merosporangial wall. Arrows in (A.81) indicate abstriction scars of sporiferous branches. ×5,500, 5,300, 1,400, 3,000, 3,200 and 3,300, respectively.

Monosporous merosporangia

Coemansia van Tieghem & Le Monnier (Figs. A.82–A85) and *Linderina* Raper and Fennell (Figs. A.86–A.90), members of the family Kickxellaceae, are characterized by formation of erect sporangiophores bearing specialized sporiferous branchlets (sporocladia) which produce one-spored sporangiola on small, ovoid or flask-shaped basal cells (pseudophialides). Prior to the discovery of zygospores in *Coemansia*, many workers placed these fungi in the Hyphomycetes (Benjamin, 1959).

The septate sporocladia of *Coemansia aciculifera* Linder are produced in acropetal succession and sympodially as the sporangiophore elongates (Fig. A.82A). Numerous short, obovoid cells develop from the concave surface of each sporocladium (Figs. A.82B, A.83), and subsequently bud holoblastically and asynchronously at their apices (Figs. A.82C, A.84). These new outgrowths develop into monosporous sporangiola (Fig. A.85), which secede (Fig. A.82D), then shed their outer, merosporangial wall as shown in Fig. A.82E (Young, 1968a, b).

Development of the aseptate sporocladia of *Linderina macrospora* Chang is similar to that of species of *Coemansia*. The merosporangia are arranged in a dense cluster, each emerging from a pseudophialide which has arisen from the sporocladium (A.86A, A.87). The merosporangia are enveloped in slime at maturity and commonly secede together, each leaving a circumscissile abstriction scar at the apex of its pseudophialide suggestive of a collarette produced by a true phialide (Fig. A.89). A striking morphological similarity is evident between the mature sporocladium of *L. macrospora* and the conidiophore of *Aspergillus* Mich. ex Fr. (Chapter 4, Fig. 4.15). The ultrastructure of sporangiospores of *Coemansia* and *Linderina* has been examined by Young (1968a, 1970a, b, 1971, 1973). Benny and Aldrich (1975) recently presented a well-illustrated account of merosporangial ontogeny in *Linderina pennispora* Raper & Fennell. Of particular interest is their description of septal formation between the pseudophialide and the merosporangiospore. The location of these septa in *L. macrospora* is indicated by the arrow in Figs. A.86B and A.88. Septal growth is centripetal, beginning from the inner margin adjacent to the plasmalemma (Fig. A.91). The crosswall remains incomplete while the merosporangium develops, presumably allowing for translocation of organelles between the two cells (Fig. A.92). Later, the septum 'forms a lenticular cavity which contains the septal plug', blocking the cytoplasmic connection between merosporangium and pseudophialide (Fig. A.93). A similar septal complex occurs at the base of the pseudophialide (Fig. A.94). Several aspects of septation here are similar to those in conidial fungi (Chapter 3). Figure A.95A–F summarizes merosporangial development in *Linderina*.

In an ultrastructural study of the trichospores of *Genistellospora homothallica* Lichtwardt, a member of the Harpellales (Zygomycotina), Moss and Lichtwardt (1976)

A.86

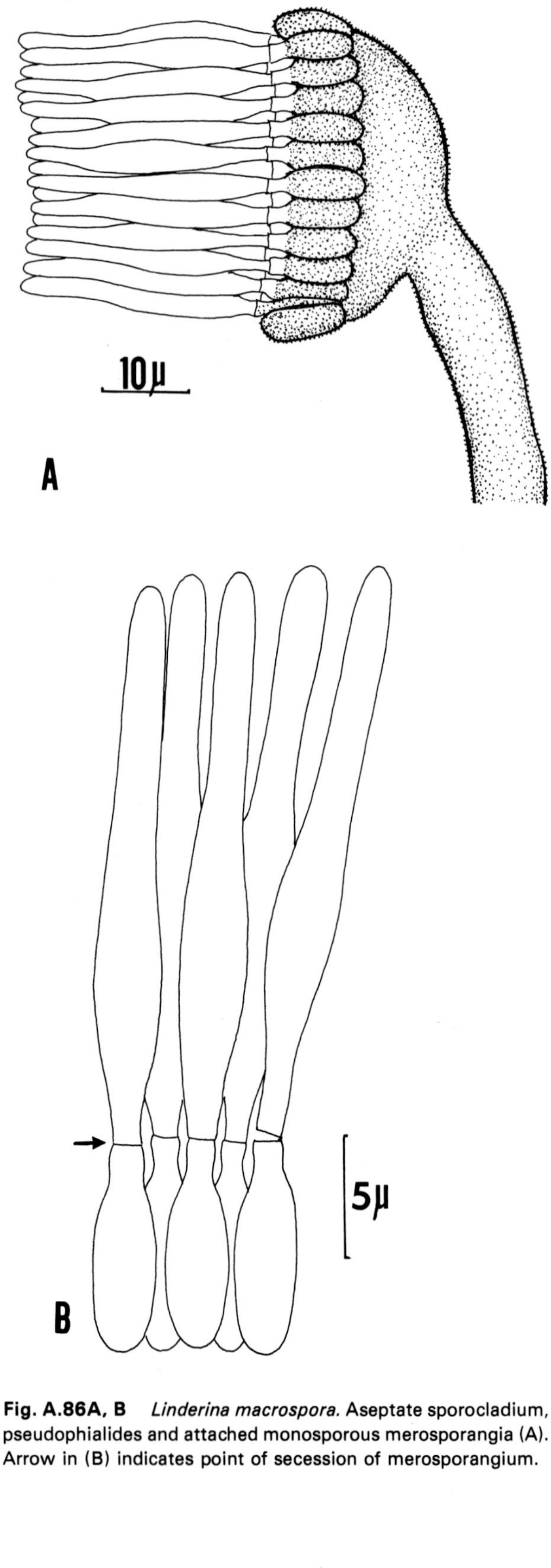

Fig. A.86A, B *Linderina macrospora.* Aseptate sporocladium, pseudophialides and attached monosporous merosporangia (A). Arrow in (B) indicates point of secession of merosporangium.

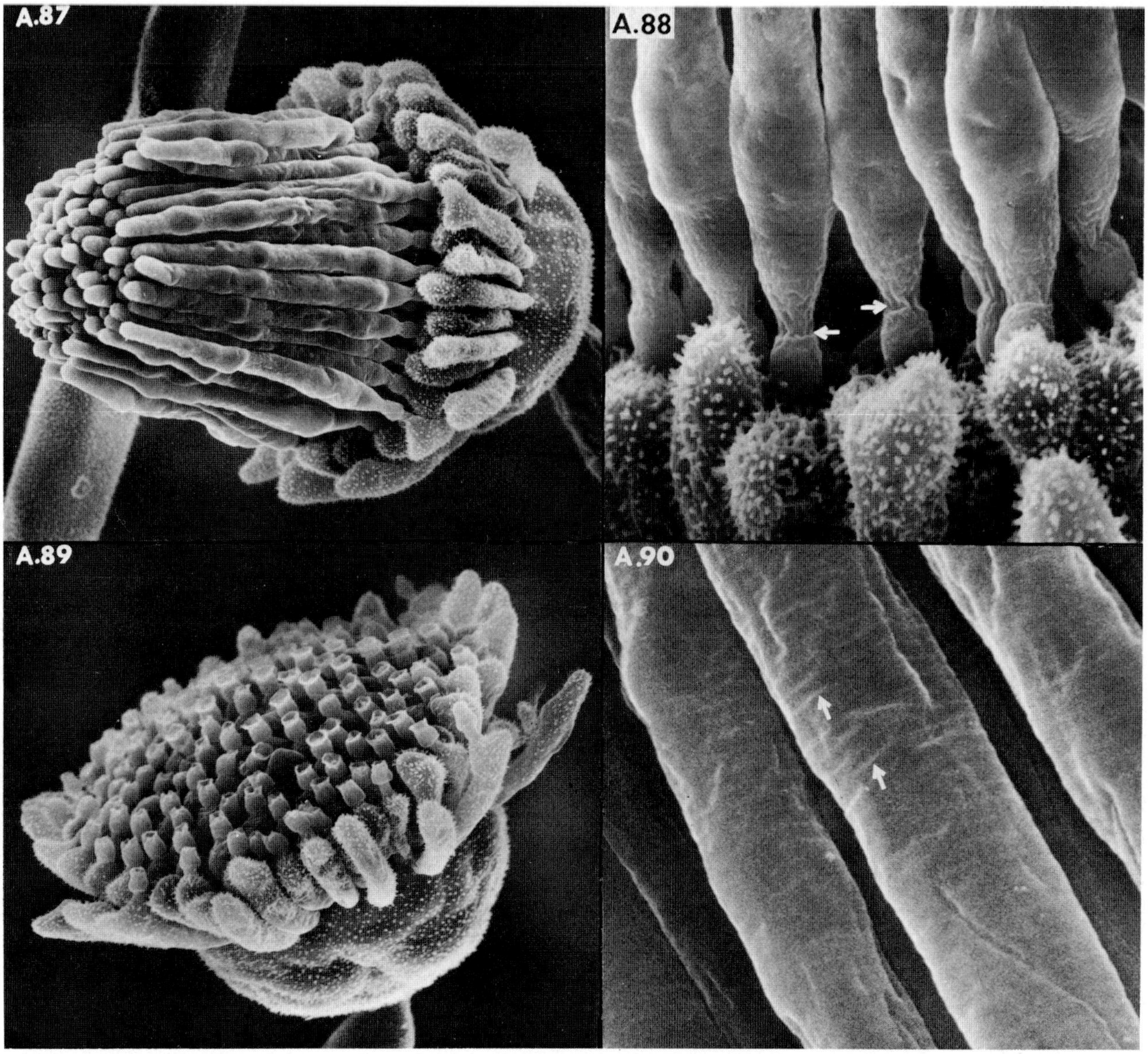

demonstrated that developmental similarities exist between these asexual propagules and the merosporangia of the Kickxellales. This provides additional evidence for a close phylogenetic affinity between the Harpellales and the Kickxellales (Kreisel, 1969; Moss and Young, 1977) and further clarifies the distinction between the trichospore and the conidium.

Figs. A.87–A.90 *Linderina macrospora.* Monosporous merosporangia at apices of pseudophialides. Secession of merosporangia (arrows in A.88) leaves collarette-like scar at apex of pseudophialide (A.89). Annular thickenings of sporangiospore wall are visible within merosporangial wall (arrows in A.90). ×1,600, 6,000, 1,500 and 7,300, respectively.

Figs. A.91–A.94 *Linderina pennispora.* Stages in the formation of a septum and septal plug at the base of the merosporangium (A.91–A.93) and pseudophialide (A.94). ×64,000, 66,000, 66,000 and 18,400, respectively.

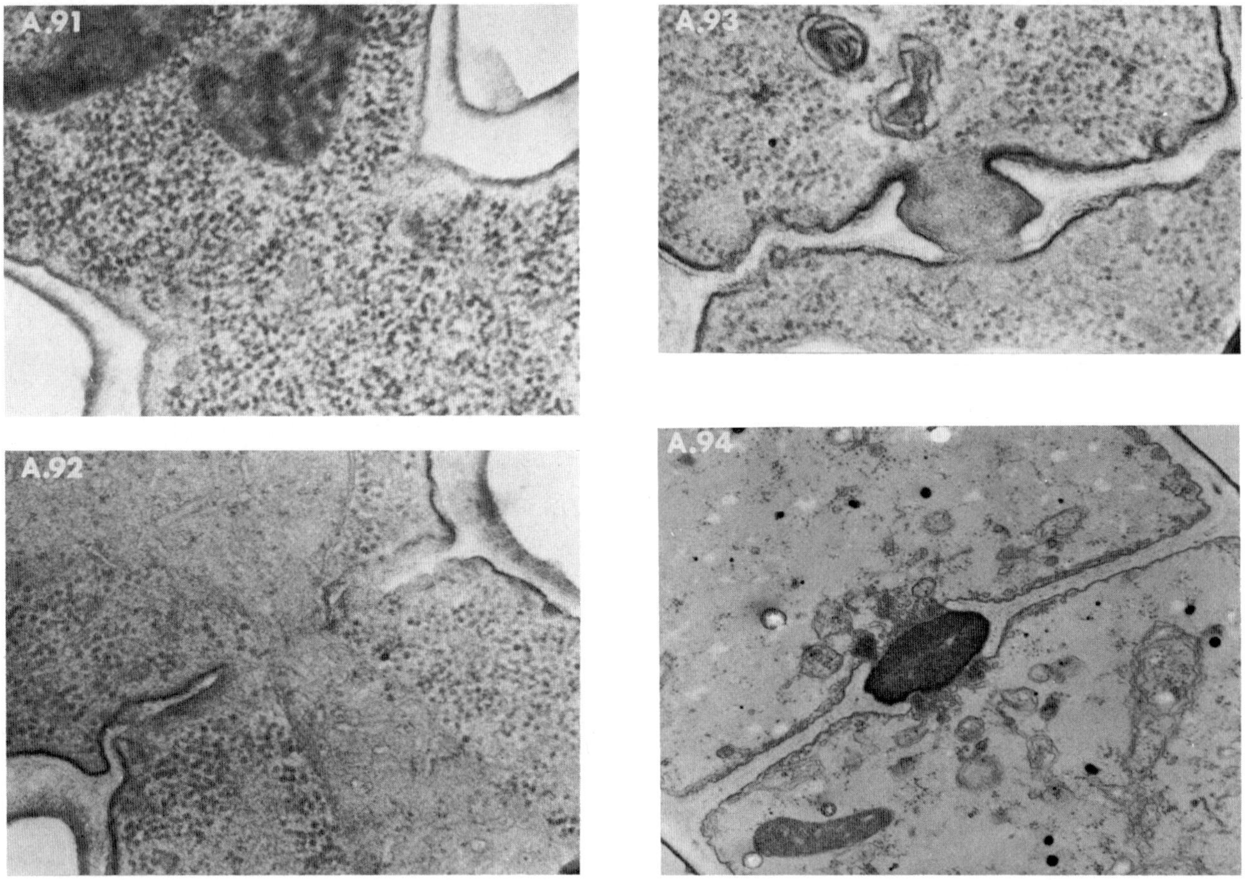

A.95

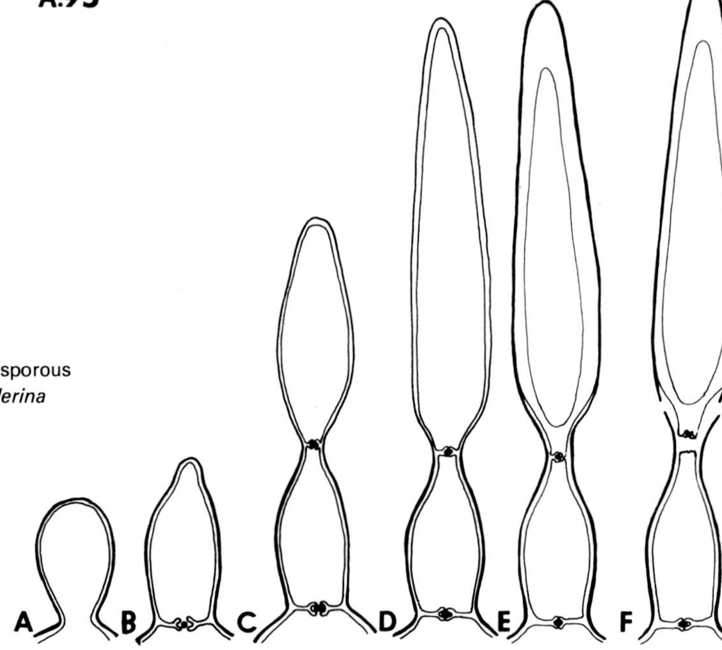

Fig. A.95A–F Diagrammatic interpretation of monosporous merosporangial ontogeny and septal formation in *Linderina pennispora.*

Spore development in the Entomophthorales and Zoopagales

Entomophthorales

Waterhouse (1973) defined three principal mechanisms of vegetative reproduction in this group of fungi: (1) fragmentation of the mycelium into *hyphal bodies* which may later bud; (2) formation of terminal spores which are usually forcibly discharged from the apex of a simple or branched sporangiophore; and (3) formation of secondary spores directly from primary spores, also forcibly discharged. In addition, *chlamydospores* may be formed. The sporangium of the Entomophthorales often functions as a ballistospore in *Entomophthora muscae* (Cohn) Fres., a parasite of houseflies (Fig. A.96A). Although these propagules are still referred to as conidia (Müller and Loeffler, 1976), Thaxter (1888) showed that the sporangial wall actually encloses a differentiated sporangiospore. Batko (1966) similarly showed that the seceded spores of *Zoophthora* Batko possess a double wall and interpreted these structures as monosporous sporangiola. *Basidiobolus* Eidam typically produces an initial swelling of the sporangiophore apex which subsequently produces a secondary swelling at its tip. The latter differentiates into the terminal, holoblastic sporangium containing an endogenous sporangiospore, while the subapical swelling is associated with the mechanism of active spore discharge (Fig. A.96B–D).

Basidiobolus microsporus Benjamin (Fig. A.97A–E) demonstrates two modes of sporangiospore development. The first is the ability of primary sporangiospores, after discharge from the apices of sporangiophores, to function as sporangia, their protoplasm cleaving into secondary spores (Fig. A.97A–C). This cleavage process is sequential, unlike the synchronous development of multispored sporangia in the Mucorales. The other developmental variation in *B. microsporus* is the formation of *microspores* from secondary sporangiospores (Fig. A.97D, E). The latter give rise synchronously to short pedicels which swell apically (Fig. A.97D), each differentiating into a single microspore (Fig. A.97E). The microspores are then passively dispersed. No analogous mechanism is known to us in the Deuteromycetes.

Basidiobolus ranarum Eidam (Fig. A.97F, G) also forms secondary spores, which in this case are either terminal on long slender stalks arising from a seceded primary sporangiospore (Fig. A.97F), or develop from another secondary spore (Fig. A.97G). The secondary spores are obclavate, each forming a short apical beak which is surrounded by a mass of adhesive material (arrows in Fig. A.97F). These propagules are not forcibly discharged but are easily detached from their stalk (e.g., by foraging insects) and the adhesive material readily binds the spores to a substrate (Drechsler, 1947; Benjamin, 1962). Similar dispersal mechanisms have evolved in the deuteromycetous fungi.

A.96

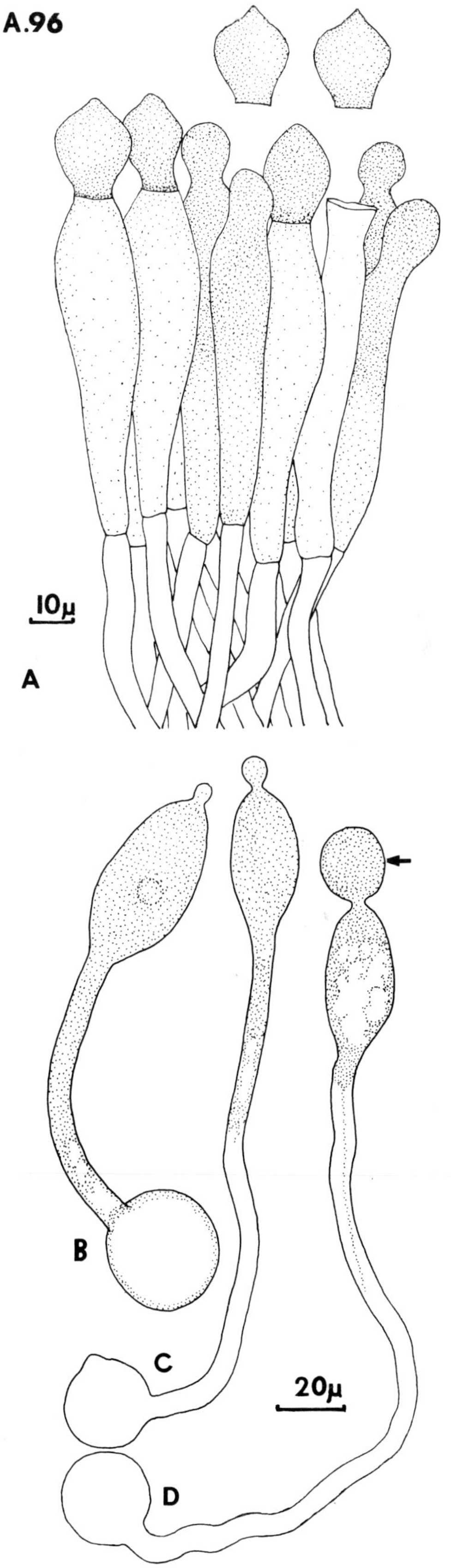

Fig. A.96A *Entomophthora muscae.*

Fig. A.96B–D *Basidiobolus ranarum.* Arrow indicates terminal sporangium.

Development of secondary conidia is also fairly common among the Hyphomycetes and, as in *Basidiobolus* spp., may occur directly from seceded primary conidia (Goos, 1975; Pirozynski, 1966).

Drechsler (1955b) illustrated the synchronous development of botryose clusters of secondary spores (*microconidia*) from forcibly discharged, primary sporangiospores in *Conidiobolus polytocus* Drechsler (Fig. A.97H, I). Each secondary spore is borne on a short denticle on the surface of the primary spore (Fig. A.97I). In *C. chlamydosporus* Drechsler, a similar proliferation of secondary spores from discharged primary spores occurs (Fig. A.97J–L) in addition to the formation of intercalary chlamydospores (Fig. A.97M, N). A comparable degree of pleomorphism occurs in the Hyphomycetes (S. J. Hughes, 1953) in which many species produce chlamydospores as well as characteristic conidia and conidiogenous cells (e.g., species of *Humicola, Acrospeira* Berk. & Br., *Chlamydomyces* Bain., and *Mycogone* Link).

Drechsler (1946) also illustrated cleavage of sporangiospores within cylindrical sporangia of the nematophagous species *Gonimochaete horridula* Drechsler (Fig. A.98). The ellipsoidal spores are produced in a chain, reported to differentiate basipetally and later escape through the ruptured apex of the sporangium. As previously demonstrated in certain primitive species of the Mucorales (e.g., species of *Mortierella, Saksenaea* Saksena), striking similarities exist between such sporangia forming endogenous spores released through an apical rupture and the phialides of several hyphomycetous fungi (e.g., *Chalara* spp.), even though clear differences exist between the development of these two kinds of reproductive structures.

A.97

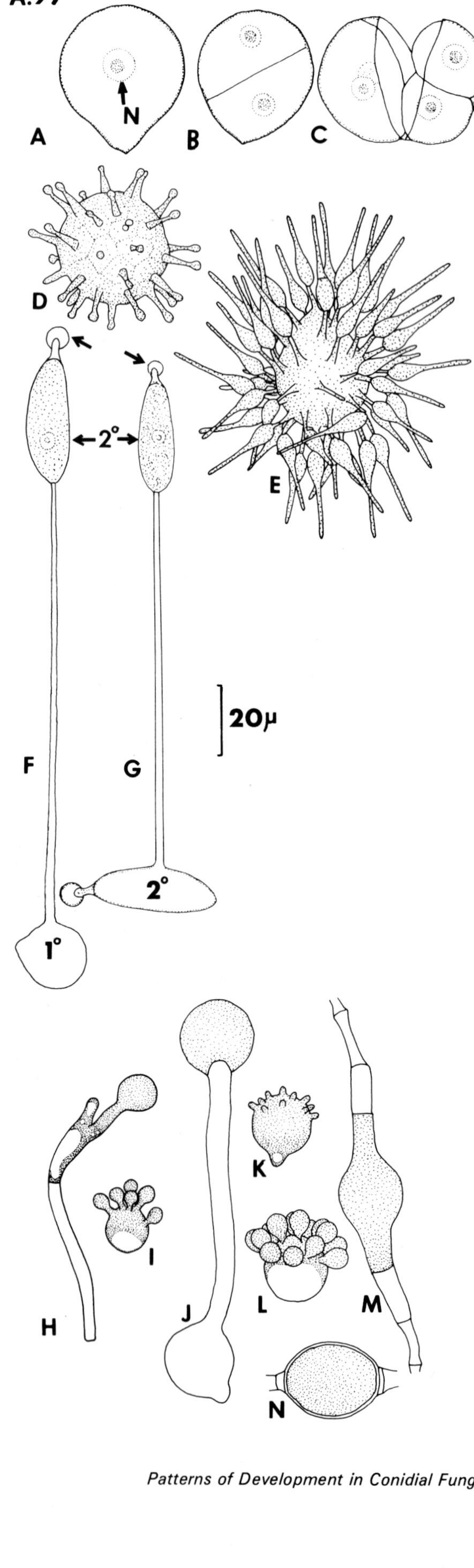

Fig. A.97A–E *Basidiobolus microsporus.* (A–C) Stages of cleavage of primary sporangiospore during formation of secondary spores. (D, E) Stages of synchronous development of microspores from secondary sporangiospores. N, nucleus (after Benjamin, 1962).

Fig. A.97F, G *Basidiobolus ranarum.* Secondary spore formation. Arrows indicate adhesive material. 1°, primary spore; 2°, secondary spore.

Fig. A.97H–N *Conidiobolus* sp. (H–I) *C. polytocus.* Primary spore attached to sporangiophore (H) and discharged 1° spore (I) which has formed 2° spores. (J–N) *C. chlamydosporus.* Proliferated 1° spore (J), stages of 2° spore formation (K, L) and intercalary chlamydospores (M, N) (after Drechsler, 1955b).

A.98

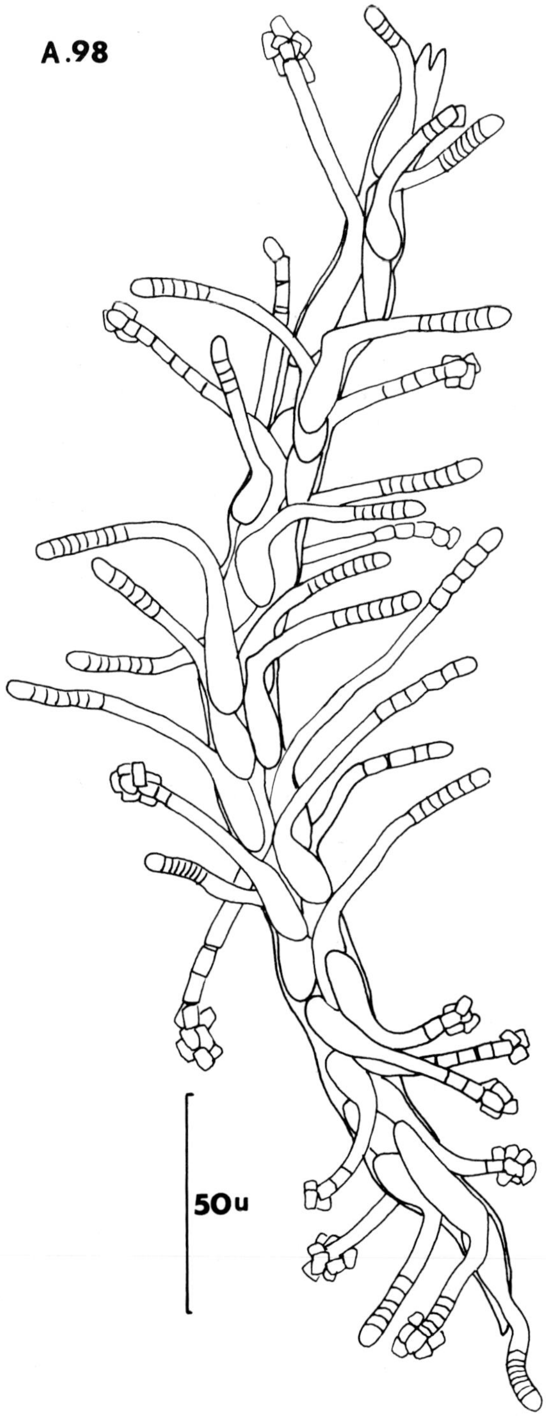

Fig. A.98 *Gonimochaete horridula.* An infected nematode (*Acrobeloides* spp.) from which sporangia have formed (after Drechsler, 1946).

Zoopagales

The Zoopagales includes a small group (approx. 13 gen., 60 spp.) of mostly predacious and endoparasitic species associated with protozoa and nematodes (Duddington, 1973). The taxonomy of this order depends, to a large extent, on the shape, size and mode of development of the asexual spores (S. J. Hughes, 1971a). Five kinds of spores are illustrated in Fig. A.99A–E, which include sessile or stalked, solitary spores (Fig. A.99A), spores formed by fragmentation of the fertile cells (Fig. A.99B), spores produced in acropetal succession in chains (Fig. A.99C) and on sympodially proliferating sporangiophores (Fig. A.99 D), and spores produced in clusters from the swollen, fertile vesicle of the sporangiophore (Fig. A.99E). The formation of both terminal and intercalary chlamydospores has also been reported. The convergent evolution of asexual reproductive mechanisms in the Zoopagales and conidial fungi is well illustrated here.

A.99

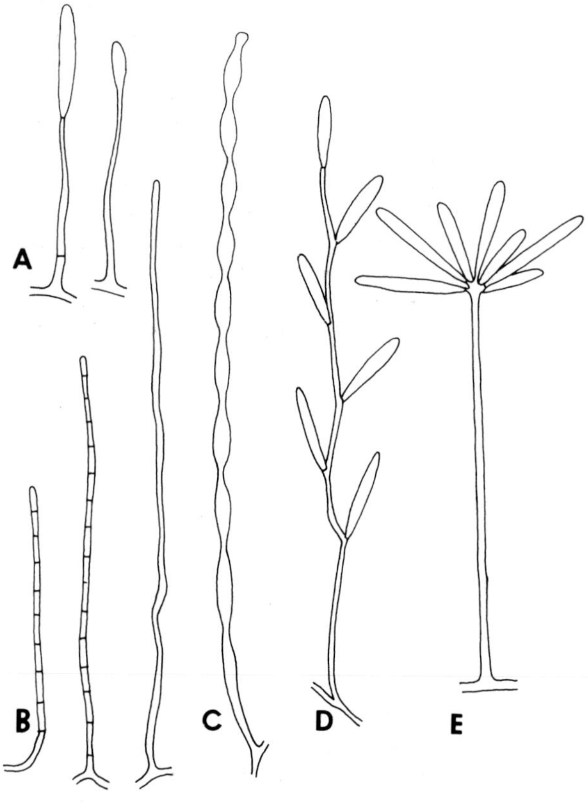

Fig. A.99A–E Modes of sporangiospore development in the Zoopagales. (A) *Stylopage haploe* Drechsler, (B) *Cochlonema pumilum* Drechsler, (C) *Zoopage thamnospira* Drechsler, (D) *Stylopage lepte* Drechsler and (E) *Stylopage cephalote* Drechsler (after Drechsler, 1936, 1938, 1939).

Glossary[1]

acervulus: a fructification consisting of an immersed (subcuticular, subepidermal or deeper) aggregation of hyphae (pseudoparenchyma) from the upper layers of which conidia are produced (Class: Coelomycetes; Order: Melanconiales); no ostiole or specialized dehiscence zone occurs.

acropetal (adj., < Greek, acro = topmost, tip; Latin, pet = seek): describes the succession of conidia which arise in chains with the youngest conidium at the apex of the chain (i.e., acropetal chain).

acropleurogenous (adj.): describes arrangement of conidia at apex and sides of fertile hypha; conidia form in basipetal succession (i.e., the apical conidium is oldest).

amerospore: a one-celled (aseptate) spore (Saccardo, 1886).

ampulla: the swollen, fertile apex of a conidiogenous cell which may give rise to botryose, holoblastic conidia.

ampulliform: swollen or flask-like in form.

anamorph: any asexual, mitotic diasporic expression of a fungus (imperfect state), as opposed to the teleomorph, which is the form producing sexual, meiotic diaspores (perfect state); the whole fungus in all its forms is called the holomorph.

annellation: ring-like scar left at the apex of an annellide after conidial secession.

annellide (noun, < Latin, annellus = a little ring): a conidiogenous cell which produces a basipetal succession of conidia; the first conidium is holoblastic, while subsequent ones are enteroblastic, each produced through the scar of the previously formed conidium, and each leaving a ring-like scar (annellation) at the fertile apex after it secedes; the conidiogenous cell elongates during conidiogenesis.

annellidic (adj.): describes mechanism of conidiogenous cell (annellide) development.

apiculate (adj.): describes cells with short projections at one or both ends.

apposition: wall growth by deposition of successive layers upon those already present.

Ascomycotina: a subdivision of fungi (Division: Mycota) in which meiospores (ascospores) are produced within a meiosporangium (ascus) by *free cell formation.*

autolysis: breakdown of cellular material by action of hydrolytic enzymes contained in the cell affected; self-digestion.

azygospore: parthenogenetic spore resembling a zygospore.

1. Comprehensive treatments of the terminology used in the taxonomy of Deuteromycetes (Fungi Imperfecti) have been presented by Kendrick (1971a) and Cooke (1974).

basauxic (adj., < Latin, basi = base, foundation; Greek, aux = grow, enlarge): describes mechanism of conidiogenous cell development in which elongation occurs at a basal growing point of a fertile cell after formation of a blastic conidium at its apex; the conidiogenous cell may continue to elongate at basal growing point after primary conidium formation in species which produce a basipetal succession of conidia.

Basidiomycotina: a subdivision of fungi (Division: Mycota) in which meiospores (basidiospores) are produced exogenously from the apex or along the sides of a meiosporangium (basidium).

basipetal (adj., < Latin, basi = base, foundation; pet = seek): describes the succession of conidia in which the youngest conidium is at the base of a chain (i.e., basipetal chain) and closest to the fertile apex of the conidiogenous cell (the latter is definitive in the case of droplets of conidia).

blastic (adj.): one of the two basic modes of conidium ontogeny—the conidium differentiates from a fertile hypha (i.e., conidiogenous cell) by the blowing-out and *de novo* growth of part of the hyphal element and is delimited by a septum; derived from budding (cf. thallic).

Blastomycetes: see Deuteromycotina.

botryose (adj.): describes kind of conidia which are produced in clusters, usually at the swollen apex of the conidiogenous cell, and arise synchronously or asynchronously.

catenulate (adj., < Latin, caten = chain): describes the arrangement of conidia in chains, end to end; use of term in this text restricted to acropetal (*true*) chains of blastic conidia and determinate chains formed during thallic-arthric conidium development (cf. *false chain*).

cervix (noun, < Latin, = neck): recurved fertile apex of conidiogenous cell of some Hyphomycetes (e.g., *Costantinella*).

chitosome: small (40–70 nm diam.) spheroidal particles present in growing fungal cells (e.g., hyphal tips, conidial initials) containing the enzyme chitin synthetase.

chlamydospore: a thick-walled, thallic, terminal or intercalary resting propagule (conidium); usually nondeciduous.

cisternal ring: ring-like arrangement of smooth, hypertrophied and sacculate endoplasmic reticulum which appears to bud and give rise to vesicles; considered to be functional equivalent of a dictyosome (Bracker, 1966, 1968).

cleavage (of sporangial cytoplasm): process of sporangiospore differentiation within sporangium by sequence of vesicle fusions and deposition of new wall material.

Coelomycetes: see Deuteromycotina.

collarette: ruptured apex of phialide (conidiogenous cell) through which a succession of enteroblastic conidia emerges.

columella: extension of the axis of a sterile stalk within mature fruit body, as in sporangium of many Mucorales.

conidial initial: a cell or part of a cell which will undergo differentiation to become a conidium.

conidiogenesis: process of conidium formation.

conidiogenous cell: a fertile cell from which, or within which a conidium (conidia) is (are) differentiated. Conidiogenous cells (fertile hyphae) may be morphologically indistinguishable from vegetative cells (vegetative hyphae) or may exhibit varying degrees of morphological differentiation.

conidiogenous locus: fertile region of conidiogenous cell from which one or more conidia are produced.

conidiophore: specialized hypha, simple or branched, on which conidiogenous cells are borne.

conidiophore mother cell: the residual cup-shaped to cylindrical structure enclosing the base of the elongated conidiogenous cell of some Hyphomycetes (e.g., *Spegazzinia, Dictyoarthrinium*).

conidium: asexual, nonmotile, usually deciduous propagule produced by deuteromycetous fungi.

critical point drying: a drying technique used in preparation of soft, biological material for the scanning electron microscope (SEM) involving conversion of a liquid phase (CO_2 or Freon), in which the sample is immersed under pressure, to the gaseous phase, thus avoiding the destructive effects on cells of the solid phase boundary.

crosswall: septum.

deciduous (of conidia, etc.): falling off or shed at maturity.

deliquescent: becoming liquefied; undergoing autolysis.

dematiaceous (adj.): darkly pigmented (mycelium, conidia, etc.).

denticle: small tooth-like projection on which propagules (conidia, sporangiospores) are borne.

dermatophytes: hyphomycetous fungi of the genera *Microsporum, Trichophyton* and *Epidermophyton* which parasitize keratinized tissue (hair, skin, nails) of man and animals.

determinate (adj.): describes conidiogenous cell (fertile hypha) in which extension growth ceases at or before the onset of conidiogenesis and does not resume during or between the formation of successive conidia.

Deuteromycotina: a subdivision of fungi (Division: Mycota) which has been formulated for convenience (i.e., a Form-Subdivision, not part of a natural classification) to include conidial states of perfect fungi (Subdivisions: Ascomycotina and Basidiomycotina) as well as forms which have no known perfect stage. Three Form-Classes are recognized: Blastomycetes (imperfect yeasts and yeast-like fungi), Hyphomycetes (mycelial forms in which reproductive structures are borne on hyphae or on aggregations of hyphae but not enclosed in fructifications), and Coelomycetes (conidia and conidiogenous cells produced in, or on specialized fructifications called pycnidia, acervuli and stromata).

dictyospore: a spore with longitudinal and transverse septa; muriform spore (Saccardo, 1886).

didymospore: a two-celled (one-septate) spore (Saccardo, 1886).

dimorphism: two forms or phases (mycelial and yeast) produced by some fungi (e.g., certain species of *Phialophora*).

disjunctor: a connective cell or wall material which binds adjacent conidia together in chains (e.g., in *Amblyosporium*).

distoseptate (adj.): describes kind of conidial septation in which individual cells are formed by ingrowth of the inner wall layer; the latter is distinct from the outer, continuous conidial wall layer(s).

doliiform (adj.): barrel-like in form.

dolipore septum: a barrel-shaped septum of dikaryotic Basidiomycete hyphae as shown by electron microscopy (Moor and McAlear, 1962).

double septum: a septum (crosswall) which delimits a deciduous conidium from the apex of a conidiogenous cell; the upper layer of the septum forms the base of the conidium, the lower layer forms the apex of the fertile cell and the abscission zone between them is apparently the region of lytic enzyme activity which occurs during separation of the two halves of the septum.

echinulate (adj.): describes cell wall surfaces (e.g., of conidia or conidiogenous cells) which have small pointed processes or spines.

Endogonales: see Zygomycetes.

enteroarthric (adj.): describes a mode of thallic-arthric conidium development in which the outer wall of the fertile hypha usually separates from the newly formed wall of the endogenous conidium.

Entomophthorales: see Zygomycetes.

euseptate (adj.): describes kind of conidial septation in which the outer wall layer(s) of the mature conidium is (are) discontinuous, while the innermost wall layer forms the crosswall(s) of the conidium.

'false' chain: basipetal chain of conidia produced by the fertile apex of a conidiogenous cell; conidia held together by wall connectives (e.g., *Penicillium, Aspergillus*) but not united cytoplasmically.

freeze-etching: a technique designed to examine ultrastructural features of frozen, fractured (cryofractured) and replicated cells; the metal replica (carbon-platinum) of the freeze-fractured cell is examined in the transmission electron microscope; a modification of this basic technique permits observation of both fracture faces of the biological sample (i.e., complementary or double replica technique).

germ tube: germination hypha which emerges from a conidium.

Gymnoascaceae: family of saprophytic or parasitic ascomycetous fungi (Subdivision: Ascomycotina) which includes the teleomorphs of dermatophytes.

helicospore (helicoconidium): a corkscrew-like, cylindrical conidium, usually septate (Saccardo, 1886).

Hemiascomycetes: class of Ascomycetes (Subdivision: Ascomycotina) in which asci are not enclosed in a differentiated fruiting body; thallus is unicellular or consists of poorly differentiated mycelium.

holoarthric (adj.): describes a mode of thallic-arthric conidium development in which all wall layers of the fertile hypha are involved in the formation of the conidial wall.

holoblastic (adj.): mode of blastic conidium ontogeny in which all wall layers of the conidiogenous cell are involved in the formation of the conidial wall.

holothallic (adj.): mode of thallic conidium formation which involves conversion of a terminal or intercalary segment of a fertile hypha into a thick-walled propagule; all wall layers of the fertile hypha are involved in formation of the conidial wall.

hypha (vegetative): the filament of a mycelium representing the vegetative unit of filamentous fungi; hyphae of Deuteromycetes, Ascomycetes and Basidiomycetes have crosswalls (i.e., are septate).

Hyphomycetes: see Deuteromycotina.

Hypocreales: order of ascomycetous fungi (Subdivision: Ascomycotina) with brightly coloured, sexual fruiting bodies (perithecia) whose anamorph is typically phialidic.

imbricate (adj.): overlapping in a regular pattern.

intercalary (adj.): among or between cells (e.g., non-apical meristems; intercalary, thallic conidia).

intrahyphal hypha (endohypha): vegetative or fertile hyphal element which is initiated within the wall of a pre-existing hypha; the innermost wall layer of the determinate, pre-existing hypha differentiates to become the elongating wall of the endohypha.

intussusception: wall growth by incorporation of new components among existing wall components; primary intussusception occurs at growing hyphal tips or apices of cells (e.g., blastic conidial initials); secondary intussusception is associated with intercalary wall growth.

karyology (adj.): refers to behaviour of nuclei.

Kickxellales: see Zygomycetes.

lignicolous (adj.): wood-inhabiting.

lomasome (noun, < Greek, loma = border; soma = body): a paramural body formed by invagination of the plasmalemma and containing numerous vesicles thought to be associated with secondary wall growth.

macroconidium (of dermatophytes): large, terminal, holothallic conidium, usually composed of two or more cells; usually spindle-shaped and thick-walled, or blunt at ends and thin-walled (e.g., *Microsporum* and *Trichophyton).*

merosporangium: cylindrical outgrowth of swollen sporangiophore apex within which sporangiospores arranged in chains are differentiated by cytoplasmic cleavage.

metulae: apical branches of conidiophore bearing phialides (e.g., *Penicillium*).

microbody: organelle bounded by single membrane enclosing specific enzymes and often containing proteinaceous crystalline bodies which may represent an orderly array of some of the enzymes present; microbodies are believed to arise from evaginations of the endoplasmic reticulum (Swanson and Webster, 1977).

microconidium (of dermatophytes): small, holothallic, usually one-celled conidium produced by species of *Microsporum* and *Trichophyton.*

monilioid (= moniliform, adj.): describes swelling at regular intervals, like string of beads, along length of hypha.

monoblastic (adj.): describes a conidiogenous cell which produces a blastic conidium at only one point.

Mucorales: see Zygomycetes.

multiperforate septum: crosswall perforated by many pores (micropores, plasmadesmal canals) produced by some Hyphomycetes (e.g., *Geotrichum candidum*).

mycelium (vegetative): mass of hyphae; thallus (vegetative body) of the fungus.

nematophagous: nematode-trapping.

Oomycetes: a class of primitive (lower) aquatic fungi (Subdivision: Mastigomycotina) producing biflagellate zoospores; oogamous with nonmotile gametes; aseptate vegetative hyphae.

pedicel: a narrow stalk supporting a sporangiolum (Class: Zygomycetes).

percurrent (adj.): describes a mechanism of conidiogenous cell proliferation in which each new axis grows through the previous apex, as in annellides.

Pezizales: the cup fungi, including morels (Subdivision: Ascomycotina).

phialide (noun, < Greek, phial = shallow cup): a conidiogenous cell producing a basipetal succession of conidia (phialoconidia); the fertile apex ruptures after or during differentiation of the primary, blastic conidium leaving a collarette; subsequent conidia are enteroblastic and produced in basipetal succession without any apparent change in length of the conidiogenous cell.

phialidic (adj.): describes mechanism of conidiogenous cell (phialide) development.

phialoconidium: a blastic conidium produced by a phialide.

phragmospore: a spore with two or more transverse septa (Saccardo, 1886).

pleomorphism: condition in which more than one independent mode of conidiogenesis (i.e., in Deuteromycetes) or sporangiosporogenesis (i.e., in Zygomycetes) occurs on the same mycelium.

polarized (tip growth): describes common mechanism of hyphal elongation by intussusception of new wall material, in region of apical pole, and associated protoplasmic differentiation (e.g., cluster of secretory vesicles at hyphal tip).

polyblastic (adj.): describes a conidiogenous cell which produces blastic conidia at several points, synchronously or asynchronously.

polyphialide: conidiogenous cell with more than one open end (collarette) from each of which a basipetal succession of conidia is produced; the new conidiogenous loci develop by sympodial proliferation of the fertile cell; one active locus or several simultaneously active loci may occur.

poroconidium: kind of blastic conidium produced by porogenous fertile cell.

porogenous (adj.): describes kind of conidiogenous cell development in which a pore is formed at junction between base of conidium and apex of fertile cell during conidiogenesis.

proliferous (adj.): reproducing by vegetative means (e.g., sympodial or percurrent proliferation).

pseudophialide: specialized monosporous merosporangium produced by the Kickxellaceae (Class: Zygomycetes).

pycnidium: flask-shaped to globose fructification with slightly tapered apex, single ostiole and containing a hymenium which bears conidia (Class: Coelomycetes; Order: Sphaeropsidales).

rachis (noun, < Greek, = spine, backbone): conidium-bearing extension of sympodially proliferated conidiogenous cell (e.g., *Tritirachium, Acrodontium*, etc.).

ramoconidium: fertile branch of conidiophore which proliferates to produce conidia and later secedes to function itself as a conidium (e.g., *Cladosporium*).

retrogressive (adj.): describes mechanism of conidiogenous cell development which involves a gradual shortening of the fertile cell during basipetal formation of successive conidia.

rhexolytic (adj. < Greek, rhexi = a break, rupture; lysis = a loosening): describes mechanism of conidial secession which occurs by the circumscissile rupture of the periclinal wall of the cell below the basal septum of the conidium; the rupture may occur as a result of mechanical stress, lytic enzyme activity, or both.

rodlet: structural unit of conidial and some hyphal walls composed of linear arrays of approximately 50 Å diameter particles which form bands with a centre to centre spacing of about 100 Å between rodlets (Hess *et al.*, 1968); clusters of parallel rodlets (fascicles) are located just below the outermost amorphous wall layer of conidia; rodlets are apparently composed mainly of protein, and some glucomannan.

schizolytic (adj. < Greek, schiz = split, cleave; lysis = a loosening): describes mechanism of conidial secession in which two layers of the septum between a conidium and conidiogenous cell, or between adjacent conidia, separate centripetally due to lytic enzyme activity.

sclerotium: a resting body produced by aggregation of hyphae forming a hardened mass with or without host tissue; usually with a darkened rind composed of pigmented, thick-walled, compact hyphae from which fruit bodies (e.g., stromata, conidiophores) or mycelium may develop.

scolecospore: a long thread-like spore (Saccardo, 1886).

secession (of conidia): mechanism of release of mature conidium from conidiogenous cell (fertile hypha).

septal pore: opening through crosswall (septum) in hypha or conidium; crosswall may be perforated by a single pore (uniperforate septum) or many pores (multiperforate septum).

septum: a crosswall in fungal hyphae and propagules; regularly occurring in members of the Deuteromycotina, Ascomycotina and Basidiomycotina.

seta: a stiff hair or modified bristle-like, thick-walled, often darkly pigmented hypha.

solitary (adj.): describes kind of conidia which are produced singly at the apex or sides of conidiogenous cells, or in basipetal succession (individually) from the fertile region of specially differentiated conidiogenous cells (i.e., phialides, annellides and retrogressive conidiogenous cells).

Spitzenkörper (noun < German, spitze = apex, körper = body): refractile body at growing hyphal tips of many septate fungi which has been correlated with a cluster of apical vesicles, as shown by electron microscopy (S. N. Grove and Bracker, 1970).

sporangiolum (pl. sporangiola): small sporangium, lacking a columella, and usually containing a small number of spores.

sporangiophore: a specialized fertile hyphal branch which supports one or more sporangia.

sporangiospore: an asexual spore produced in a sporangium.

sporangium: unicellular, asexual reproductive structure within which spores are produced by cytoplasmic cleavage.

sporocladium: specialized fertile (spore-producing) branchlet produced by the Kickxellaceae (Class: Zygomycetes).

sporodochium: a cushion-like mass of conidiogenous cells produced by certain Hyphomycetes (Family: Tuberculariaceae); conidia and conidiogenous cells produced above the surface of substratum.

stauroconidium: star-like conidium having three or more arms.

stroma (pl. stromata): a mass or matrix of vegetative hyphae, with or without host tissue, in which or on which fructifications are produced (e.g., Class: Coelomycetes; Order: Sphaeropsidales).

subulate (adj.): slender and tapering to a point.

sympodial (adj.): describes a mechanism of conidiogenous cell proliferation in which each new growing point appears just behind and to one side of the previous apex, often resulting in a geniculate cell configuration.

sympodulophialide: conidiogenous cell which produces a basipetal succession of conidia; each conidium (after the first) produced from alternate sides of the slightly proliferated fertile cell apex, above the collarette (e.g., *Chloridium virescens*).

synnema: compact aggregation of erect hyphae and conidiogenous cells bearing conidia; produced by some Hyphomycetes.

tetraradiate (adj.): describes four-armed conidium, characteristic of many species of aquatic Hyphomycetes.

thallic (adj.): one of two basic modes of conidium ontogeny—the conidium differentiates by conversion of a segment of a fertile hypha which may involve some enlargement and secondary wall growth; the conidium is delimited by a septum or septa (cf. blastic).

thallic-arthric (adj.): describes one kind of thallic conidium development which occurs by fragmentation and disarticulation of determinate fertile hyphae into chains of conidia; septation of the fertile hypha is at first acropetal, and cytoplasmic continuity between cells of the chain is maintained (i.e., catenulate or *true* chain formation), but later septation is random as the fertile hypha disarticulates.

thermophilic (adj.): describes those fungi which are capable of active growth at high temperatures (40–50°C), and sporulation at 45°C and above.

translocation: transport of water, minerals or food; most often used to refer to food transport in plants.

trichospore: thick-walled, asexual reproductive propagule produced by specialized fungi parasitic on arthropods (Class: Trichomycetes; Subdivision: Zygomycotina).

Uredinales: the rust fungi (Subdivision: Basidiomycotina).

Ustilaginales: the smut fungi (Subdivision: Basidiomycotina).

Woronin body: rounded or elongated oval, highly refractive, cytoplasmic organelle bounded by a single membrane and found in cells of ascomycetous and deuteromycetous fungi.

Zoopagales: see Zygomycetes.

Zygomycetes: class of lower, terrestrial fungi (Subdivision: Zygomycotina), mostly saprophytic, some parasitic forms; no motile cells; asexual reproduction by multisporous sporangia, few-spored or monosporous sporangiola, and few-spored or monosporous merosporangia; sexual reproduction is by zygospores; recognized orders include the Mucorales (usually saprobic, reproduce asexually by sporangia, sporangiola and few-spored merosporangia), Kickxellales (reproduce asexually by monosporous merosporangia), Entomophthorales and Zoopagales (parasites of insects, protozoa and nematodes) and Endogonales (form endotrophic mycorrhizal associations: reproduce by zygospores, sporangia and chlamydospores).

zygospore: thick-walled, sexual spore produced by Zygomycetes; formed by fusion of a pair of often morphologically similar gametangia.

Bibliography

Adelman, M. R. and Taylor, E. W. (1969). Further purification and characterization of slime mold myosin and slime mold actin. *Biochemistry*, **8**:4976–4988.

Ainsworth, G. C. (1971). *Ainsworth and Bisby's Dictionary of the Fungi*, (6th edn). Commonwealth Mycol. Inst. Kew, Surrey, England.

Al-Doory, Y. (1971). The ultrastructure of *Cryptococcus neoformans*. *Sabouraudia*, **9**:113–118.

Allen, R. D. and Kimiya, N. (Eds). (1964). *Primitive Motile Systems in Cell Biology*. Academic Press, New York.

Anderson, T. F. (1951). Techniques for the preservation of three-dimensional structure in preparing specimens for the electron microscope. *N.Y. Acad. Sci.*, **13**:130–134.

Apinis, A. E. (1963). Occurrence of thermophilous microfungi in certain alluvial soils near Nottingham. *Nova Hedwigia*, **5**:57–78.

Arx, J. A. von. (1967). *Pilzkunde. Ein kurzer Abriss der Mykologie unter besonderer Berücksichtigung der Pilze in Reinkultur*. J. Cramer, Lehre.

Arx, J. A. von. (1970). *The Genera of Fungi Sporulating in Pure Culture*. J. Cramer, Lehre.

Arx, J. A. von. (1974). *The Genera of Fungi Sporulating in Pure Culture*. (2nd edn). J. Cramer, Lehre.

Arx, J. A. von. (1977). Ascomycetes as Fungi Imperfecti. *Second Int. Mycol. Congress*, Tampa, Florida. Abstract, p. 23.

Avers, C. J. (1971). Peroxisomes of yeast and other fungi. *Sub-Cell. Biochem.*, **1**:25–37.

Bailey, A. and Bisalputra, T. (1970). A preliminary account of the application of thin-sectioning, freeze-etching and scanning electron microscopy to the study of coralline algae. *Phycology*, **9**:83–101.

Baker, K. L., Hooper, G. R and Beneke, E. S. (1977). Ultrastructural development of merosporangia in the mycoparasite *Syncephalis sphaerica* (Mucorales). *Can. J. Bot.*, **55**:2207–2215.

Barron, G. L. (1961). Studies on species of *Oidiodendron, Helicodendron*, and *Stachybotrys* from soil. *Can. J. Bot.*, **39**:1563–1571.

Barron, G. L. (1962). New species and new records of *Oidiodendron*. *Can. J. Bot.*, **40**:589–607.

Barron, G. L. (1968). *The Genera of Hyphomycetes from Soil*. Williams and Wilkins Co., Baltimore.

Barstow, W. E. and Lovett, J. S. (1974). Apical vesicles and microtubules in rhizoids of *Blastocladiella emersonii*: effects of actinomycin D and cycloheximide on development during germination. *Protoplasma*, **82**:103–117.

Bartnicki-Garcia, S. (1968). Cell wall chemistry, morphogenesis and taxonomy of fungi. *Ann. Rev. Microbiol.*, **22**:87–108.

Bartnicki-Garcia, S. (1969). Cell wall composition and other biochemical markers in fungal phylogeny. In *Phytochemical Phylogeny. Proc. Phytochem. Soc. Symp., Bristol*. J. B. Harborne (Ed.). Academic Press, London, pp. 81–103.

Bartnicki-Garcia, S. (1973). Fundamental aspects of hyphal morphogenesis. In *Microbiol Differentiation*. J. M. Ashworth and J. E. Smith (Eds). 23rd Symp. Soc. Gen. Microbiol, pp. 245–267.

Bartnicki-Garcia, S. and Lippman, E. (1969). Fungal morphogenesis: cell wall construction in *Mucor rouxii*. *Science (New York)*, **165**:302–304.

Bartnicki-Garcia, S., Bracker, C. E. and Ruiz-Herrera, J. (1977). Reassembly of solubilized chitin synthetase into functional structures resembling chitosomes. *Second Int. Mycol. Congress*, Tampa, Florida. Abstract, p. 41.

Batko, A. (1966). On the subgenera of the fungus genus *Zoophthora* Batko 1964 (Entomophthoraceae). *Acta Mycologia*, **2**:15–21.

Beckett, A. (1976). Ultrastructural studies on germinating ascospores of *Daldinia concentrica*. *Can. J. Bot.*, **54**:698–705.

Beckett, A., Heath, J. B. and McLaughlin, D. J. (1974). *An Atlas of Fungal Ultrastructure*. Longman, London.

Begres, F. M., Horner, H. T. Jr. and Dodd, J. D. (1970). A simple technique for integrating transmission electron microscopy and light microscopy for diatom taxonomy. *J. Phycol. (Suppl.)*, **6**:10.

Benjamin, R. K. (1959). The merosporangiferous Mucorales. *Aliso*, **4**:321–433.

Benjamin, R. K. (1960). Two new members of the Mucorales. *Aliso*, **4**:523–530.

Benjamin, R. K. (1961). Addenda to 'The merosporangiferous Mucorales'. *Aliso*, **5**:11–19.

Benjamin, R. K. (1962). A new *Basidiolobus* that forms microspores. *Aliso*, **5**:223–233.

Benjamin, R. K. (1963). Addenda to 'The merosporangiferous Mucorales' II. *Aliso*, **5**:273–288.

Benjamin, R. K. (1965). Addenda to 'The merosporangiferous Mucorales' III. *Dimargaris*. *Aliso*, **6**:1–10.

Benjamin, R. K. (1966). The merosporangium. *Mycologia*, **58**:1–42.

Benjamin, R. K. and Mehrotra, B. S. (1963). Obligate azygospore formation in two species of *Mucor*

(Mucorales). *Aliso,* **5**:235–245.
Benny, G. L. (1972). Histochemistry of the cell wall and septum of vegetative cells of selected species of Dimargaritaceae and Kickxellaceae. *Mycologia,* **64**:854–862.
Benny, G. L. and Aldrich, H. C. (1975). Ultrastructural observations on septal and merosporangial ontogeny in *Linderina pennispora* (Kickxellales; Zygomycetes). *Can. J. Bot.,* **53**:2325–2335.
Benny, G. L. and Benjamin, R. K. (1975). Observations on Thamnidiaceae (Mucorales). New taxa, new combinations, and notes on selected species. *Aliso,* **8**:301–351.
Benny, G. L. and Benjamin, R. K. (1976). Observations on Thamnidiaceae (Mucorales). II. *Chaetocladium, Cokeromyces, Mycotypha* and *Phascolomyces. Aliso,* **8**:391–424.
Bibel, D. J., Crumrine, D. A., Yee, K. and King, R. D. (1977). Development of arthrospores of *Trichophyton mentagrophytes. Infect. Immun.,* **15**:958–971.
Boedijn, K. B. (1958). Notes on the Mucorales of Indonesia. *Sydowia,* **12**:321–362.
Boerema, G. H. (1965). Spore development in the form-genus *Phoma. Persoonia,* **3**:413–417.
Boerema, G. H. and Bollen, G. J. (1975). Conidiogenesis and conidial septation as differentiating criteria between *Phoma* and *Ascochyta. Persoonia,* **8**:111–144.
Bojović-Cvetić, D. and Vuyicić, R. (1974). Ultrastructure of conidiophores in *Aspergillus flavus. Trans. Br. Mycol. Soc.,* **63**:131–135.
Bourne, M. (1968). Frontispiece. In *Fundamentals of Mycology.* J. H. Burnett (Ed.). Edward Arnold, London.
Boyde, A. and Echlin, P. (1973). Freezing and freeze-drying—a preparative technique for SEM. In *Scanning Electron Microscopy/1973.* IITRI, Chicago, pp. 759–765.
Boyer, M. G. and Isaac, P. K. (1964). Some observations on white pine blister rust as compared by light and electron microscopy. *Can J. Bot.,* **42**:1305–1309.
Bracker, C. E. (1966). Ultrastructural aspects of sporangiospore formation in *Gilbertella persicaria.* In *The Fungus Spore. Proc. Symp. Colston Res. Soc., 18, Bristol, England.* M. F. Madelin (Ed.). Butterworths, London, pp. 39–60.
Bracker, C. E. (1967). Ultrastructure of fungi. *Ann. Rev. Phytopath.,* **5**:343–374.
Bracker, C. E. (1968). The ultrastructure and development of sporangia in *Gilbertella persicaria. Mycologia,* **60**:1016–1067.
Bracker, C. E. (1971). Cytoplasmic vesicles in germinating spores of *Gilbertella persicaria. Protoplasma,* **72**:381–397.
Bracker, C. E. (1977). Structure and transformation of chitosomes during chitin microfibril synthesis. *2nd Int. Mycol. Congr.,* Tampa, Florida, Abstract, p. 64.
Bracker, C. E. and Butler, E. E. (1963). The ultrastructure and development of septa in hyphae of *Rhizoctonia solani. Mycologia,* **55**:35–58.
Bracker, C. E. and Butler, E. E. (1964). Function of the septal pore apparatus in *Rhizoctonia solani* during protoplasmic streaming. *J. Cell Biol.,* **21**:152–157.
Bracker, C. E., Ruiz-Herrera, J. and Bartnicki-Garcia, S. (1976). Structure and transformation of chitin synthetase particles (chitosomes) during microfibril synthesis *in vitro. Proc. Natl. Acad. Sci.,* **73**:4570–4574.
Bracker, C. E., Grove, S. N., Heintz, C. E. and Morré, D. J. (1971). Continuity between endomembrane components in hyphae of *Pythium* spp. *Cytobiologie,* **4**:1–8.
Brefeld, O. (1872). *Botanische Untersuchungen über Schimmelpilze. VIII, Heft 1.* Verlag von Arthur Felix, Leipzig.
Brewer, J. G. and Boerema, G. H. (1965). Electron microscope observations on the development of pycnidiospores in *Phoma* and *Ascochyta* spp. Proc. *Konikl. Nederl. Akademie van Wetenschappen-Amsterdam,* **68**:82–97.
Brotzman, H. G., Calvert, O. H., Brown, M. F. and White, J. A. (1975). Holoblastic conidiogenesis in *Helminthosporium maydis. Can J. Bot.,* **53**:813–817.
Brown, W. (1923). Experiments on the growth of fungi on culture media. *Ann. Bot. (London),* **37**:105–129.
Brushaber, J. A. and Jenkins, S. F. Jr. (1971). Lomasomes and vesicles in *Poria monticola. Can. J. Bot.,* **49**:2075–2079.
Buckley, P. M., Sjaholm, V. E. and Sommer, N. F. (1966). Electron microscopy of *Botrytis cinerea* conidia. *J. Bacteriol.,* **91**:2037–2044.
Buckley, P. M., Wyllie, T. D. and Devay, J. E. (1969). Fine structure of conidia and conidium formation in *Verticillium albo-atrum* and *V. nigrescens. Mycologia,* **61**:240–250.
Buller, A. H. R. (1933). *Researches on Fungi. V.* Univ. of Toronto Press, Toronto, Ontario.
Burdett, I. D. J. and Murray, R. G. E. (1974). Electron microscope study of septum formation in *Escherichia coli* Strains B and B/r during synchronous growth. *J. Bacteriol.,* **119**:1039–1056.
Burnett, J. H. (1968). *Fundamentals of Mycology.* Edward Arnold, London.
Calley, W. A. (1976). Morphological investigations of several human pathogenic fungi under variable conditions of temperature and light in a defined liquid medium. M.A. Thesis, Univ. Texas, Austin, Texas, 147 pp.
Calonge, F. D. (1968). Origin and development of intrahyphal hyphae in *Scerotinia fructigera. Mycologia,* **60**:932–942.
Campbell, R. (1968). An electron microscope study of spore structure and development in *Alternaria brassicicola. J. Gen. Microbiol.,* **54**:381–392.
Campbell, R. (1969). Further electron microscope studies of the conidium of *Alternaria brassicicola. Arch. Mikrobiol.,* **69**:60–68.
Campbell, R. (1970). An electron microscope study of exogenously dormant spores, spore germination, hyphae and conidiophores of *Alternaria brassicicola. New Phytol.,* **69**:287–293.
Campbell, R. (1972). Ultrastructure of conidium ontogeny in the deuteromycete fungus *Stachybotrys atra* Corda. *New Phytol.,* **71**:1143–1149.
Campbell, R. (1975). The ultrastructure of the *Arthrinium*

state of *Apiospora montagnei* Sacc. *Protoplasma,* **83**:51–60.
Carmichael, J. W. (1971). Blastospores, aleuriospores, chlamydospores. In *Taxonomy of Fungi Imperfecti.* B. Kendrick (Ed.). Univ. Toronto Press, Toronto, Ontario, pp. 50–56.
Carroll, F. E. (1972). A fine-structural study of conidium initiation in *Stemphylium botryosum* Wallroth. *J. Cell Sci.,* **11**:33–47.
Carroll, F. E. and Carroll, G. C. (1973). Senescence and death of the conidiogenous cell in *Stemphylium botryosum* Wallroth. *Arch. Mikrobiol.,* **94**:109–124.
Carroll, F. E. and Carroll, G. C. (1974a). The fine structure of conidium initiation in *Ulocladium atrum. Can. J. Bot.,* **52**:443–446.
Carroll, G. C. (1966). A study of ascosporogenesis in *Saccobolus kerverni* and *Ascodesmis sphaerospora.* Ph.D. Thesis, Univ. Texas, Austin, Texas.
Carroll, G. C. (1967). The ultrastructure of ascospore delimitation in *Saccobolus kerverni. J. Cell Biol.,* **33**:218–224.
Carroll, G. C. and Carroll, F. E. (1974b). The fine structure of conidium development in *Phialocephala dimorphospora. Can. J. Bot.,* **52**:2119–2128.
Chan, C. and Stephen, R. C. (1967). Intrahyphal hyphae in the genus *Linderina. Can. J. Bot.,* **45**:1995–1998.
Chesters, C. G. C. (1969). Morphology as a taxonomic criterion. In *The Fungi,* vol. III. G. C. Ainsworth and A. S. Sussman (Eds). Academic Press, New York and London, pp. 517–542.
Chesters, C. G. C. and Greenhalgh, G. N. (1964). *Geniculosporium serpens* gen. et sp. nov., the imperfect stage of *Hypoxylon serpens. Trans. Br. Mycol. Soc.,* **47**:393–401.
Clements, F. E. and Shear, C. L. (1931). *The Genera of Fungi.* Wilson, New York.
Clough, K. S. and Patrick, Z. A. (1972). Naturally occurring perforations in chlamydospores of *Thielaviopsis basicola* in soil. *Can. J. Bot.,* **50**:2251–2253.
Coffey, M. D., Palevitz, B. A. and Allen, P. J. (1972). The fine structure of two rust fungi, *Puccinia helianthi* and *Melamspora lini. Can. J. Bot.,* **50**:231–240.
Cole, G. T. (1969). An exploration of the complementary functions of mature morphology and conidiophore-conidium ontogeny in the taxonomy of hyphomycetes. Ph.D. Thesis. Univ. of Waterloo, Waterloo, Canada.
Cole, G. T. (1971). The sympodula and the sympodioconidium. In *Taxonomy of Fungi Imperfecti.* B. Kendrick (Ed.). Univ. of Toronto Press, Toronto, pp. 141–159.
Cole, G. T. (1972). Microfibrils in the cytoplasm of fertile hyphae of the imperfect fungus, *Drechslera sorokiniana. J. Ultrastruct. Res.,* **41**:563–571.
Cole, G. T. (1973a). Ultrastructure of conidiogenesis in *Drechslera sorokiniana. Can. J. Bot.,* **51**:629–638.
Cole, G. T. (1973b). Ultrastructural aspects of conidiogenesis in *Gonatobotryum apiculatum. Can. J. Bot.,* **51**:1677–1684.
Cole, G. T. (1973c). A correlation between rodlet orientation and conidiogenesis in Hyphomycetes. *Can. J. Bot.,* **51**:2413–2422.
Cole, G. T. (1974). Conidiophore and conidium ontogeny in *Spegazzinia tessarthra. Can. J. Bot.,* **52**:1259–1264.
Cole, G. T. (1975a). The thallic mode of conidiogenesis in the Fungi Imperfecti. *Can. J. Bot.,* **53**:2983–3001.
Cole, G. T. (1975b). A preparatory technique for examination of imperfect fungi by scanning electron microscopy. *Cytobios,* **12**:115–121.
Cole, G. T. (1976a). Conidiogenesis in pathogenic hyphomycetes. I. *Sporothrix, Exophiala, Geotrichum* and *Microsporum. Sabouraudia,* **14**:81–98.
Cole, G. T. (1976b). Conidium ontogeny in marine hyphomycetous fungi: *Asteromyces cruciatus* and *Zalerion maritimum. Marine Biology,* **38**:147–158.
Cole, G. T. (1976c). Application of scanning electron microscopy to developmental and taxonomic studies of the Fungi Imperfecti. In *Scanning Electron Microscopy/1976.* IITRI, Chicago, pp. 527–534.
Cole, G. T. (1977). *Conidial development in the Fungi Imperfecti. I. Holoblastic conidia and poroconidia.* 16 mm time-lapse film, Inst. für den Wissenschaftlichen Film, Göttingen.
Cole, G. T. (1978). Conidiogenesis in the black yeasts. *Proc. Int. Conf. on Mycoses. Brazilia.* Pan Am. Health Organiz. Pub., Washington, Publ. no. 356, pp. 66–78.
Cole, G. T. and Aldrich, H. C. (1971a). Ultrastructure of conidiogenesis in *Scopulariopsis brevicaulis. Can. J. Bot.,* **49**:745–755.
Cole, G. T. and Aldrich, H. C. (1971b). Demonstration of myelin figures in unfixed, freeze-etched fungus spores. *J. Cell Biol.,* **51**:873–874.
Cole, G. T. and Aldrich, H. C. (1971c). Scanning and transmission electron microscopy and freeze-etching techniques used in ultrastructural studies of Hyphomycetes. In *Taxonomy of Fungi Imperfecti.* B. Kendrick (Ed.). Univ. of Toronto Press, Toronto, pp. 292–300.
Cole, G. T. and Kendrick, W. B. (1968a). Conidium ontogeny in Hyphomycetes. The imperfect state of *Monascus ruber* and its meristem arthrospores. *Can. J. Bot.,* **46**:987–992.
Cole, G. T. and Kendrick, W. B. (1968b). A thin culture chamber for time-lapse photomicrography of fungi at high magnifications. *Mycologia,* **60**:340–344.
Cole, G. T. and Kendrick, W. B. (1969a). Conidium ontogeny in Hyphomycetes. The phialides of *Phialophora, Penicillium,* and *Ceratocystis. Can. J. Bot.,* **47**:779–789.
Cole, G. T. and Kendrick, W. B. (1969b). Conidium ontogeny in Hyphomycetes. The annellophores of *Scopulariopsis brevicaulis. Can. J. Bot.* **47**:925–929.
Cole, G. T. and Kendrick, W. B. (1969c). Conidium ontogeny in Hyphomycetes. The arthrospores of *Oidiodendron* and *Geotrichum* and the endoarthrospores of *Sporendonema. Can. J. Bot.,* **47**:1773–1780.
Cole, G. T. and Kendrick, W. B. (1969d). Time-lapse studies of conidiophore and conidium ontogeny in Hyphomycetes. *XIth International Botanical Congress,* Seattle, Abstract, p. 34.

Cole, G. T. and Kendrick, W. B. (1971). Conidium ontogeny in Hyphomycetes. Development and morphology of *Cladobotryum*. *Can. J. Bot.*, **49**:595–599.

Cole, G. T. and Kendrick, W. B. (1973). Taxonomic studies of *Phialophora*. *Mycologia*, **65**:661–688

Cole, G. T., Nag Raj, T. R. and Kendrick, W. B. (1969). A simple technique for time-lapse photomicrography of microfungi in plate culture. *Mycologia*, **61**:726–730.

Collinge, A. J. and Trinci, A. P. J. (1974). Hyphal tips of wild-type and spreading colonial mutants of *Neurospora crassa*. *Arch. Mikrobiol.*, **99**:353–368.

Cook, B. E. (1972). The distribution and role of cytoplasmic membrane-bounded vesicles during the development of botryose solitary blastospores of two fungi. *New Phytol.*, **71**:1135–1141.

Cook, B. E. (1974). The development of conidiophores and conidia in the imperfect fungus *Oedocephalum roseum*. *New Phytol.*, **73**:115–130.

Cooke, W. B. (1974). Terminology of the Fungi Imperfecti. *Mycopath. Mycol. Appl.*, **53**:45–67.

Cooney, D. G. and Emerson, R. (1964). *Thermophilic Fungi: An Account of their Biology, Activities and Classification*. W. H. Freeman and Co., San Francisco.

Cooper, B. H., Grove, S., Mims, C. and Szaniszlo, P. J. (1973). Septal ultrastructure in *Phialophora pedrosoi, Phialophora verrucosa* and *Cladosporium carrionii*. *Sabouraudia*, **11**:127–130.

Corlett, M., Chong, J. and Kokko, E. G. (1976). The ultrastructure of the *Spilocaea* state of *Venturia inaequalis in vivo*. *Can. J. Microbiol*, **22**:1144–1152.

Correns, C. (1899). *Untersuchungen über die Vermehrung der Laubmoose durch Brutorgane und Stecklinge*. Gustav Fischer, Jena.

Cortat, M. (1971). Localisation intracellulaire et fonctions des β-1, 3-glucanases lors du bourgeonnement chez *Saccharomyces cerevisiae*. Ph.D. Thesis, l'Ecole Polytechnique Federale, Zurich, Switzerland, 87 pp.

Da Riva Ricci, D. and Kendrick, W. B. (1972). Computer modelling of hyphal tip growth in fungi. *Can. J. Bot.*, **50**:2455–2462.

De Bary, A. (1886). *Morphologie und Physiologie der Pilze, Flechten und Myxomyceten. XII*. Wilhelm Engelmann, Leipzig.

De Bary, A. (1887). *Comparative Morphology and Biology of Fungi, Mycetozoa and Bacteria*. (Translated by H. E. F. Garnsey.) Clarendon Press, Oxford.

De Hoog, G. S. (1972). The genera *Beauveria, Isaria, Tritirachium* and *Acrodontium* gen. nov. *C. B. S. Studies in Mycology*, **1**:1–41.

De Hoog, G. S. (1974). The genera *Blastobotrys, Sporothrix, Calcarisporium* and *Calcarisporiella* gen. nov. *C. B. S. Studies in Mycology*, **7**: 1–84.

Decker, S. and Maier, S. (1975). Fine structure of mesosomal involvement during *Bacillus macerans* sporulation. *J. Bacteriol*, **121**:363–372.

Delvecchio, V. G., Corbaz, R. and Turian, G. (1969). An ultrastructural study of the hyphae, endoconidia and chlamydospores of *Thielaviopsis basicola*. *J. Gen. Microbiol.*, **58**:23–27.

Demoulin, V. (1974). The origin of Ascomycetes and Basidiomycetes. The case for a red algal ancestry. *Bot. Rev.*, **40**:315–345.

Dobbs, C. G. and English, M. P. (1954). *Piptocephalis xenophila* sp. nov. parasitic on non-mucorine hosts. *Trans. Br. Mycol. Soc.*, **37**:375–389.

Dodge, C. W. (1935). *Medical Mycology*. C. V. Mosby, St. Louis.

Domsch, K. H. and Gams, W. (1972). *Fungi in Agricultural Soils*. Longman, London.

Dowding, E. S. (1955). *Endogone* in Canadian rodents. *Mycologia*, **47**:51–57.

Drechsler, C. (1936). New conidial Phycomycetes destructive to terricolous amoebae. *Mycologia*, **28**:363–389.

Drechsler, C. (1938). New Zoopagaceae capturing and consuming soil amoebae. *Mycologia*, **30**:137–157.

Drechsler, C. (1939). Five new Zoopagaceae destructive to rhizopods and nematodes. *Mycologia*, **31**:388–415.

Drechsler, C. (1946). A nematode-destroying Phycomycete forming immotile spores in aerial evacuation tubes. *Bull. Torrey Bot. Club*, **73**:1–17.

Drechsler, C. (1947). A *Basidiobolus* producing elongated secondary conidia with adhesive beaks. *Bull. Torrey Bot. Club*, **74**:403–413.

Drechsler, C. (1955a). A new species of *Rhopalomyces* occurring in Florida. *Bull. Torrey Bot. Club*, **82**:473–479.

Drechsler, C. (1955b). Two species of *Conidiobolus* that produce microconidia. *Amer. J. Bot.*, **42**:793–802.

Duddington, C. L. (1973). Zoopagales. In *The Fungi*. vol. IVB. G. C. Ainsworth, F. K. Sparrow and A. S. Sussman (Eds). Academic Press, New York. pp. 231–234.

Dykstra, M. J. (1974a). An ultrastructural examination of the structure and germination of asexual propagules of four mucoralean fungi. *Mycologia*, **66**:477–489.

Dykstra, M. J. (1974b). Some ultrastructural features in the genus *Septobasidium*. *Can. J. Bot.*, **52**:971–972.

Ehrlich, M. A. and Ehrlich, H. G. (1969). Urediospore development in *Puccinia graminis*. *Can. J. Bot.*, **47**:2061–2064.

Ehrlich, M. A., Ehrlich, H. G. and Schafer, J. F. (1968). Septal pores in the Heterobasidiomycetidae, *Puccinia graminis* and *P. recondita*. *Amer. J. Bot.*, **55**:1020–1027.

Ellis, D. H. and Griffiths, D. A. (1975a). The fine structure of conidial development in the genus*Torula*. I. *T. herbarum* (Pers.) Link ex S. F. Gray and *T. herbarum* f. *quaternella* Sacc. *Can. J. Microbiol.*, **21**:1661–1675.

Ellis, D. H. and Griffiths, D. A. (1975b). The fine structure of conidial development in the genus *Torula*. II. *T. caligans* (Batista and Upadhyay) M. B. Ellis and *T. terrestris* Misra. *Can. J. Microbiol.*, **21**:1921–1929.

Ellis, D. H. and Griffiths, D. A. (1977). The fine structure of conidiogenesis in *Alysidium resinae* (= *Torula ramosa*). *Can. J. Bot.*, **55**:676–684.

Ellis, J. J., Rhodes, L. J. and Hesseltine, C. W. (1976). The genus *Amylomyces*. *Mycologia*, **68**:131–143.

Ellis, M. B. (1971a). Porospores. In *Taxonomy of Fungi Imperfecti*. B. Kendrick (Ed.). Univ. of Toronto Press, Toronto, pp. 71–74.

Ellis, M. B. (1971b). *Dematiaceous Hyphomycetes*. Commonw. Mycol. Inst. Publ., Kew, Surrey.

Ellis, T. T., Rogers, M. A. and Mims, C. W. (1972). The fine structure of the septal pore cap in *Coprinus stercorarius*. *Mycologia*, **64**:681–688.

Embree, C. W. (1963). The status of *Gliocephalis*. *Mycologia*, **55**:127–128.

Farr, D. F. and Lichtwardt, R. W. (1967). Some cultural and ultrastructural aspects of *Smittium culisetae* (Trichomycetes) from mosquito larvae. *Mycologia*, **59**:172–182.

Févre, M. (1977). Subcellular localization of glucanase and cellulase in *Saprolegnia monoica* Pringsheim. *J. Gen. Microbiol.*, **103**:287–295.

Févre, M. and Dumas, C. (1977). β-glucan synthetases from *Saprolegnia monoica*. *J. Gen. Microbiol.*, **103**:297–306.

Fischer, A. (1892). Die Pilze Deutschlands, Oesterreichs und der Schweiz. IV. Abteilung: Phycomycetes. In Rabenhorst, L. *Kryptogamen-Flora* (2nd Edn), vol. 1, pp.1–505.

Fletcher, J. (1971). Conidium ontogeny in *Penicillium*. *J. Gen. Microbiol.*, **67**:207–214.

Fletcher, J. (1972). Fine structure of developing merosporangia and sporangiospores of *Syncephalastrum racemosum*. *Arch. Mikrobiol.*, **87**:269–284.

Frederick, S. E., Gruber, P. J. and Newcomb, E. H. (1975). Plant microbodies. *Protoplasma*, **84**:1–29.

Fromme, H. G., Pfautsch, M., Pfefferkorn, G. and Bystricky, V. (1972). Critical point drying—a preparation method for scanning electron microscopy. *Microsc. Acta*, **73**:29–37.

Galgoczy, J. (1975). Dermatophytes: conidium ontogeny and classification. *Acta Microbiol. Acad. Sci.*, **22**:105–136.

Gams, W. (1969). Gliederrungsprinzipien in der Gattung *Mortierella*. *Nova Hedwigia*, **18**:30–43.

Gams, W. (1973). Phialides with solitary conidia? Remarks on conidium ontogeny in some hyphomycetes. *Persoonia*, **7**:161–169.

Gams, W., Chien, C.-Y. and Domsch, K. H. (1972). Zygospore formation by the heterothallic species *Mortierella elongata* and related homothallic species *Mortierella epigama*. *Trans. Br. Mycol. Soc.*, **58**:5–13.

Gams, W., van der Aa, H. A., van der Plaats-Niterink, A. J., Samson, R. A. and Stalpers, J. A. (1975). *CBS Course of Mycology*. Centraalbureau voor Schimmelcultures, Baarn.

Garrison, R. G., Lard, J. W. and Field, M. F. (1970). Ultrastructural changes during the yeastlike to mycelial-phase conversion of *Blastomyces dermatitidis* and *Histoplasma capsulatum*. *J. Bacteriol.*, **101**:628–635.

Garrison, R. G., Lane, J. W. and Johnson, O. R. (1971). Electron microscopy of the transitional cell of *Histoplasma capsulatum*. *Mycopath. Mycol. Appl.*, **44**:121–129.

Gerdemann, J. W. and Trappe, J. M. (1974). The Endogonaceae in the Pacific Northwest. *Mycol. Mem.*, **5**:1–76.

Ghiorse, W. C. and Edwards, M. R. (1973). Ultrastructure of *Aspergillus fumigatus* conidia. Development and maturation. *Protoplasma*, **76**:49–59.

Giesy, R. M. and Day, P. R. (1965). The septal pores of *Coprinus lagopus* in relation to nuclear migration. *Amer. J. Bot.*, **52**:287–293.

Gilleland, H. E. and Murray, R. G. E. (1975). Demonstration of cell division by septation in a variety of gram-negative rods. *J. Bacteriol.*, **121**:721–725.

Girbardt, M. (1957). Der Spitzenkörper von *Polystictus versicolor* (L.). *Planta (Berlin)*, **50**:47–59.

Girbardt, M. (1958). Über die Substrukture von *Polystictus versicolor* (L.). *Arch. Mikrobiol.*, **28**:255–269.

Girbardt, M. (1961). Licht- und elektronenmikroskopische Untersuchungen an *Polystictus versicolor*, V. Zur Frage der Kernfärbung mit Eisen–Hämatoxylin. *Flora*, **150**:427–440.

Girbardt, M. (1968). Ultrastructure and dynamics of the moving nucleus. In *Aspects of Cell Motility*. *Symb. Soc. Exp. Biol., 22nd*. P. L. Miller (Ed.). University Press, Cambridge, England, pp. 249–259.

Girbardt, M. (1969). Die Ultrastruktur der Apikalregion von Pilzhyphen. *Protoplasma*, **67**:413–441.

Girbardt, M. (1973). Die Pilzzelle. In *Grundlagen der Cytologie*. G. C. Hirsch, H. Ruska and P. Sitte (Eds). Fischer-Verlag, Jena, pp. 441–460.

Gooday, G. W. (1971). An autoradiographic study of hyphal growth of some fungi. *J. Gen Microbiol.*, **67**:125–133.

Gooday, G. W., Fawcett, P., Green, D. and Shaw, G. (1973). The formation of fungal sporopollenia in the zygospore wall of *Mucor mucedo*: a role for the sexual carotenogenesis in the Mucorales. *J. Gen. Microbiol.*, **74**:233–239.

Gooday, G. W. and Hunsley, D. (1971). Cellulose wall ingrowths in *Phytophthora parasitica*. *Trans. Br. Mycol. Soc.*, **57**:178–179.

Goos, R. D. (1956). Classification of the Fungi Imperfecti. *Proc. Iowa Acad. Sci.*, **63**:311–320.

Goos, R. D. (1969). Conidium ontogeny in *Cacumisporium capitulatum*. *Mycologia*, **61**:52–56.

Goos, R. D. (1975). Production of secondary conidia by *Helicoma proliferans*. *Mycologia*, **67**:888–891.

Goos, R. D. and Pirozynski, K. A. (1975). Fungi of Barro Colorado Island: new and interesting Hyphomycetes. *Can. J. Bot.*, **53**:2927–2932.

Goos, R. D. and Tubaki, K. (1973). Conidium ontogeny in *Riessia semiophora*. *Can. J. Bot.*, **51**:1439–1442.

Greenhalgh, G. N. and Chesters, C. G. C. (1968). Conidiophore morphology in some British members of the Xylariaceae. *Trans. Br. Mycol. Soc.*, **51**:57–82.

Griffiths, D. A. (1973a). The fine structure of conidial development in *Epicoccum nigrum*. *J. Microscopie*, **17**:55–64.

Griffiths, D. A. (1973b). Fine structure of the chlamydospore wall in *Fusarium oxysporum*. *Trans. Br. Mycol. Soc.*, **61**:1–6.

Griffiths, D. A. (1974). Development and structure of the aleuriospores of *Humicola grisea* Traaen. *Can. J. Microbiol.*, **20**:55–58.

Grove, S. N. (1972). Apical vesicles in germinating conidia of *Aspergillus parasiticus*. *Mycologia*, **64**:638–641.

Grove, S. N. and Bracker, C. E. (1970). Protoplasmic

organization of hyphal tips amongst fungi: vesicles and Spitzenkörper. *J. Bact.*, **104**:989–1009.

Grove, S. N., Bracker, C. E. and Morré, D. J. (1968). Cytomembrane differentiation in the endoplasmic reticulum-Golgi apparatus-vesicle complex. *Science*, **161**:171–173.

Grove, S. N., Bracker, C. E. and Morré, D. J. (1970). An ultrastructural basis for hyphal tip growth in *Pythium ultimum*. *Amer. J. Bot.*, **57**:245–266.

Grove, S. N., Marlow, J. D. and Szaniszlo, P. J. (1972). Isolation and partial chemical characterization of apical secretory vesicles from growing hyphae of the fungus, *Gilbertella persicaria*. *J. Cell Biol.*, **55**:99.

Grove, S. N., Morré, D. J. and Bracker, C. E. (1967). Dictyosomes in vegetative hyphae of *Pythium ultimum*. *Proc. Indiana Acad. Sci.*, **76**:210–214.

Grove, S. N., Oujezdsky, K. B. and Szaniszlo, P. J. (1973). Budding in the dimorphic fungus *Phialophora dermatitidis*. *J. Bacteriol.*, **115**:323–329.

Grove, W. B. (1935). *British Stem- and Leaf-Fungi (Coelomycetes)*, vol. I. Cambridge Univ. Press, Cambridge, England.

Grove, W. B. (1937). *British Stem- and Leaf-Fungi (Coelomycetes)*, vol. II. Cambridge Univ. Press, Cambridge, England.

Gull, K. (1975). Cytoplasmic microfilament organization in two basidiomycete fungi. *J. Ultrastruct. Res.*, **50**:226–232.

Gustafson, R. A., Hardcastle, R. V. and Szaniszlo, P. J. (1975). Budding in the dimorphic fungus *Cladosporium werneckii*. *Mycologia*, **67**:942–951.

Hammill, T. M. (1971). Fine structure of annellophores. I. *Scopulariopsis brevicaulis* and *S. koningii*. *Amer. J. Bot.*, **58**:88–97.

Hammill, T. M. (1972a). Electron microscopy of phialoconidiogenesis in *Metarrhizium anisopliae*. *Amer. J. Bot.*, **59**:317–326.

Hammill, T. M. (1972b). Fine structure of annellophores. V. *Stegnosporium pyriforme*. *Mycologia*, **64**:654–657.

Hammill, T. M. (1972c). Electron microscopy of conidiogenesis in *Chloridium chlamydosporis*. *Mycologia*, **64**:1054–1065.

Hammill, T. M. (1972d). Fine structure of annellophores. II. *Doratomyces nanus*. *Trans. Br. Mycol. Soc.*, **59**:249–253.

Hammill, T. M. (1972e). Fine structure of annellophores. III. *Acrogenospora sphaerocephala*. *Can. J. Bot.*, **50**:581–585.

Hammill, T. M. (1973a). Fine structure of annellophores. IV. *Spilocaea pomi*. *Trans. Br. Mycol. Soc.*, **60**:65–68.

Hammill, T. M. (1973b). Fine structure of conidiogenesis in the holoblastic, sympodial *Tritirachium roseum*. *Can. J. Bot.*, **51**:2033–2036.

Hammill, T. M. (1974a). Electron microscopy of phialides and conidiogenesis in *Trichoderma saturnisporum*. *Amer. J. Bot.*, **61**:15–24.

Hammill, T. M. (1974b). Septal pore structure in *Trichoderma saturnisporum*. *Amer. J. Bot.*, **61**:767–771.

Hammill, T. M. (1976). Light and electron microscopy of sporangial development in *Mucor* sp. *Session 10. Contributed papers. Fungal Ultrastructure*. A.I.B.S., New Orleans.

Hammill, T. M. (1977a). Light microscopic observations of karyology during conidiogenesis in *Scopulariopsis koningii*. *Mycologia*, **69**:417–421.

Hammill, T. M. (1977b). Transmission electron microscopy of annellides and conidiogenesis in the synnematal hyphomycete *Trichurus spiralis*. *Can. J. Bot.* **55**:233–244.

Hammill, T. M. (1977c). Cytological observations of conidiogenesis by phialides. *Second Int. Mycol. Congress*, Tampa, Florida. Abstract, p. 264.

Hammill, T. M. (1977d). Karyology during conidiogenesis in *Gliomastix murorum*: light microscopy. *Amer. J. Bot.*, **64**:1140–1151.

Hanlin, R. T. (1971). Morphology of *Nectria haematococca*. *Amer. J. Bot.*, **58**:105–116.

Hanlin, R. T. (1976). Phialide and conidium development in *Aspergillus clavatus*. *Amer. J. Bot.*, **63**:144–155.

Happ, G. M., Happ, C. M. and Barras, S. J. (1976). Bark beetle-fungal symbiosis. II. Fine structure of a basidiomycetous ectosymbiont of the southern pine beetle. *Can. J. Bot.*, **54**:1049–1062.

Harvey, I. C. (1974a). Striated fibers in the cytoplasm of the imperfect fungus *Pleiochaeta setosa* (Kirchn.) Hughes. *Protoplasma*, **80**:371–380.

Harvey, I. C. (1974b). Light and electron microscope observations of conidiogenesis in *Pleiochaeta setosa* (Kirchn.) Hughes. *Protoplasma*, **82**:203–221.

Hashimoto, T., Morgan, J. and Conti, J. S. F. (1973). Morphogenesis and ultrastructure of *Geotrichum candidum septa*. *J. Bacteriol.*, **116**:447–455.

Hashimoto, T., Wu-Yuan, C. D. and Blumenthal, H. J. (1976). Isolation and characterization of the rodlet layer of *Trichophyton mentagrophytes* microconidial wall. *J. Bacteriol.*, **127**:1543–1549.

Hashioka, Y. (1975). Conidia liberation in *Pyricularia*. In *Proc. First Congr., Int. Assoc. Microbiol. Soc., Tokyo. Vol. 2. Developmental Microbiol. Ecol.* T. Hasegawa (Ed.). Science Council of Japan Pub., pp. 141–144.

Hashmi, M. H., Kendrick, W. B. and Morgan-Jones, G. (1972). Conidium ontogeny in Hyphomycetes. The genera *Torulomyces* Delitsch and *Monocillium* Saksena. *Can. J. Bot.*, **50**:1461–1463.

Hashmi, M. H. and Morgan-Jones, G. (1973). Conidium ontogeny in Hyphomycetes. The meristem arthrospores of *Wallemia sebi*. *Can. J. Bot.*, **51**:1669–1671.

Hashmi, M. H., Morgan-Jones, G. and Kendrick, W. B. (1973). Conidium ontogeny in Hyphomycetes. The blastoconidia of *Cladosporium herbarum* and *Torula herbarum*. *Can. J. Bot.*, **51**:1089–1091.

Haskins, R. H. (1975). Septa ultrastructure and hyphal branching in the pleomorphic imperfect fungus *Trichosporonoides oedocephalis*. *Can. J. Bot.*, **53**:1139–1148.

Haskins, R. H., Brushaber, J. A., Child, J. J. and Holtby, L. B. (1976). The ultrastructure of sexual reproduction in *Pythium acanthicum*. *Can. J. Bot.*, **54**:2193–2203.

Havelkova, M. and Mensik, P. (1966). The Golgi apparatus

in the regenerating protoplasts of *Schizosaccharomyces. Naturwissenschaften,* **53**:562.

Hawes, C. R. and Beckett, A. (1977a). Conidium ontogeny in the *Chalara* state of *Ceratocystis adiposa.* I. Light microscopy. *Trans. Br. Mycol. Soc.,* **68**:259–265.

Hawes, C. R. and Beckett, A. (1977b). Conidium ontogeny in the *Chalara* state of *Ceratocystis adiposa.* II. Electron microscopy. *Trans. Br. Mycol. Soc.,* **68**:267–276.

Hawker, L. E. (1965). Fine structure of fungi as revealed by electron microscopy. *Biol. Rev.,* **40**:52–92.

Hawker, L. E. (1966). Germination: morphological and anatomical changes. In *The Fungus Spore.* M. F. Madelin (Ed.). Butterworths, London, England, pp. 151–161.

Hawker, L. E., Gooday, M. A. and Bracker, C. E. (1966). Plasmodesmata in fungal cell walls. *Nature,* **212**:635.

Hawker, L. E. and Hendy, R. J. (1963). An electron microscope study of germination of conidia of *Botrytis cinerea. J. Gen. Microbiol.,* **33**:43–46.

Hawker, L. E., Thomas, B. and Beckett, A. (1970). An electron microscope study of structure and germination of conidia of *Cunninghamella elegans* Lendner. *J. Gen. Microbiol.,* **60**:181–189.

Heath, I. B. (1977). Morphology and ontogeny of microbodies in the oomycete fungus *Sapromyces elongatus. Can. J. Bot.,* **55**:556–562.

Heath, I. B. and Greenwood, A. D. (1970). The structure and formation of lomasomes. *J. Gen. Microbiol.,* **62**:129–137.

Heath, I. B., Gay, J. L. and Greenwood, A. D. (1971). Cell wall formation in the Saprolegniales: cytoplasmic vesicles underlying developing walls. *J. Gen. Microbiol.,* **65**:225–232.

Heath, M. C. (1975). Structure and function of fungal septa. In *Fine Structure of Fungi. Proc. First Intersectional Congr. of Inter. Assoc. Microbiol. Soc., Tokyo. Vol. 2. Developmental Microbiol. Ecol.* H. Takeo (Ed.). Science Council of Japan Pub., pp. 117–124.

Hemmes, D. E. and Hohl, H. R. (1973). Mitosis and nuclear degeneration: simultaneous events during secondary sporangia formation in *Phytophthora palmivora. Can. J. Bot.,* **51**:1673–1675.

Hemmes, D. E. and Wong, L. D. S. (1975). Ultrastructure of chlamydospores of *Phytophthora cinnamomi* during development and germination. *Can. J. Bot.,* **53**:2945–2957.

Hennebert, G. L. (1962). *Wardomyces* and *Asteromyces. Can. J. Bot.,* **40**:1203–1216.

Hess, W. M. (1973). Ultrastructure of fungal spore germination. *Shokubutsu Byogai Kenkyu,* **8**:71–84.

Hess, W. M. (1975). Spore germination of fungi. In *Fine Structure of Fungi. Proc. First Intersectional Congr. Int. Assoc. Microbiol. Soc., Tokyo. Vol. 2. Developmental Microbiol. Ecol.* H. Takeo (Ed.). Science Council of Japan Pub., pp. 83–91.

Hess, W. M. and Weber, D. J. (1973). Ultrastructure of dormant and germinated sporangiospores of *Rhizopus arrhizus. Protoplasma,* **77**:15–33.

Hess, W. M., Sassen, M. M. A. and Remsen, C. C. (1968). Surface characteristics of *Penicillium* conidia. *Mycologia,* **60**:290–303.

Hesseltine, C. W. (1952). A survey of the Mucorales. *Trans. N.Y. Acad. Sci.,* Ser. 2, **14**:210–214.

Hesseltine, C. W. (1961). Carotenoids in the fungi Mucorales: special reference to Choanephoraceae. *U.S. Dept. Agric. Tech. Bull.* no. 1245:1–33.

Hesseltine, C. W. and Benjamin, C. R. (1957). Notes on the Choanephoraceae. *Mycologia,* **49**:723–733.

Hesseltine, C. W. and Ellis, J. J. (1973). Mucorales. In *The Fungi,* vol. IVB. G. C. Ainsworth, F. K. Sparrow and A. S. Sussman (Eds). Academic Press, New York, pp. 187–217.

Heunert, H.-H. (1972). Microtechnique for the observation of living microorganisms. *Zeiss Information,* **20**:40–49.

Hoch, H. C. and Maxwell, D. P. (1974). Proteinaceous hexagonal inclusions in hyphae of *Whetzelinia sclerotiorum* and *Neurospora crassa. Can. J. Microbiol.,* **20**:1029–1035.

Horridge, G. A. and Tamm, S. L. (1969). Critical point drying for scanning electron microscopic study of ciliary motion. *Science,* **163**:817–818.

Huang, H. C. and Patrick, Z. A. (1974). Karyology of conidiogenesis and endoconidium germination in *Thielaviopsis basicola. Can. J. Bot.,* **52**:2263–2267.

Hughes, G. C. (1975). Studies of fungi in oceans and estuaries since 1961. I. Lignicolous, caulicolous and foliicolous species. *Oceanogr. Mar. Biol. Ann. Rev.,* **13**:69–180.

Hughes, G. C. and Bisalputra, A. A. (1970). Ultrastructure of Hyphomycetes. Conidium ontogeny in *Peziza ostracoderma. Can. J. Bot.,* **48**:361–366.

Hughes, S. J. (1951). Studies on microfungi. XI. Some hyphomycetes which produce phialides. *Commonw. Mycol. Papers,* **45**:1–36.

Hughes, S. J. (1953). Conidiophores, conidia and classification. *Can. J. Bot.,* **31**:577–659.

Hughes, S. J. (1970). Ontogeny of spore forms in Uredinales. *Can. J. Bot.,* **48**:2147–2157.

Hughes, S. J. (1971a). Phycomycetes, Basidiomycetes, and Ascomycetes as Fungi Imperfecti. In *Taxonomy of Fungi Imperfecti.* W. B. Kendrick (Ed.). Univ. of Toronto Press, pp. 7–33.

Hughes, S. J. (1971b). Percurrent proliferations in fungi, algae, and mosses. *Can. J. Bot.,* **49**:215–231.

Hughes, S. J. (1971c). On conidia of fungi and gemmae of algae, bryophytes, and pteridophytes. *Can. J. Bot.,* **49**:1319–1339.

Hughes, S. J. (1976). Sooty moulds. *Mycologia,* **68**:693–820.

Hughes, S. J. and Kendrick, W. B. (1968). New Zealand Fungi. 12. *Menispora, Codinaea, Menisporopsis. N.Z. Jour. Bot.,* **6**:323–375.

Hunsley, D. (1973). Apical wall structure in hyphae of *Phytophthora parasitica. New Phytol.,* **72**:985–990.

Hunsley, D. and Burnett, J. H. (1968). Dimensions of microfibrillar elements in fungal walls. *Nature,* **218**:462–463.

Hunsley, D. and Burnett, J. H. (1970). The ultrastructural architecture of the walls of some hyphal fungi. *J. Gen. Microbiol.,* **62**:203–218.

Hunsley, D. and Gooday, G. W. (1974). The structure and development of septa in *Neurospora crassa*. *Protoplasma*, **82**: 125–146.
Ingold, C. T. (1956). The conidial apparatus of *Trichothecium roseum*. *Trans. Br. Mycol. Soc.*, **39**:460–464.
Ingold, C. T. (1975). Hooker Lecture 1974. Convergent evolution in aquatic fungi: the tetraradiate spore. *Biol. J. Linn. Soc.*, **7**: 1–25.
Ingold, C. T., Dann, V. and McDougall, P. J. (1968). *Tripospermum camelopardus* sp. nov. *Trans. Br. Mycol. Soc.*, **51**:51–56.
Jeffries, P. and Young, T. W. K. (1975). Ultrastructure of the sporangiospore of *Piptocephalis unispora* (Mucorales). *Arch. Mikrobiol.*, **105**:329–333.
Jeffries, P. and Young, T. W. K. (1976). Physiology and fine structure of sporangiospore germination in *Piptocephalis unispora* prior to infection. *Arch. Mikrobiol.*, **107**:99–107.
Jersild, R., Mishkin, S. and Niederpruem, D. J. (1967). Origin and ultrastructure of complex septa in *Schizophyllum commune* development. *Arch. Mikrobiol.*, **57**:20–32.
Jones, D. R. (1973). Ultrastructure of septal pore in *Uromyces dianthi*. *Trans. Br. Mycol. Soc.*, **61**:227–235.
Katz, P. and Rosenberger, R. F. (1970). The utilization of galactose by an *Aspergillus nidulans* mutant lacking galactose phosphate-UDP glucose transferase and its relation to cell wall synthesis. *Arch. Mikrobiol.*, **74**:41–51.
Katz, P. and Rosenberger, R. F. (1971). Hyphal wall synthesis in *Aspergillus nidulans*: effect of protein synthesis inhibition and osmotic shock on chitin insertion and morphogenesis. *J. Bacteriol.*, **108**: 184–190.
Kelley, R. O., Decker, A. F. and Bluemink, J. G. (1973). Ligand-mediated osmium binding: its application in coating biological specimens for scanning electron microscopy. *J. Ultrastruct. Res.*, **45**:254–258.
Kendrick, W. B. (1958). *Sympodiella*, a new Hyphomycete genus. *Trans. Br. Mycol. Soc.*, **41**:519–521.
Kendrick, W. B. (1962). The *Leptographium* complex. *Verticicladiella* Hughes. *Can. J. Bot.*, **40**: 771–797.
Kendrick, W. B. (Ed.). (1971a). *Taxonomy of Fungi Imperfecti.* Univ. of Toronto Press, Toronto.
Kendrick, W. B. (1971b). Conclusions and recommendations. In *Taxonomy of Fungi Imperfecti.* B. Kendrick (Ed.). Univ. of Toronto Press, Toronto, pp. 253–262.
Kendrick, W. B. (1971c). Arthroconidia and meristem arthroconidia. In *Taxonomy of Fungi Imperfecti.* B. Kendrick (Ed.). Univ. of Toronto Press, Toronto, pp. 160–171.
Kendrick, W. B. (1971d). Conidium shape. In *Taxonomy of Fungi Imperfecti.* W. B. Kendrick (Ed.). Univ. of Toronto Press, Toronto, pp. 301–304.
Kendrick, W. B. and Carmichael, J. W. (1973). Hyphomycetes. In *The Fungi*, Vol. IVA. G. C. Ainsworth, F. K. Sparrow and A. S. Sussman (Eds). Academic Press, New York, pp. 323–509.
Kendrick, W. B. and Chang, M. G. (1971). Karyology of conidiogenesis in some Hyphomycetes. In *Taxonomy of Fungi Imperfecti.* W. B. Kendrick (Ed.). Univ. Of Toronto Press, Toronto, pp. 279–291.
Kendrick, W. B. and Cole, G. T. (1968). Conidium ontogeny in Hyphomycetes. The sympodulae of *Beauveria* and *Curvularia*. *Can. J. Bot.*, **46**: 1297–1301.
Kendrick, W. B. and Cole, G. T. (1969). Conidium ontogeny in Hyphomycetes. *Trichothecium roseum* and its meristem arthrospores. *Can. J. Bot.*, **47**:345–350.
Kendrick, W. B. and Watling. R. (1977). Mitospores in Basidiomycetes. *Second Int. Mycol. Congress, Tampa, Florida.* Abstract. p. 341.
Kendrick, W. B., Cole, G. T. and Bhatt, G. C. (1968). Conidium ontogeny in Hyphomycetes. *Gonatobotryum apiculatum* and its botryose blastospores. *Can. J. Bot.*, **46**:591–596.
Khan, S. R. (1975). Wall structure and germination of spores in *Cunninghamella echinulata*. *J. Gen. Microbiol.*, **90**: 115–124.
Khan, S. R. and Aldrich, H. C. (1973). Conidiogenesis in *Termitaria synderi* (Fungi Imperfecti). *Can. J. Bot.*, **51**:2307–2314.
Khan, S. R. and Talbot, P. H. B. (1975). Monosporous sporangiola in *Mycotypha* and *Cunninghamella*. *Trans. Br. Mycol. Soc.*, **65**:29–39.
King, D. S. and Jong, S. C. (1976). Induction of arthroconidia in *Trichosporon*. *Mycopathologia*, **59**: 61–63.
Kitajima, Y., Sekiya, T., Nozawa, Y. and Ito, Y. (1976). Ultrastructure of the *Trichophyton mentagrophytes* septum: visualization of the lamellar structure. *Japan J. Med. Mycol.*, **17**:31–36.
Kohlmeyer, J. (1969). Marine fungi of Hawaii including the new genus *Heliascus*. *Can. J. Bot.*, **47**: 1469–1487.
Kohlmeyer, J. (1975). New clues to the possible origin of Ascomycetes. *Bioscience*, **25**:86–93.
Kreger-van Rij, N. J. W. and Veenhuis, M. (1969). Septal pores in *Endomycopsis platypodis* and *Endomycopsis monospora*. *J. Gen. Microbiol.*, **57**:91–96.
Kreger-van Rij, N. J. W. and Veenhuis, M. (1971a). Septal pores in *Trichosporon cutaneum*. *Sabouraudia*, **9**:36–38.
Kreger-van Rij, N. J. W. and Veenhuis, M. (1971b). Bipolar budding in yeasts—an electron microscope study. *Antonie van Leeuwenhoek, J. Microbiol. Serol.*, **37**: 125–136.
Kreger-van Rij, N. J. W. and Veenhuis, M. (1972). Some features of vegetative and sexual reproduction in *Endomyces* species. *Can. J. Bot.*, **50**: 1691–1695.
Kreger-van Rij, N. J. W. and Veenhuis, M. (1974). Spores and septa in the genus *Dipodascus*. *Can. J. Bot.*, **52**: 1335–1338.
Kreisel, H. (1969). *Grundzüge eines natürlichen systems der Pilze.* J. Cramer, Lehre.
Kuhlman, E. G. (1972). Variation in zygospore formation among species of *Mortierella*. *Mycologia*, **64**:325–341.
Kuhlman, E. G. (1975). Zygospore formation in *Mortierella alpina* and *M. spinosa*. *Mycologia* **67**:678–681.
Kurtzman, C. P., Baker, F. L. and Smiley, M. J. (1974). Specimen holder to critical point dry microorganisms for scanning electron microscopy. *Appl. Microbiol.*, **28**:708–712.
Laane, M. M. (1969). Further cytological studies in the genus *Penicillium*. *Hereditas*, **62**: 153–184.

Lane, J. W. and Garrison, R. G. (1970). Electron microscopy of the yeast to mycelial phase conversion of *Sporotrichum schenckii. Can. J. Microbiol.,* **16**: 747–749.

Lendner, A. (1908). Les Mucorinées de la Suisse. In *Beiträge zur Kryptogamenflora der Schweiz,* Vol. 3. K. J. Wyss (Ed.). Berne, pp. 1–180.

Lewis, E. R. and Nemanic, M. K. (1973). Critical point drying techniques. *Scanning Electron Microscopy/1973.* IITRI. Chicago, pp. 767–773.

Lichtwardt, R. W. (1973). The Trichomycetes: What are their relationships? *Mycologia,* **65**: 1–20.

Lin, L. P., Lee, Y. Y., Fong, J. C. and Hsieh, S. I. (1975). Ultrastructure of membrane-complex systems in *Agaricus bisporus.* In *Fine Structure of Fungi. Proc. First Intersectional Congr. Int. Assoc. Microbiol. Soc., Tokyo. Vol. 2. Developmental Microbiol. Ecol.* H. Takeo (Ed.). Science Council of Japan Pub., pp. 136–140.

Linder, D. H. (1929). A monograph of the helicosporous Fungi Imperfecti. *Ann. Missouri Bot. Gard.,* **16**: 227–388.

Littlefield, L. J. and Bracker, C. E. (1971). Ultrastructure of septa in *Melampsora lini. Trans. Br. Mycol. Soc.,* **56**: 181–188.

Lowry, R. J., Durkee, T. L. and Sussman, A. S. (1967). Ultrastructural studies of micro-conidium formation in *Neurospora crassa. J. Bacteriol.,* **94**: 1757–1763.

Lung, B. (1974). The preparation of small particulate specimens by critical point drying: application for scanning electron microscopy. *J. Microsc.,* **101**: 77–80.

Lurie, H. I., Shadomy, H. J. and Still, W. J. S. (1971). An electron microscopic study of *Cryptococcus neoformans* (Coward Strain). *Sabouraudia,* **9**: 15–16.

Luttrell, E. S. (1963). Taxonomic criteria in *Helminthosporium. Mycologia,* **55**: 643–674.

Madelin, M. F. (Ed.). (1966). *The Fungus Spore.* Butterworths. London.

Madelin, M. F. and Dorabjee, S. (1974). Conidium ontogeny in *Wallemia sebi. Trans. Br. Mycol. Soc.,* **63**: 121–130.

Mangenot, F. (1952). Recherches méthodiques sur les champignons de certain bois en décomposition. Ph.D. Thesis. Faculté des Sciences, Nancy, France.

Mangenot, F. and Reisinger, O. (1976). Form and function of conidia as related to their development. In *The Fungal Spore.* D. J. Weber and W. M. Hess (Eds). John Wiley, New York, pp. 789–847.

Manocha, M. S. and Shaw, M. (1967). Electron microscopy of uredospores of *Melampsora lini* and of rust-infected flox. *Can. J. Bot.,* **45**: 1575–1582.

Marchant, H. J. (1973). Processing small delicate biological specimens for scanning electron microscopy. *J. Microsc.,* **97**: 369–371.

Marchant, H. J. (1976). Plasmodesmata in algae and fungi. In *Intercellular Communication in Plants: Studies on Plasmodesmata.* B. E. S. Gunning and A. W. Robards (Eds). Springer-Verlag, Berlin, pp. 59–80.

Marchant, R. (1975). An ultrastructural study of 'phialospore' formation in *Fusarium culmorum* grown in continuous culture. *Can. J. Bot.,* **53**: 1978–1987.

Marchant, R. and Moore, R. T. (1973). Lomasomes and plasmalemmasomes in fungi. *Protoplasma,* **76**: 235–247.

Marchant, R. and Robards, A. W. (1968). Membrane systems associated with the plasmalemma of plant cells. *Ann. Bot., New Series,* **32**: 457–471.

Marchant, R. and Wessels, J. G. H. (1973). Septal structure in normal and modified strains of *Schizophyllum commune* carrying mutations affecting septal dissolution. *Arch. Mikrobiol.,* **90**: 35–45.

Marchant, R., Peat, A. and Banbury, G. H. (1967). The ultrastructural basis of hyphal growth. *New Phytol.,* **66**: 623–629.

Martin, E. (1940). The morphology of *Taphrina deformans. Amer. J. Bot.,* **27**: 743–751.

Mason, E. W. (1933). Annotated account of fungi received at the Imperial Mycological Institute, List II, Fasc. **2**: 1–67.

Mason, E. W. (1937). Annotated account of fungi received at the Imperial Mycological Institute. List II. Fasc. **3**: 68–99.

Matruchot, M. L. (1899). Notes mycologiques. I. *Gliocephalis hyalina. Bull. Soc. Mycol. Fr.,* **15**: 254–262.

Matsushima, T. (1971). *Microfungi of the Solomon Islands and Papua-New Guinea.* Shionogi Research Lab., Kobe, Japan. 78 pp., 169 figs., 48 plates.

Matsushima, T. (1975). *Icones Microfungorum a Matsushima Lectorum.* Kobe, Japan.

Mayfield, J. E. (1974). Septal involvement in nuclear migration in *Schizophyllum commune. Arch. Mikrobiol.,* **95**: 115–124.

McClure, W. K., Park, D. and Robinson, P. M. (1968). Apical organization in the somatic hyphae of fungi. *J. Gen. Microbiol.,* **50**: 177–182.

McCully, E. K. and Bracker, C. E. (1972). Apical vesicles in growing bud cells of heterobasidiomycetous yeasts. *J. Bacteriol.,* **109**: 922–926.

McKeen, W. E. (1971). Woronin bodies in *Erysiphe graminis* D. C. *Can. J. Microbiol.,* **17**: 1557–1560.

McKeen, W. E. (1972). Nuclear movement in *Erysiphe graminis hordei. Can. J. Microbiol.,* **18**: 1333–1336.

Meredith, D. S. (1961). Spore discharge in *Deightoniella torulusa* (Syd.) Ellis. *Ann. Bot., New Series,* **25**: 271–278.

Meredith, D. S. (1962). Spore discharge in *Cordana musae* (Zimm.) Höhnel and *Zygosporium oscheoides* Mont. *Ann. Bot. New Series,* **26**: 233–241.

Meredith, D. S. (1963). Violent spore release in some Fungi Imperfecti. *Ann. Bot., New Series,* **27**: 39–47.

Migula, W. (1910). Pilze. In *Thome's Kryptogamen-Flora von Deutschland, Deutsch-Osterreich und der Schweiz,* Band III, Teil 1. Verlag Friedrich von Zezschwitz, Gera. pp. 58–242.

Miller, C. V. and Anderson, N. A. (1961). Proliferations of conidiophores and intrahyphal hyphae in *Aspergillus niger. Mycologia,* **53**: 433–436.

Mollenhauer, H. H. and Morré, D. J. (1966). Golgi apparatus and plant secretion. *Ann. Rev. Plant Physiol.,* **17**: 27–46.

Moor, H. and Mühlethaler, K. (1963). Fine structure in frozen-etched yeast cells. *J. Cell Biol.,* **17**: 609–628.

Moore, R. T. (1955). Index to the Helicosporae. *Mycologia,* **47**: 90–103.

Moore, R. T. (1957). Index to the Helicosporae: Addenda. *Mycologia,* **49**: 580–587.

Moore, R. T. (1963). Fine structure of mycota. 10. Thallus formation in *Puccinia podophylli* aecia. *Mycologia,* **55**:633–642.
Moore, R. T. (1965). The ultrastructure of fungal cells. In *The Fungi,* vol. I. G. C. Ainsworth and A. Sussman (Eds). Academic Press, New York, pp. 95–118.
Moore, R. T. (1971). An alternative concept of the fungi based on their ultrastructure. In *Recent Advances in Microbiology.* A. Pérez-Miravete and D. Peláez (Eds). Asociacion Mexicana de Microbiologia, Mexico, pp. 49–64.
Moore, R. T. (1972). Ustomycota, a new division of higher fungi. *Antonie van Leeuwenhoek J. Microbiol. Serol.,* **38**:567–584.
Moore, R. T. (1977). Dolipore disjunction in *Antromycopsis broussonetiae* Pat. *Expt. Mycol.,* **1**:92–101.
Moore, R. T. and Marchant, R. (1972). Ultrastructural characterization of the basidiomycete septum of *Polyporus biennis. Can. J. Bot.,* **50**:2463–2469.
Moore, R. T. and McAlear, J. H. (1961). Fine structure of mycota. 5. Lomasomes, previously uncharacterized hyphal structures. *Mycologia,* **53**:194–200.
Moore, R. T. and McAlear, J. H. (1962). Fine structure of mycota. 7. Observations on septa of Ascomycetes and Basidiomycetes. *Amer. J. Bot.,* **49**:86–94.
Moreau, F. (1953). Les champignons. *Encycl. Mycol.* 13. Lechevalier, Paris.
Morgan-Jones, G., Nag Raj, T. R. and Kendrick, W. B.(1972). Conidium ontogeny in Coelomycetes. IV. Percurrently proliferating phialides. *Can. J. Bot.,* **50**:2009–2014.
Morré, D. J., Mollenhauer, H. H. and Bracker, C. E. (1971). Origin and continuity of Golgi apparatus. In *Results and Problems in Cell Differentiation. Vol. 2. Origin and Continuity of Cell Organelles.* J. Reinert and H. Ursprung (Eds). Springer-Verlag, Berlin, p. 82.
Moss, S. T. and Lichtwardt, R. W. (1976). Development of trichospores and their appendages in *Genistellospora homothallica* and other Harpellales and fine-structural evidence for the sporangial nature of trichospores. *Can. J. Bot.,* **54**:2346–2364.
Moss, S. T. and Young, T. W. K. (1977). Comparative wall structure in Kickxellales and Harpellales. *Second Int. Mycol. Congress,* Tampa, Florida. Abstract, p. 456.
Müller, E. and Loeffler, W. (1976). *Mycology.* (Translated by B. Kendrick and F. Bärlocher.) G. Thieme, Stuttgart.
Murray, G. M. and Maxwell, D. P. (1974). Ultrastructure of conidium germination of *Cochliobolus carbonus. Can. J. Bot.,* **52**:2335–2340.
Nachmias, V. T. and Ingram, W. C. (1970). Actomyosin from *Physarum polycephalum*: Electron microscopy of myosin-enriched preparations. *Science,* **170**:743–745.
Nachmias, V. T., Huxley, H. E. and Kessler, D. (1970). Electron microscope observations on actomyosin and actin preparations from *Physarum polycephalum* and on their interaction with heavy meromyosin subfragment I from muscle myosin. *J. Mol. Biol.,* **50**:83–90.
Nag Raj, T. R. and Kendrick, B. (1975). *A Monograph of* Chalara *and Allied Genera.* Wilfred Laurier University Press, Waterloo, Ontario.
Nannfeldt, J. A. and Eriksson, J. (1952). On the genus *Costantinella* Matruchot (Hyphomycetes). *Svensk Botan. Tidskr.,* **46**:109–128.
Newman, H. N. and Tailby, P. W. (1973). A simple apparatus for the freeze drying of biological specimens preparatory to scanning electron microscopy. *Stain Technol.,* **48**:145–149.
Nicot, J. and Leduc, A. (1957). Mise en evidence d'un mucilage dans la paroi des spores du *Trichothecium roseum* Link ex Fr. *C. R. Acad. Sci., Paris,* **244**:1403–1405.
Nicot, J. and Mariat, F. (1973). Caractères morphologiques et position systématique de *Sporothrix schenckii,* agent de la sporotrichose humaine. *Mycopath. Mycol. Appl.,* **49**:53–65.
Niederpruem, D. J. and Wessels, J. G. H. (1969). Cytodifferentiation and morphogenesis in *Schizophyllum commune. Bacteriol. Rev.,* **33**:505–535.
O'Brien, T. P. (1972). The cytology of cell-wall formation in some eukaryotic cells. *Bot. Rev.,* **38**:87–118.
O'Donnell, K. L., Fields, W. G. and Hooper, G. R. (1976). Light and electron microscopic observations of conidiogenesis in *Peziza quelepidotia. Can. J. Bot.,* **54**:1084–1093.
Oláh, G. M. (1970). Présence vraisemblable de dictyosome simple dans le mycelium de *Paecilomyces berlinensis* Ola'h et Ackerman. *C. R. Acad. Sci. Ser. D, Sci. Nat. (Paris),* **271**:1504–1507.
Oláh, G. M. (1971). Présence vraisemblable de dictyosome simple dans le mycelium de *Paecilomyces berlinensis* et particularités ultrastructurales de cette espèce. *Communication présentée au 39e Congrès de l'A.C.F.A.S. (Assoc. Can. Fr. Av. Sci.). Sherbrooke.*
Oláh, G. M. (1973). Activités enzymatiques et infrastructure du dictyosome simple. *C. R. Acad. Sci., Paris,* **276**:2261–2263.
Oláh, G. M. (1974). Localisation infrastructural des activité enzymatiques et ontogénie du dictyosome simple de *Paecilomyces berlinensis* (Fungi Imperfecti). Travaux mycologiques dédiés à R. Kühner, numéro spécial du *Bulletin de la Société Linnéenne de Lyon,* pp. 292–312.
Oláh, G. M. and Reisinger, O. (1974). Etude ultrastructurale et cytochimique de l'appareil sporifère chez *Phialophora richardsiae. Can. J. Bot.,* **52**:2473–2480.
Oliver, P. T. P. (1972). Conidiophore and spore development in *Aspergillus nidulans. J. Gen. Microbiol.,* **73**:45–54.
Oso, B. A. (1969). Electron microscopy of ascus development in *Ascobolus. Ann. Bot., New Series,* **33**:205–209.
Oujedsky, K. B., Grove, S. N. and Szaniszlo, P. J. (1973). Morphological and structural changes during the yeast-to-mold conversion of *Phialophora dermatitidis. J. Bacteriol.,* **113**:468–477.
Oulevey-Matikian, N. and Turian, G. (1968). Contrôle metabolique et aspects ultrastructuraux de la conidiation (macro-microconidies) de *Neurospora crassa. Arch. Mikrobiol.,* **60**:35–58.
Park, D. and Robinson, P. M. (1966). Internal pressure of hyphal tips of fungi and its significance in morphogenesis. *Ann. Bot.,* **30**:425–439.

Park, D. and Robinson, P. M. (1969). Sporulation in *Geotrichum candidum. Trans. Br. Mycol. Soc.*, **52**: 213–222.

Park, D. and Robinson, P. M. (1970). Germination studies with *Geotrichum candidum. Trans. Br. Mycol. Soc.*, **54**: 83–92.

Parthasarathy, M. V. and Mühlethaler, K. (1972). Cytoplasmic microfilaments in plant cells. *J. Ultrastruct. Res.*, **38**: 46–62.

Pidoplichko, N. M. and Milko, A. A. (1971). *Atlas of Mucoralean Fungi.* Academy of Sciences of the Ukranian S. S. R., Kiev, (In Russian).

Pirozynski, K. A. (1966). The genus *Xenosporium. Commonw. Mycol. Papers*, **105**: 21–35.

Pirozynski, K. A. (1969). Reassessment of the genus *Amblyosporium. Can. J. Bot.*, **47**: 325–334.

Plumb, R. T. and Turner, R. H. (1972). Scanning electron microscopy of *Erysiphe graminis. Trans. Br. Mycol. Soc.*, **59**: 149–178.

Pollack, F. G. and Miller, O. K. (1976). *Antromycopsis broussonetiae* found to be the name of the imperfect state of *Pleurotus cystidiosus. Mem. N. Y. Bot, Gdn.*, **28**: 174–178.

Pontecorvo, G., Roper, J. A., Hemmons, L. M., Macdonald, K. D. and Bufton, A. W. J. (1953). The genetics of *Aspergillus nidulans. Adv. Genetics* **5**: 141–238.

Porter, K. R., Kelley, D. and Andrews, P. M. (1972). The preparation of cultured cells and soft tissues for scanning electron microscopy. In *Proc. of the Fifth Annual Stereoscan Colloquium.* Kent Cambridge Scientific Company, pp. 1–19.

Raudaskoski, M. (1970). Occurrence of microtubules and microfilaments, and origin of septa in dikaryotic hyphae of *Schizophyllum commune. Protoplasm*, **70**: 415–422.

Raudaskoski, M. (1972). Occurrence of microtubules in the hyphae of *Schizophyllum commune* during intercellular nuclear migration. *Arch. Mikrobiol.*, **86**: 91–100.

Raudaskoski, M. and Koltin, Y. (1973). Ultrastructural aspects of a mutant of *Schizophyllum commune* with continuous nuclear migration. *J. Bacteriol.*, **116**: 981–988.

Reeves, F. (1967). The fine structure of ascospore formation in *Pyronema domesticum. Mycologia*, **59**: 1018–1033.

Reichle, R. E. and Alexander, J. V. (1965). Multiperforate septations, Woronin bodies and septal plugs in *Fusarium. J. Cell Biol.*, **24**: 489–496.

Reichle, R. E. and Lichtwardt, R. W. (1972). Fine structure of the Trichomycete, *Harpella melusinae* from black-fly guts. *Arch. Mikrobiol.*, **81**: 103–125.

Reisinger, O. (1968). Remarques sur les genres *Dendryphiella* et *Dendryphion. Bull. Soc. Mycol. Fr.*, **84**: 27–51.

Reisinger, O. (1972). *Contribution a l' ètude ultrastructurale de l' appareil sporifère chez quelques Hyphomycètes à paroi mèlanisée.* Thèse Sciences Naturelles, Nancy.

Reisinger, O. and Guedenet, J. C. (1968). Morphologie ultrastructurale et critères taxonomiques chez les Deutèromycètes. I. Les paroi sporales chez *Dendryphiella vinosa* (Berk. et Curt.) Reisinger. *Bull. Soc. Mycol. Fr.*, **84**: 19–26.

Reisinger, O. and Kiffer, E. (1974). Contribution to the fungal microflora of the Congo. IV. *Polyschema congolensis* sp. nov.: Taxonomy and ultrastructure. *Trans. Br. Mycol. Soc.*, **62**: 289–294.

Reisinger, O. and Oláh, G. M. (1974). Etude ultrastructurale et cytochimique de la conidiogénèse chez *Beauveria bassiana* (Bals.) Vuill. *Can. J. Microbiol.*, **20**: 1387–1392.

Reynolds, E. S. (1963). The use of lead citrate at high pH as an electron-opaque stain in electron microscopy. *J. Cell Biol.*, **17**: 208–212.

Robb, J., Harvey, A. E. and Shaw, M. (1973). Ultrastructure of hyphal walls and septa of *Cronartium ribicola* on tissue cultures of *Pinus monticola. Can. J. Bot.*, **51**: 2301–2305.

Robertson, N. F. (1958). Observations on the effect of water on the hyphal apices of *Fusarium oxysporum. Ann. Bot., New Series*, **22**: 159–173.

Robertson, N. F. (1959). Experimental control of hyphal branching and branch form in hyphomycetous fungi. *J. Linn. Soc. (Bot.)*, **56**: 207–211.

Robertson, N. F. (1965a). The fungal hypha. *Trans. Br. Mycol. Soc.*, **48**: 1–8.

Robertson, N. F. (1965b). The mechanism of cellular extension and branching. In *The Fungi*, vol. I. G. C. Ainsworth and A. S. Sussman (Eds). Academic Press, New York, pp. 613–623.

Robertson, N. F. (1968). The growth process in fungi. *Ann. Rev. Phytopathol.*, **6**: 115–136.

Robinow, C. F. and Caten, C. E. (1969). Mitosis in *Aspergillus nidulans. J. Cell Sci.*, **5**: 403–431.

Roelofsen, P. A. (1950). Cell wall structure in the growth-zone of *Phycomyces* sporangiophores. I. Model experiments and microscopical observations. *Biochim. Biophys. Acta*, **6**: 340–356.

Roelofsen, P. A. (1959). *The plant cell-wall. Encyclopedia of plant anatomy*, Vol. 3. Gebrüder Borntraeger, Berlin-Nikolasee.

Rogers, J. D. (1966). Notes on the conidial stage of *Hypoxylon fuscum. Mycologia*, **58**: 459–464.

Rogerson, C. T. (1970). The Hypocrealean fungi (Ascomycetes, Hypocreales). *Mycologia*, **62**: 865–910.

Roquebert, M. F. and Abadie, M. (1973). Etude ultrastructurale de la sporogenèse chez un micromycète: *Stilbothamnium nudipes* Haum. *C. R. Acad. Sci. Paris*, **276**: 2883–2885.

Ruiz-Herrera, J., Sing, V. O., Van der Woude, W. J. and Bartnicki-Garcia, S. (1975). Microfibril assembly by granules of chitin synthetase. *Proc. Nat. Acad. Sci.*, **72**: 2706–2710.

Russ, J. C. and Kabaya, A. (1970). Preparation of samples for scanning electron microscopy. *Proc. 28th Ann. EMSA meeting*, Claitor's Pub. Co. New Orleans, La, p. 380.

Saccardo, P. A. (1886). Hyphomyceteae. *Sylloge Fungorum*, **4**: 1–807.

Saksena, S. B. (1953). A new genus of the Mucorales. *Mycologia*, **45**: 426–436.

Samson, R. A. (1974). *Paecilomyces* and some allied Hyphomycetes. *C.B.S. Studies in Mycology*, **6**: 1–119.

Schipper, M. A. A., Samson, R. A. and Stalpers, J. A.

(1975). Zygospore ornamentation in the genera *Mucor* and *Zygorhynchus*. *Persoonia,* **8**:321–328.
Schol-Schwarz, M. B. (1959). The genus *Epicoccum* Link. *Trans. Br. Mycol. Soc.,* **42**:149–173.
Schol-Schwarz, M. B. (1968). *Rhinocladiella* its synonym *Fonsecaea* and its relation to *Phialophora. Antonie van Leeuwenhoek, J. Microbiol. Serol.* **34**:119–152.
Schol-Schwarz, M. B. (1970). Revision of the genus *Phialophora* (Moniliales). *Persoonia,* **6**:59–94.
Schramm, K.-H. and Petzold, H. (1973). Significance of the methods for preparing fungi on agar for critical examination with the scanning electron microscope. *Microsc. Acta,* **75**:128–136.
Schroeter, J. (1897). Mucorineae. In Engler and Prantl, *Die Natürlichen Pflanzenfamilien,* vol. 1, pp. 119–134.
Shearer, C. A. (1973). Fungi of the Chesapeake Bay and its tributaries. II. The genus *Conioscypha. Mycologia,* **65**:128–136.
Shearer, C. A. and Motta, J. J. (1973). Ultrastructure and conidiogenesis in *Conioscypha* (Hyphomycetes). *Can. J. Bot.,* **51**:1747–1751.
Simmons, E. G. (1966). Symposium on stability in dynamic microbial systems. 5. The theoretical bases for classification of the Fungi Imperfecti. *Quart. Rev. Biol.,* **41**:113–123.
Smith, G. (1961). *Polypaecilum* gen. nov. *Trans. Br. Mycol. Soc.,* **44**:437–440.
Smith, J. E., Gull, K., Anderson, J. G. and Deans, S. G. (1976). Organelle changes during fungal spore germination. In *The Fungal Spore: Form and Function.* D. J. Weber and W. M. Hess (Eds). Wiley, New York, pp. 301–352.
Snider, P. J. (1968). Nuclear movements in *Schizophyllum,* In *Aspects of Cell Motility. Symp. Soc. Exp. Biol. 22nd.* P. L. Miller (Ed.). University Press, Cambridge, England, pp. 261–283.
Spurr, A. R. (1969). A low-viscosity epoxy resin embedding medium for electron microscopy. *J. Ultrastruct. Res.,* **26**:31–43.
Stalpers, J. A. (1974a). Revision of the genus *Oedocephalum* (Fungi Imperfecti). *Proc. Koninkl. Nederl. Akademie van Wetenschappen-Amsterdam,* **77**:383–401.
Stalpers, J. A. (1974b). *Spiniger,* a new genus for imperfect states of Basidiomycetes. *Proc. Koninkl. Nederl. Akademie van Wetenschappen-Amsterdam,* **77**:402–407.
Stevens, R. B. (Ed.) (1974). *Mycology Guidebook.* University of Wash. Press, Seattle.
Stevens, R. T. (1975). Fine structure of sporogenesis and septum formation in *Micromonospora globosa* Kriss and *M. fusca* Jensen. *Can. J. Microbiol.,* **21**:1081–1088.
Streiblová, E. (1971). Cell division in yeasts. In *Recent Advances in Microbiology.* A. Pérez-Miravete and D. Peláez (Eds). *Proc. Symp. 10th Int. Congr. Microbiol.,* pp. 131–140.
Streiblová, E. and Wolf, A. (1972). Cell wall growth during the cell cycle of *Schizosaccharomyces pombe. Zeitschrift für Allg. Mikrobiologie,* **12**:673–684.
Subramanian, C. V. (1956). Hyphomycetes I and II. *J. Indian Bot. Soc.,* **35**:53–91 and 446–494.
Subramanian, C. V. (1962a). The classification of the Hyphomycetes. *Bull. Bot. Surv. India,* **4**:249–259.
Subramanian, C. V. (1962b). A classification of the Hyphomycetes. *Curr. Sci.,* **31**:409–411.
Subramanian, C. V. (1965). Spore types in the classification of the Hyphomycetes. *Mycopathol. Mycol. Appl.,* **26**:373–384.
Subramanian, C. V. (1971a). Conidial ontogeny in Fungi Imperfecti with special reference to cell wall relationships. *J. Indian Bot. Soc.,* **50**A:51–59.
Subramanian, C. V. (1971b). The phialide. In *Taxonomy of Fungi Imperfecti.* B. Kendrick (Ed.). Univ. of Toronto Press, Toronto, pp. 92–119.
Subramanian, C. V. (1972a). Conidial chains, their nature and significance in the taxonomy of Hyphomycetes. *Curr. Sci.,* **41**:43–49.
Subramanian, C. V. (1972b). Conidium ontogeny. *Curr. Sci.,* **41**:619–624.
Subramanian, C. V. (1973). Reflections on the conidium. *J. Indian Bot. Soc.,* **52**:1–16.
Subramanian, C. V. (Ed.) (1978). *Proceedings of the International Symposium on Taxonomy of Fungi.* Univ. of Madras. In press.
Sugimura, M. (1973). Scanning and transmission electron microscopy of individual cells. *J. Electron Microsc.,* **22**:367–369.
Sutton, B. C. (1973). Coelomycetes. In *The Fungi,* vol. IVA. G. C. Ainsworth, F. K. Sparrow and A. S. Sussman (Eds). Academic Press, New York, pp. 513–582.
Sutton, B. C. and Sandhu, D. K. (1969). Electron microscopy of conidium development and secession in *Cryptosporiopsis* sp., *Phoma fumosa, Melanconium bicolor,* and *M. apiocarpum. Can. J. Bot.,* **47**:745–749.
Swanson, C. P. and Webster. P. L. (1977). *The Cell* (4th Edn). Prentice-Hall, Englewood Cliffs, N.J.
Syrop, M. (1973). The ultrastructure of the growing regions of aerial hyphae of *Rhizopus sexualis* (Smith) Callen. *Protoplasma,* **76**:309–314.
Takada, H., Yagi, T. and Hiraoka, J. (1965). Elektronoptische Untersuchungen an *Endomycopsis fibuliger* auf festen Nährböden. *Protoplasma,* **59**:494–505.
Talbot, P. H. B. (1971). *Principles of Fungal Taxonomy.* Macmillan, London.
Tanaka, K. and Iino, A. (1974). Critical point drying method using dry ice. *Stain Technol.,* **49**:203–206.
Tanaka, K. and Yanagita, T. (1963). Electron microscopy of ultrathin sections of *Aspergillus niger.* II. Fine structure of the conidia-bearing apparatus. *J. Gen. Appl. Microbiol.,* **9**:189–203.
Thaxter, R. (1888). The Entomophthoraceae of the United States. *Mem. Boston Soc. Nat. Hist.,* **4**:133–201.
Thaxter, R. (1922). A revision of the Endogonaceae. *Proc. Amer. Acad. Arts Sci.,* **57**:291–351.
Thielke, C. (1972). Zisternenaggregate bei hoheren Pilzen. *Protoplasma,* **75**:335–339.
Tokunaga, M., Tokunaga, J. and Harada, K. (1973). Scanning and transmission electron microscopy of sterigma and conidiospore formation in *Aspergillus* group. *J. Electron Microsc.,* **22**:27–38.

Trinci, A. P. J. and Collinge, A. J. (1973). Structure and plugging of septa of wild type and spreading colonial mutants of *Neurospora crassa*. *Arch. Mikrobiol.*, **91**:355–364.

Trinci, A. P. J. and Collinge, A. J. (1974a). Occlusion of the septal pore of damaged hyphae of *Neurospora crassa* by hexagonal crystals. *Protoplasma*, **80**:57–67.

Trinci, A. P. J. and Collinge, A. J. (1974b). Spore formation in nitrogen and carbon starved cultures of *Geotrichum candidum* and *Mucor racemosus*. *Trans. Br. Mycol. Soc.*, **62**:351–358.

Trinci, A. P. J. and Saunders, P. T. (1977). Tip growth of fungal hyphae. *J. Gen. Microbiol.*, **103**:243–248.

Trinci, A. P. J., Peat, A. and Banbury, G. H. (1968). Fine structure of phialide and conidiospore development in *Aspergillus giganteus* 'Wehmer'. *Ann. Bot. (London)*, **32**:241–249.

Tsukahara, T. (1970). Electron microscopy of conidiospore formation in *Aspergillus niger*. *Sabouraudia.*, **8**:93–97.

Tubaki, K. (1958). Studies on Japanese Hyphomycetes. V. Leaf and stem group with a discussion of the classification of Hyphomycetes and their perfect stages. *J. Hattori Bot. Lab.*, **20**:142–244.

Tubaki, K. (1963). Taxonomic study of Hyphomycetes. *Ann. Rep. Inst. Ferment. Osaka*, **1**:25–54.

Tubaki, K. (1967). An undescribed species of *Heleococcum* from Japan. *Trans. Mycol. Soc. Japan*, **8**:5–10.

Tubaki, K. (1975). *Hypomyces* and the conidial states in Japan. *Rept. Tottori Mycol. Inst. (Japan)*, **12**:161–169.

Turner, R. H. and Green, C. D. (1973). Preparation of biological material for scanning electron microscopy by critical point drying from water miscible solvents. *J. Microsc.*, **97**:357–363.

van Emden, J. H. and Veenbaas-Rijks, J. W. (1973). *Cylicognone*, a new genus of Hyphomycetes with unusual conidium ontogeny. *Acta Bot. Neerl.*, **22**:637–640.

Vian, B. (1972). Aspects, en cryodécapage, de la fusion des membranes des vésicules cytoplasmiques et du plasmalemme lors des phénomènes de sécrétion végétale. *C. R. Acad. Sci., Paris*, **275**:2471–2474.

Vigil, E. L. (1973). Structure and function of plant microbodies. *Sub-Cell. Biochem.*, **2**:237–285.

Vuillemin, P. (1910a). Matériaux pour une classification rationelle des Fungi Imperfecti. *C. R. Acad. Sci., Paris*, **150**:882–884.

Vuillemin, P. (1910b). Les conidiosporés. *Bull. Soc. Sci. Nancy*, **2**:129–172.

Vuillemin, P. (1911). Les aleuriosporés. *Bull. Soc. Sci. Nancy*, **3**:151–175.

Wakefield, E. M. and Bisby, G. R. (1941). List of Hyphomycetes recorded for Britain. *Trans. Br. Mycol. Soc.*, **25**:49–126.

Wang, C. J. K. (1964). Studies on *Trichosporium heteromorphum* Nannfeldt. *Can J. Bot.*, **42**:1011–1016.

Wang, C. J. K. (1966). Annellophores in *Torula jeanselmei*. *Mycologia*, **58**:614–621.

Waterhouse, G. M. (1973). Entomophthorales. In *The Fungi*, vol. IVB. G. C. Ainsworth, F. K. Sparrow and A. S. Sussman (Eds). Academic Press, New York, pp. 219–229.

Webster, J. (1952). Spore projection in the hyphomycete *Nigrospora sphaerica*. *New Phytol.*, **51**:229–235.

Webster, J. (1966). Spore projection in *Epicoccum* and *Arthrinium*. *Trans. Br. Mycol. Soc.*, **49**:339–343.

Webster, J. (1970). *Introduction to Fungi*. The University Press, Cambridge.

Wells, K. (1964). The basidia of *Exidia nucleata*. 1. Ultrastructure. *Mycologia*, **56**:327–341.

Wergin, W. P. (1973). Development of Woronin bodies from microbodies in *Fusarium oxysporum* f. sp. *lycopersici*. *Protoplasma*, **76**:249–260.

Weisburg, S. H. and Turian, G. (1971). Ultrastructure of *Aspergillus nidulans* conidia and conidial lomasomes. *Protoplasma*, **72**:55–67.

Wetmore, C. M. (1973). Multiperforate septa in lichens. *New Phytol.*, **72**:535–538.

White, J. A., Calvert, O. H. and Brown, M. F. (1973). Ultrastructure of the conidia of *Helminthosporium maydis*. *Can. J. Bot.*, **51**:2006–2008.

Willets, H. J. and Calonge, F. D. (1969). Spore development in the brown rot fungi (*Sclerotinia* spp.). *New Phytol.*, **68**:123–131.

Wilsenach, R. and Kessel, M. (1965a). On the function and structure of the septal pore of *Polyporus rugulosus*. *J. Gen. Microbiol.*, **40**:397–400.

Wilsenach, R. and Kessel, M. (1965b). The role of lomasomes in wall formation in *Penicillium vermiculatum*. *J. Gen. Microbiol.*, **40**:401–404.

Wilsenach, R. and Kessel, M. (1965c). Micropores in the cross-wall of *Geotrichum candidum*. *Nature*, **207**:545–546.

Winborn, W. B. and Guerrero, D. L. (1974). The use of a single tissue specimen for both transmission and scanning electron microscopy. *Cytobios*, **10**:83–91.

Yamashita, K. and Yamashita, T. (1972). *Polypaecilum insolitum* (= *Scopulariopsis divaricata*) isolated from cases of otomycosis. *Sabouraudia*, **10**:128–131.

Young, T. W. K. (1968a). Electron microscopic study of asexual spores in Kickxellaceae. *New Phytol.*, **67**:823–836.

Young, T. W. K. (1968b). Electron microscopic study of the asexual structures in Mucorales. *Proc. Linn. Soc., London*, **179**:1–9.

Young, T. W. K. (1969a). Ultrastructure of aerial hyphae in *Linderina pennispora*. *Ann. Bot., New Series.*, **33**:211–216.

Young, T. W. K. (1969b). Electron and phase-contrast microscopy of spores in two species of the genus *Mycotypha* (Mucorales). *J. Gen. Microbiol.*, **55**:243–249.

Young, T. W. K. (1970a). Ultrastructure of the spore wall of *Linderina*. *Trans. Br. Mycol. Soc.*, **54**:15–25.

Young, T. W. K. (1970b). Arrangement of the microfibrils in walls of aerial hyphae of *Linderina*. *Trans. Br. Mycol. Soc.*, **55**:29–35.

Young, T. W. K. (1971). Ultrastructure of the wall of the germinating sporangiospore of *Linderina pennispora* (Mucorales). *Ann. Bot.*, **35**:183–191.

Young, T. W. K. (1973). Ultrastructure of the sporangiospore of *Coemansia reversa* (Mucorales). *Trans.*

Br. Mycol. Soc., **60**: 57–63.

Zachariah, K. and Fitz-James, P. C. (1967). The structure of phialides in *Penicillium claviforme*. *Can. J. Microbiol.*, **13**: 249–256.

Zalokar, M. (1959). Growth and differentiation of *Neurospora* hyphae. *Amer. J. Bot.*, **46**: 602–610.

Zycha, H. (1935). Mucorinae. *Kryptogamenflora der Mark Brandenburg*, **6a**: 1–264.

Zycha, H., Siepmann, R. and Linnemann, G. (1969). *Mucorales*. Cramer, Weinheim.

Index

Note: the page numbers shown in italics indicate figure captions.